International Conference on SDGs and Bibliometric Studies (ICoSBi 2023)

E3S Web of Conferences Volume 450 (2023)

Surabaya, Indonesia
3 - 4 October 2023

Editors:

Nadi Suprapto
Binar Kurnia Prahani
Muhammad Satriawan
Hapsari Peni Agustin Tjahyaningtijas

ISBN: 978-1-7138-8449-1

Printed from e-media with permission by:

Curran Associates, Inc.
57 Morehouse Lane
Red Hook, NY 12571

Some format issues inherent in the e-media version may also appear in this print version.

For additional information, please contact EDP Sciences – Web of Conferences
at the address below.

EDP Sciences – Web of Conferences
17, Avenue du Hoggar
Parc d'Activité de Courtabœuf
BP 112
F-91944 Les Ulis Cedex A
France

Phone: +33 (0) 1 69 18 75 75
Fax: +33 (0) 1 69 28 84 91

contact-edps@webofconferences.org

Additional copies of this publication are available from:

Curran Associates, Inc.
57 Morehouse Lane
Red Hook, NY 12571 USA
Phone: 845-758-0400
Fax: 845-758-2633
Email: curran@proceedings.com
Web: www.proceedings.com

TABLE OF CONTENTS

- **Scientific editor (s) board:**

ICoSBi
International Conference SDGs Bibliometrics Studies

Kampus Merdeka
INDONESIA JAYA

GreenMetric

SIER

PELINDO

Organized By
Directorate of Innovation, Ranking, and Scientific Publication – State University of Surabaya (UNESA), Indonesia

The 1st International Conference on SDGs and Bibliometrics Studies (ICoSBi)

Theme

Empowering the Sustainability of Higher Education Institution Toward Green Digital University

Opening Remarks

Prof. Dr. Nurhasan, M.Kes.
Rector of UNESA

Keynote Speaker

Prof. Dr. Ir. Arjen Wals
Professor of Transformative Learning for Socio-Ecological Sustainability
Wagheningen University
UNESCO Chair

#1 UI GreenMetric

Dr. H. Emil Elestianto Dardak, B.Bus., M.Sc.
Vice Governor of East-Java, Indonesia

Prof. Dr. Ir. Riri Fitri Sari MM MSc.
Chairperson of UI GreenMetric, Indonesia

Assoc. Prof. Dr. Muhammad Faizal A. Ghani
Head of Department Educational Management, Planning, and Policy University of Malaya, Malaysia

#50 UI GreenMetric

Edi Priyanto, S.KM., MM
Director of Human Resources PT Pelindo Multi Terminal, Indonesia

*in confirmation

Dr. Bagus Muljadi
University of Nottingham, UK

#3 UI GreenMetric

Scope

a. 6 Indicators UI Green Metric:

1. *Setting and Infrastructure*
2. *Energy and Climate Change*
3. *Waste*
4. *Water*
5. *Transportation*
6. *Education and Research*

b. 17 SDGs:

1. *No poverty*
2. *Zero Hunger*
3. *Good Health and Well Being*
4. *Quality Education*
5. *Gender Equality*
6. *Clean Water and Sanitation*
7. *Affordable and Clean Energy*
8. *Decent Work and Economic Growth*
9. *Industry Innovation and infrastructure*
10. *Reduced inequalities*
11. *Sustainable cities and communities*
12. *Responsible consumption and production*
13. *Climate Action*
14. *Life Below Water*
15. *Life on Land*
16. *Peace, Justice, and Strong Institutions*
17. *Partnerships for the Goals*

c. Bibliometric studies in all subjects

Publication

All submitted papers will be reviewed by our internal and external reviewers. Papers declared to be of good quality will proceed to the selected proceedings: "E3S Web of Conference" which is indexed by Scopus.

Important Dates

Abstract Submission
September 10, 2023

Abstract Acceptance
September 13, 2023

Full Paper Submission Deadline
September 26, 2023

Registration Deadline:
September 27, 2023

Conference Day
October 3-4, 2023

Revised Paper Deadline (FOR SELECTED PAPER)
October 6, 2023

Registration Link

https://icosbi.unesa.ac.id

Date

3-4th OCT 2023
Ciputra World Hotel Surabaya

Fee

Offline participants:
Early bird: IDR 2,250,000 Regular: IDR 2,500,000
*Payment above includes: seminar kit, certificate, proceeding, fullboard meeting facilities at Ciputra World Hotel Surabaya

Online participants:
Early bird: IDR 1,000,000 Regular: IDR 1,250,000
*Payment above includes: proceeding and certificate

Payment via Bank Transfer:
9422091923079190022

Contact

Shelly Andari
+62 812-3575-2696

Riska
+62 812-3265-2200

Website: https://icosbi.unesa.ac.id
email: icosbi@unesa.ac.id

Statement of Peer review

In submitting conference proceedings to *Web of Conferences*, the editors of the proceedings certify to the Publisher that

1. They adhere to its **Policy on Publishing Integrity** in order to safeguard good scientific practice in publishing.
2. All articles have been subjected to peer review administered by the proceedings editors.
3. Reviews have been conducted by expert referees, who have been requested to provide unbiased and constructive comments aimed, whenever possible, at improving the work.
4. Proceedings editors have taken all reasonable steps to ensure the quality of the materials they publish and their decision to accept or reject a paper for publication has been based only on the merits of the work and the relevance to the journal.

Title, date and place of the conference:

International Conference on SDGs and Bibliometric Studies (ICoSBi) 2023
(Surabaya, Indonesia, October 3-4, 2023)

Proceedings editor(s):

Prof. Nadi Suprapto, Ph.D.
Assoc. Prof. Dr. Binar Kurnia Prahani
Dr. Muhammad Satriawan
Dr. Hapsari Peni Agustin Tjahyaningtijas

Date and editor's signature:

November 03, 2023

The role of universities in realizing sustainability: Analysis of UI GreenMetric contributions and publications related to sustainability

Riri Fitri Sari[1*], *Yogi Andrian* Sidiyanto[2], and *Jauzak Hussaini* Windiatmaja[1]

[1]Department of Electrical Engineering, Universitas Indonesia, Depok, 16424, Indonesia
[2]UI GreenMetric, Universitas Indonesia, Depok, 16424, Indonesia

Abstract. This paper discusses how UI GreenMetric and sustainability publications interact to strengthen universities' contributions to global sustainability. It examines data from six universities, including Wageningen University & Research (Netherlands), Universitas Indonesia (Indonesia), Universitas Diponegoro (Indonesia), University of Malaya (Malaysia), Universitas Airlangga (Indonesia), and Universitas Negeri Surabaya (Indonesia), linking UI GreenMetric rankings with sustainability-related publications. Results in the 2022 UI GreenMetric Rankings show that Wageningen University & Research secured the top position, with an average of 2395 sustainability-related publications in the last three years and 3221 publications from Elsevier. Universitas Indonesia has 4725 total publications from Elsevier and 1525 publications related to sustainability from UI GreenMetric data. Other universities, such as Universitas Diponegoro, University of Malaya, Universitas Airlangga, and Universitas Negeri Surabaya, also showcased their commitment to sustainability through their publication records. This study concludes that higher-ranked universities tend to produce more sustainability-focused publications, indicating a positive relationship between sustainability commitment and research impact. Institutions with elevated UI GreenMetric rankings consistently prioritize sustainability, increasing research productivity in this field.

1 Introduction

The issues related to sustainability and the achievement of the Sustainable Development Goals (SDGs) have garnered global attention in recent decades [1]. Sustainability has emerged as one of the most pressing challenges facing humanity and our planet today [2]. The increasingly evident impacts of climate change, declining biodiversity, social inequality, and various other issues serve as reminders of the urgent need for immediate and serious action to protect our environment and enhance human well-being worldwide [3-4]. The SDGs agenda, adopted by the United Nations in 2015, established an ambitious

* Corresponding author: riri@ui.ac.id

framework consisting of 17 goals that encompass various aspects of human life, ranging from poverty eradication to environmental protection [5-8]. The seventeenth goals of SDGs are shown as shown in Figure 1. The SDGs aim to create a fairer, more sustainable, and inclusive world [9]. However, achieving these goals is not a straightforward task; it necessitates global cooperation, innovation, and in-depth research to address the diverse challenges faced in bringing about positive change [10-11].

One of elements in addressing these challenges is the crucial role played by universities worldwide. Universities have the potential to serve as centers of innovation, knowledge, and advocacy in the effort to achieve sustainable development [12-13]. Universities possess the capacity to generate in-depth and relevant scientific research in various fields related to sustainability, from environmental science to social economics [14]. Moreover, universities also play a role in educating future generations, who will be the leaders of tomorrow in the pursuit of sustainable development goals [15-16]. Integrated sustainability education in university curricula offers opportunities for students to comprehend the complexities of sustainability issues, develop critical thinking skills, and ignite their leadership spirit [17]. Universities can also serve as role models in implementing sustainable practices on their campuses, inspiring positive actions in society. Universities can become influential agents of change by bridging the academic world with the private sector, government, and civil society organizations [18]. Collaboration between universities and external parties can facilitate the transfer of knowledge, technology, and best practices that support sustainable development [19]. Hence, universities have a significant responsibility in establishing and maintaining broad and synergistic networks to achieve sustainability goals more effectively.

Fig. 1. Sustainable Development Goals [7]

One initiative for measuring and assessing universities' contributions to environmental sustainability is UI GreenMetric. This initiative aims to evaluate the sustainability performance of universities worldwide through six indicators, including setting and infrastructure, energy and climate change, waste, water, transportation, and education and research. UI GreenMetric indicators are related to the SDGs as shown in Figure 2. UI GreenMetric provides universities a platform to gauge their environmental impact and encourages enhancements in sustainable practices.

The participation of universities in UI GreenMetric creates opportunities for them to share information about their sustainable efforts and exchange experiences with universities around the world. This facilitates collaboration and learning among universities, aiding in the identification of best practices that can be applied in various contexts. Additionally, UI GreenMetric ranking provides incentives for universities to continually improve their sustainability efforts, as their ranking can impact their image and reputation both nationally and internationally.

The number of sustainability-related publications in the last three years is one of the UI GreenMetric indicators. Scientific publications related to sustainability constitute one crucial way in which universities can make substantial contributions to supporting sustainability issues. Universities often serve as hubs for scientific research and development, granting them access to in-depth knowledge across various aspects of sustainability [20]. Research and publications in the field of sustainability encompass diverse topics, ranging from climate change and renewable energy to environmental sustainability, food security, and social innovation [21]. These publications can furnish a robust scientific foundation for more effective policymaking and assist in addressing complex sustainability challenges. This paper discusses how UI GreenMetric and sustainability publications interact to strengthen universities' contributions to global sustainability.

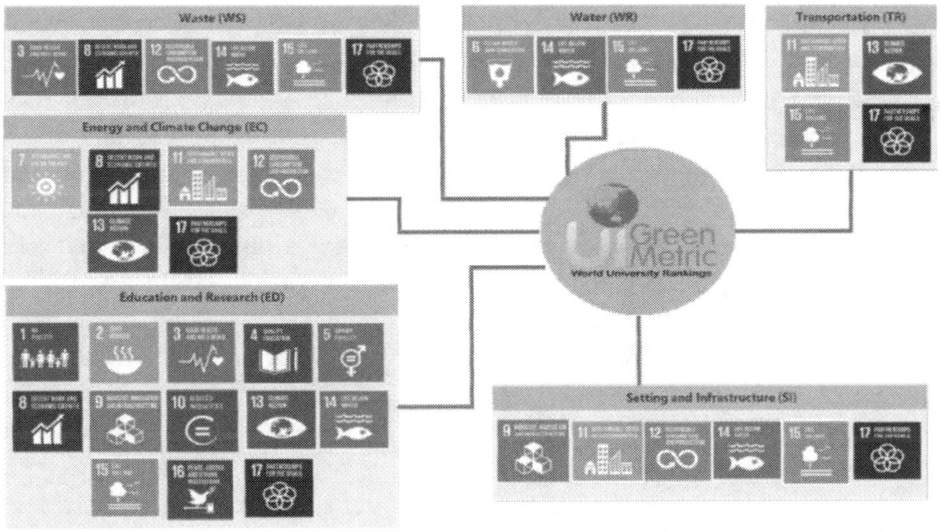

Fig. 2. UI GreenMetric and SDGs [22]

2 Methods

In this study, data were collected from six universities: Wageningen University & Research (Netherlands), Universitas Indonesia (Indonesia), Universitas Diponegoro (Indonesia), University of Malaya (Malaysia), Universitas Airlangga (Indonesia), and Universitas Negeri Surabaya (Indonesia). The data encompassed several key variables, including university rankings in the 2022 UI GreenMetric, the average number of sustainability-related publications over the past three years (2020-2022) from UI GreenMetric data, and the total query data publications from Elsevier pertaining to sustainability over the same three-year period.

The study adopted a cross-sectional design, compiling information from these universities and sources to explore potential relationships between sustainability-related metrics and university rankings. To achieve this, data was collected from the specified sources (UI GreenMetric and Elsevier) for the selected universities and subsequently subjected to analysis. The analysis aimed to unveil any correlations or patterns between university rankings, sustainability-related publications, and query data publications.

3 Result and Discussion

In Table 1, a comprehensive comparison of six prominent universities is presented, highlighting their respective positions in the 2022 UI GreenMetric Ranking alongside key metrics related to sustainability publications. These universities include Wageningen University & Research in the Netherlands, Universitas Indonesia in Indonesia, Universitas Diponegoro in Indonesia, the University of Malaya in Malaysia, Universitas Airlangga in Indonesia, and Universitas Negeri Surabaya in Indonesia. The table encompasses critical data points such as their UI GreenMetric Ranks for 2022, the average number of sustainability-related publications spanning the last three years, and the total publication figures pertaining to sustainability from Elsevier over the same period. This comparative analysis offers valuable insights into the universities' commitment to sustainability and their contributions to the field through scholarly publications

The subject areas for publications at each university, based on Scopus data, are shown in Figure 3. The figure illustrates that there is variation in the distribution of scholarly publications at each university. In general, it can be observed that all universities focus on broader fields of study, including science, technology, medicine, and social sciences. Wageningen University & Research has a strong emphasis on Agricultural and Biological Sciences (26.1%). University of Malaya, Universitas Indonesia, and Universitas Airlangga demonstrate their focus on Medicine, with respective subject area percentages of 26.1%, 25.4%, and 22.3%. Universitas Negeri Surabaya has a significant percentage in Physics and Astronomy (21.5%), while Universitas Diponegoro shows a strong commitment to Environmental Science (21.4%). Each university has a unique profile of subject areas, with some subjects being their primary focus, reflecting their specialization and commitment in particular fields of study.

The comparison between the number of publications from UI GreenMetric data and Elsevier is as shown in Figure 4, with universities arranged from left to right based on their rankings in 2022 UI GreenMetric. Wageningen University & Research from the Netherlands stands out as the top performer in the 2022 UI GreenMetric ranking, securing the first position. This suggests a strong commitment to sustainability practices and policies, reflected in its top-tier ranking. Furthermore, the university's substantial average of 2395 sustainability-related publications over the last three years underscores its dedication to advancing knowledge in the field of sustainability. The university's high position in the UI GreenMetric ranking aligns with its prolific research output in this area.

In contrast, Indonesian universities, including Universitas Indonesia (UI), Universitas Diponegoro (UNDIP), Universitas Airlangga (UNAIR), and Universitas Negeri Surabaya (UNESA), exhibit lower rankings than Wageningen University & Research in the UI GreenMetric. While they face challenges in achieving higher positions in the sustainability ranking, their contributions to sustainability-related publications are noteworthy. For example, UI and UNDIP have impressive averages of 4725 and 4491 sustainability-related publications, respectively, over the last three years, indicating their active involvement in academic research on sustainability. These universities are leveraging their research capabilities to address sustainability issues despite their lower UI GreenMetric rankings. The University of Malaya from Malaysia, with a UI GreenMetric ranking of 50, occupies an intermediate position among the universities studied. Its average of 1284 sustainability-related publications demonstrates a significant commitment to research in sustainability. However, there is room for improvement in its UI GreenMetric ranking to align more closely with its research output.

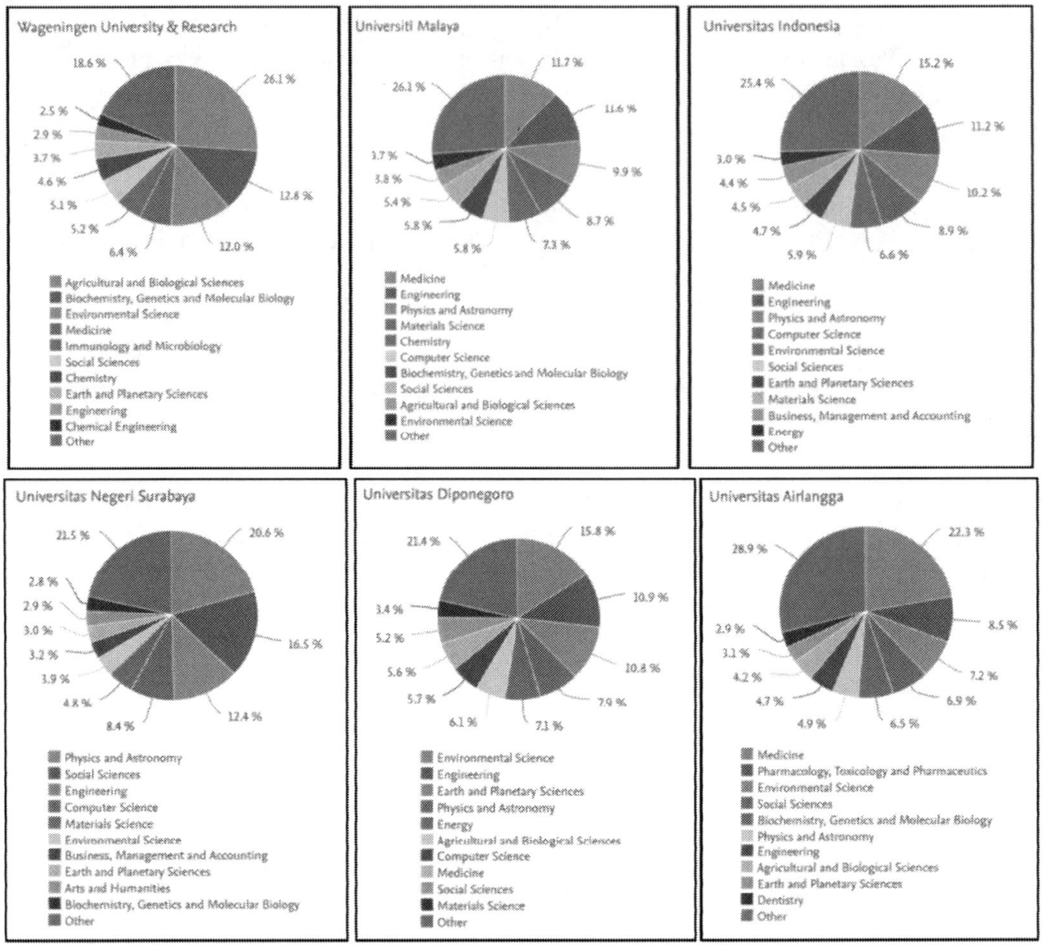

Fig. 3. University Subject Area Publications

Table 1. Comparison UI GreenMetric Rank and Universities Publications Related to Sustainability

No	University	Country	2022 UI Green Metric Rank	Average sustainability-related publications for last three years	Total publications data from Elsevier related to sustainability last 3 years
1	Wageningen University & Research	Netherlands	1	2395	3221
2	Universitas Indonesia	Indonesia	24	4725	1525
3	Universitas Diponegoro	Indonesia	29	1144	887
4	University of Malaya	Malaysia	50	4491	2000
5	Universitas Airlangga	Indonesia	83	1284	768
6	Universitas Negeri Surabaya	Indonesia	242	825	82

It's also worth noting that when considering the total publications data from Elsevier related to sustainability over the last three years, Wageningen University & Research, and the University of Malaya appear to have a more extensive presence in sustainability research globally, with 3221 and 2000 publications, respectively. In contrast, Universitas Negeri Surabaya (UNESA) lags significantly in this regard, with only 82 publications.

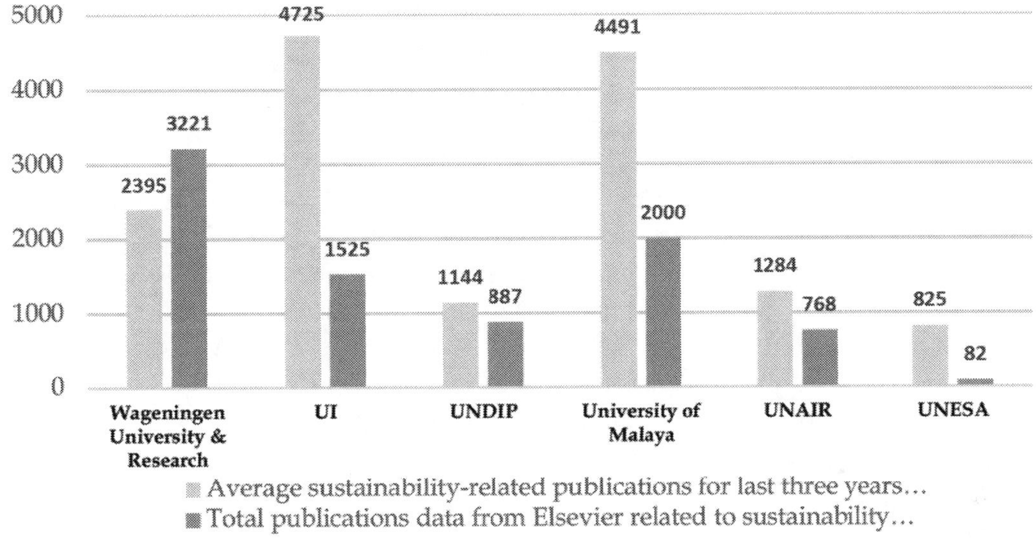

Fig. 4. Universities Publications Related to Sustainability

Universities with higher rankings have a greater number of publications related to sustainability. For instance, Wageningen University & Research had an average of 2395 sustainability-related publications in the last three years, while UNESA has only 82 publications. Therefore, there is a positive correlation between the ranking in UI GreenMetric, the number of sustainability-related publications from Elsevier, and the total sustainability-related publications in the last three years based on UI GreenMetric data. The more sustainability-related publications a university produces, the better its performance and ranking in UI GreenMetric tend to be. This indicates that universities with a stronger commitment to sustainability practices tend to have a greater impact in the field of sustainability research.

The university rankings in UI GreenMetric consistently reflect the level of focus and commitment of universities to sustainability. Universities with higher rankings in UI GreenMetric tend to have a stronger focus on sustainable initiatives and practices. This is reflected in the university's publications related to sustainability in the last three years. Universities with more publications related to sustainability tend to have a good ranking in UI GreenMetric. Universities with high rankings tend to produce more publications in the sustainability domain, indicating that they are actively involved in research and scientific publications that support sustainability practices. Therefore, there is a positive relationship between the ranking in UI GreenMetric, the number of sustainability-related publications based on Elsevier data, and the total sustainability-related publications in the last three years from UI GreenMetric data. This suggests that universities with a stronger commitment to sustainability practices tend to have a greater impact in the field of sustainability research.

4 Conclusion

The global importance of sustainability and the pursuit of Sustainable Development Goals (SDGs) cannot be overstated, given the pressing challenges of climate change, biodiversity loss, and social inequality. Universities worldwide play a pivotal role in addressing these challenges by serving as hubs of innovation, knowledge, and advocacy. It generates essential scientific research, educates future leaders, and promotes sustainable practices. Initiatives like UI GreenMetric provide a platform for universities to assess their environmental impact, foster collaboration, and incentivize sustainability efforts. The strong correlation between UI GreenMetric rankings and sustainability-related publications underscores the influence of universities committed to sustainability. In a world where sustainability is paramount, universities are key agents of change and progress in the global pursuit of a fairer, more sustainable, and inclusive future.

References

1. F. Indana, R. W. Pahlevi, Cogent Business & Management **10**, 2 (2023)
2. C. Vasconcelos, N. Orion, Sustainability **13**, 3 (2021)
3. A. Hassoun, M. Prieto, M. Rodríguez, Y. Bouzembrak, H.J.P. Marvin, N. Pallarés, F. Barba, B. Punia, Sneh, V. Chaudhary, S. Ibrahim, G. Bono, Food Research International **162** (2022)
4. P. H. Raven, D. L. Wagner, Proceedings of the National Academy of Sciences **118**, 2 (2021)
5. Å. Persson, N. Weitz, M. Nilsson, Comparative & International Environmental Law. **25**, 1 (2016)
6. B. Reyers, E. R. Selig, Nature Ecology & Evolution **4**, 8 (2020)
7. 17 Goals of Sustainability Development Goals (2015). Available online at https://sdgs.un.org/goals
8. W. L. Filho, S. K. Tripathi, J. B. Guerra, R. Giné-Garriga, V. O Lovren, J. Willats, International Journal of Sustainable Development & World Ecology **26**, 2 (2019)
9. C. Cardoso, Worldwide Hospitality and Tourism Themes **12**, 6 (2020)
10. O. Larrinaga, N. Calvo, Journal of Engineering and Technology Management **35** (2015)
11. N. Alghamdi, A. den Heijer, H. de Jonge, Int. J. Sustain. in HEб **1**, 84 (2017)
12. L. Dryjanska, J. Kostalova, D. Vidović, Higher Education Practices for Social Innovation and Sustainable Development **6** (2022)
13. W. C., Clark, L. V. Kerkhoff, L. Lebel, G. C. Gallopin, Proceedings of the National Academy of Sciences **113**, 17 (2016)
14. R. Hansmann, H. A. Mieg, P. Frischknecht, International Journal of Sustainable Development & World Ecology **19**, 5 (2012)

15. M. Dzimińska, J. Fijałkowska, Ł. Sułkowski, Sustainability **12**, 11 (2020)
16. K. Kohl, C. Hopkins, M. Barth, G. Michelsen, J. Dlouhá, D. A. Razak, Z. Abidin Bin Sanusi, I. Toman, International Journal of Sustainability in Higher Education **23**, 2 (2022)
17. A. Amran, I. Jasin, M. Perkasa, et al., Journal of Physics: Conference Series **1521**, 4 (2020)
18. R. Scheyvens, G. Banks, E. Hughes, Sustainable Development **24**, 6 (2016)
19. J. Wang, M. Yang, P. Maresova, Sustainability **12**, 6 (2020)
20. B. C. Stahl, K. Chatfield, C. T. Holter, A. Brem, Journal of Cleaner Production **239** (2019)
21. M. G. Gómez Zermeño, L. Y. Alemán de la Garza, International Journal of Sustainability in Higher Education **22**, 2 (2021)
22. UI Green Metric World University Rankings; Methodology (2023). Available online at https://greenmetric.ui.ac.id/publications/guidelines

The Influence of student understanding and concern for the environment on sustainable campus in the UI GreenMetric Ranking

Hakim Hakim[1]* and *Tri* Endangsih[1]

[1] Architecture Study Program, Faculty of Engineering, Budi Luhur University, Jakarta

Abstract. Development that ignores the concept of sustainability can result in losses and negative impacts on environmental sustainability as well as acceleration of the phenomena of climate change and global warming. Therefore, it is important to minimize environmental damage by implementing the concept of sustainable development in various aspects to meet the needs of future generations. This research aims to provide information and determine the influence of students' understanding and concern for the environment in achieving a sustainable campus. The research method uses a correlational approach with a survey type via questionnaire. The data analysis technique uses regression testing with SPSS software version 26.00. The results of this research show that there is a significant positive influence if students' understanding and concern for the environment increases, this will have an impact on campus sustainability. The highest influence on a sustainable campus is environmental awareness, followed by students' understanding of the environment.

1 Introduction

The concept of sustainable development is a development concept that pays attention to the continued availability of natural resources and environmental sustainability for the current and future generations [1, 2]. Development that ignores the concept of sustainability has the potential to result in losses and negative impacts on environmental sustainability, such as various damages that occur due to meeting needs that are not environmentally friendly. Other damages include destruction of the availability of natural resources, decreased food and energy security, natural disasters, acceleration of the phenomena of climate change and global warming. Based on the damage that has occurred, the concept of sustainable development needs to be intensified in its application in various aspects of life so as not to threaten the availability of meeting the needs of future generations [3].

Sustainable campus development is a strategic step because the campus environment has a very large mass, so the frequency of use of clean water and energy in the campus environment is also large. Apart from that, waste and waste management that is not environmentally friendly will have an impact on environmental pollution. If these problems are ignored and not followed up, they will have a negative effect on environmental sustainability in and around campus. These bad effects give rise to

* Corresponding author: hakim@budiluhur.ac.id

discomfort such as increasing temperatures in the campus environment as a result of pollution resulting from the use of private vehicles, lack of sufficient clean water in the environment and outside the campus, and environmental pollution as a result of waste management that is not environmentally sound as a result of activities. activities on campus [4]. A sustainable campus is a concept that implements sustainable development in the aspect of campus environmental management. Sustainable campuses emerge from academics' awareness of the importance of campuses contributing to maintaining the integrity of the current environment for future generations, starting in the campus environment [5]. A sustainable campus is defined as a concept that prioritizes the practice of sustainable environmental protection, management and preservation efforts in educational institutions [6].

The criteria for creating a sustainable campus include the layout and condition of campus facilities and infrastructure, energy utilization and anticipation of global warming, integrated waste management, economical use of water, use and creation of environmentally friendly transportation facilities, and environmentally sound education. In Indonesia itself, this concept has begun to be widely implemented on various campuses [7, 5]. A sustainable campus is a concept of campus management that considers ecological sustainability consisting of material, air, water, land and energy components, by creating an environmentally sound lifestyle through developing knowledge, community, health and institutions for the communities that are part of it. from the campus environment [8, 9]. A sustainable campus (green campus) is closely related to a university's high commitment to creating environmentally friendly campus infrastructure. A sustainable campus (green campus) is an integration concept in research, education and community service in environmental management. Thus, the role of campuses in environmental conservation efforts is very substantial [10, 8, 7].

UI Green-Metric World University Ranking is a movement that contributes to the discourse of sustainable development for the field of education and campus greening, becomes a promotional tool initiated by the campus as an agent of change related to sustainable achievements, becomes a basic guideline regarding campus sustainability standards for educational institutions High levels throughout the world, and socialize to the general public about sustainable programs on campus [11, 12]. A sustainable campus based on the UI Green-metric has six assessment categories including: (1) setting and infrastructure, (2) energy and climate change, (3) waste, (4) water (water), (5) transportation, and (6) education and research (education and research)[13]. The assessment criteria in the UI Green Metric include six categories, namely regulation and infrastructure, energy and climate change, waste, water, transportation, and education and research [14, 10, 15].

This research aims to obtain information and determine the influence of students' environmental understanding and concern for sustainable campuses in the UI Green Metric ranking. Students' understanding and concern regarding sustainable campuses in higher education regarding the importance of environmental sustainability in the future has increased. One of the universities that has a vision to play an active role in sustainable development at the national and global level is Budi Luhur University, Jakarta [16]. Budi Luhur University Jakarta is one of the universities participating in the UI Green-Metric World University Ranking program [13]. It is hoped that the results of the research will increase student understanding and concern, and the UI Green Metric at national and international levels at Budi Luhur University Jakarta will also increase. Therefore, it is very important that this research is carried out regarding students' environmental understanding and concern for sustainable campuses in the UI Green Metric ranking. Comprehension is a measure of a person's ability to be able to understand or understand the activities they are carrying out, so in learning, teachers must understand or comprehend what they are teaching to students [17, 18]. In understanding (comprehension) a person is asked to prove that he understands simple relationships between facts [19]. Apart from understanding that achieving a sustainable campus also requires caring for the environment. This is an expression when someone pays attention feels something that is happening and takes positive action regarding the incident. Concern does not always have to be directed at a person, but also towards all living creatures and the surrounding environment [20, 21]. Environmental concern is an attitude towards environmental consequences [22]. Environmental concern is an important attribute that can represent an individual's affection, likes and dislikes, and concern for the environment [23, 24].

2 Methods

Students' understanding and concern regarding sustainable campuses in higher education regarding the importance of environmental sustainability in the future has increased. One of the universities that has a vision to play an active role in sustainable development at the national and global level is Budi Luhur University, Jakarta. Budi Luhur University Jakarta is one of the universities participating in the UI Green-Metric World University Ranking program. This research is limited to: From the identification of the problems above, the researcher limits the research problem to: students' environmental understanding and concern for sustainable campuses in the UI Green Metric ranking. The research aims to obtain information and determine the influence of students' environmental understanding and concern for sustainable campuses in the UI Green Metric ranking. provide a contribution to the efforts made to increase students' understanding and awareness of sustainable campuses in the UI Green Metric ranking, so that it can be more applied in the campus environment. Apart from that, it can be used by campus policymakers in efforts to sustain the campus in the UI Green Metric ranking.

The focus of this research was carried out at Budi Luhur University with a sample of all students at Budi Luhur University Jakarta in 2022-2023. Implementation time is from May to July 2023. The sampling technique in this research uses simple random sampling. Simple random sampling is a sample taken from a population by not selecting individuals who are members of the sample based on certain reasons or subjective reasons. The number of samples taken was based on the formula proposed by Slovin, that is:

$$n = \frac{N}{N\,(d)^2 + 1} \quad (1)$$

Information :
n = Number of samples sought

N = Total population
e = *error* (0.05)
Known: Total population (N) = 75
e (*error*) = 0.05

Thus, the sample size can be calculated as follows:

$$n = \frac{75}{75\,(0.05)^2 + 1}$$

$$= \frac{75}{1.2}$$

= 62.5 (rounded 63)

The total sample was 63 students at Budi Luhur University.

After determining the sample size, the second stage is determining the research variables. Research variables are based on indicators of students' level of understanding and sense of concern for the environment to achieve a sustainable campus. The research variables are as follows: 1) Students' environmental understanding in this research is the students' ability to provide explanations or give more detailed descriptions of the environment through indicators of self-confidence, opportunities to communicate freely, and involve themselves actively and creatively; 2) Environmental concern in this research is the emotional involvement of students towards environmental protection through indicators of concern about the deteriorating quality of the environment, attention to the environment, emotional involvement in environmental protection issues, and playing an active role in improving the quality of the environment; 3) A sustainable campus in this research is students' awareness of the importance of the campus contributing to maintaining the integrity of the current environment for future generations, starting in the campus environment through indicators of the layout and condition of campus facilities and infrastructure, energy utilization and anticipation of global warming. , integrated waste management, economical use of water, use and creation of environmentally friendly transportation facilities, and environmentally sound education.

The third stage is to determine the data collection method based on the problems and objectives to be achieved in this research. So it was determined that the method used was a survey method to get an overview of students' level of understanding and concern for the environment using correlational techniques. This research method is concerned with determining whether or not there is a relationship between variables and how strong the relationship is (correlation coefficient). The relationship between research variables can be seen in Figure 1 below:

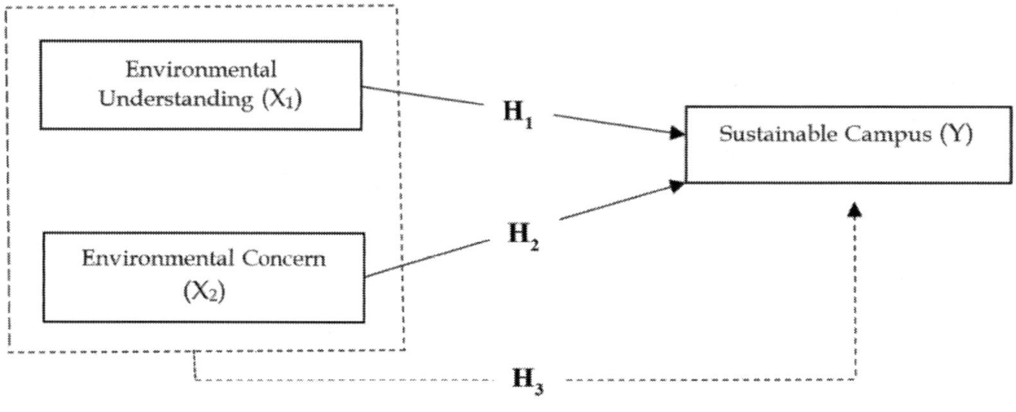

Fig. 1. Research Data Collection Methods

Information :
X1 = Environmental Understanding
X2 = Environmental Concern
Y = Sustainable Campus in the UI Green Metric Ranking

The fourth stage is to determine the validity and reliability values. Validity testing and reliability calculations were carried out on 30 students at Budi Luhur University, Jakarta. The results of testing the validity and calculating the reliability of the employee environmental knowledge variable instrument and the environmental understanding, environmental awareness and sustainable campus variable instruments in the UI Green Metric ranking are as follows:

1. Testing the validity and reliability of environmental understanding (X1), the environmental understanding variable instrument (X1) is composed of 15 items. The item validity testing technique is carried out by determining the correlation coefficient between the item score and the total score. This determination uses the product moment correlation formula. The validity test was carried out at $\alpha = 0.05$ and for r table with $n - 2 = 30 - 2 = 28$ was 0.361. If rcount > r table, then the item is declared valid. Vice versa, if rcount < r table then the item is declared dropped. Of the 15 instrument items that were tested and declared valid, there were 14 items. Meanwhile, 1 item is invalid/dropped. The reliability coefficient for the environmental understanding variable instrument (X1) is calculated using Cronbach's Alpha formula as follows: 1) The calculated reliability coefficient for 30 people and 15 items is 0.891. This means that the questionnaire given to respondents has a confidence level of (0.891)2 X 100% = 79.4%; 2) The results of calculating the reliability coefficient were 30 people and 14 valid items after removing 1 drop item, namely 0.898. This means that the questionnaire given to respondents has a confidence level of (0.898)2 X 100% = 80.6%.
2. Testing the validity and reliability of environmental concern (X2), the environmental concern variable instrument (X2) is composed of 16 items. The item validity testing technique is carried out by determining the correlation coefficient between the item

score and the total score. This determination uses the product moment correlation formula. The validity test was carried out at $\alpha = 0.05$ and for r table with $n - 2 = 30 - 2 = 28$ was 0.361. If $r_{count} > r_{table}$, then the item is declared valid. Vice versa, if $r_{count} < r_{table}$ then the item is declared dropped. Of the 16 instrument items that were tested and declared valid, there were 13 items. Meanwhile, 3 items are invalid/dropped. The reliability coefficient for the environmental concern variable instrument (X2) is calculated using the Cronbach's Alpha formula as follows: 1) The calculated reliability coefficient for 30 people and 16 items is 0.867. This means that the questionnaire given to respondents has a confidence level of (0.867)2 X 100% = 75.2%; 2) The results of calculating the reliability coefficient were 30 people and 13 valid items after removing 3 drop items, namely 0.883. This means that the questionnaire given to respondents has a confidence level of (0.883)2 X 100% = 78%.

3. Testing the validity and reliability of sustainable campus in the UI Green Metric (Y) ranking, the sustainable campus variable instrument in the UI Green Metric (Y) ranking is composed of 18 items. The item validity testing technique is carried out by determining the correlation coefficient between the item score and the total score. This determination uses the product moment correlation formula. The validity test was carried out at $\alpha = 0.05$ and for r table with $n - 2 = 30 - 2 = 28$ was 0.361. If $r_{count} > r_{table}$, then the item is declared valid. Vice versa, if $r_{count} < r_{table}$ then the item is declared dropped. Of the 18 instrument items that were tested and declared valid, there were 16 items. Meanwhile, 2 items are invalid/dropped. The reliability coefficient for the sustainable campus variable instrument in the UI Green Metric (Y) ranking is calculated using the Alpha Cronbach formula as follows: 1) The calculated reliability coefficient for 30 people and 18 items is 0.909. This means that the questionnaire given to respondents has a confidence level of (0.909)2 X 100% = 82.62%; 2) The results of calculating the reliability coefficient for 30 people and 16 valid items after removing 2 drop items, namely 0.919. This means that the questionnaire given to respondents has a confidence level of (0.919)2 X 100% = 84.46%.

The fifth stage is an analysis using data analysis techniques with classic assumption tests consisting of normality, linearity, heteroscedasticity and multicollinearity tests. Data analysis with descriptive statistics and inferential statistics. Hypothesis testing using t-test (partial) and F test / ANOVA (simultaneous). The analysis program for processing data in research used SPSS software version 26.00.

3 Result and discussion

3.1. Description of data and processing

Based on data obtained from the field, descriptive statistical data processing was then carried out, the results of this research are presented in Table 1 as follows:

Table 1. Statistical Description of Research Data

		Environmental Understanding	Environmental Concern	Sustainable Campus
N	Valid	63	63	63

	Environmental Understanding	Environmental Concern	Sustainable Campus
Missing	0	0	0
Mean	49.59	46.71	58.97
Median	50.00	48.00	62.00
Mode	50	52	65
Std. Deviation	8.368	7.821	10.565
Variance	70.020	61.175	111.612
Range	33	33	41
Minimum	31	26	37
Maximum	64	59	78
Sum	3124	2943	3715

The results of the validity testing of research variables are as follows:

1). Sustainable Campus in the UI Green Metric (Y) Ranking

From the data that has been collected regarding sustainable campuses in the UI Green Metric (Y) ranking, 16 valid items were obtained so that the value range is 16 – 80. Mean = 58.97; median = 62.00; mode = 65; standard deviation = 10.565; variance = 111.612; range = 41; minimum = 37; maximum = 78; sum = 3715.

2) Environmental Understanding (X1)

Based on the environmental understanding data (X1) collected, 14 valid items were obtained, so the value range was between 14 – 70. Mean = 49.59; median = 50.00; mode = 50; standard deviation = 8.368; variance = 70.020; range = 33; minimum = 31; maximum = 64; sum = 3124.

3). Environmental Concern (X2)

Based on the environmental awareness data (X2) collected, 13 valid items were obtained, so the value range was between 13 – 65. Mean = 46.71; median = 48.00; mode = 52; standard deviation = 7.821; variance = 61.175; range = 33; minimum = 26; maximum = 59; sum = 2943.

3.2. Classic assumption test results

3.2.1 Normality test

Normality testing uses the Kolmogorov-Smirnov test from two groups of data. The normality test results are presented in table 2 as follows:

Table 2. Data Normality Test Results

Group	N	α	*Kolmogorov-Smirnov (Sig.)*	Conclusion
Y atas X_1	63	0,05	0,200	Normal
Y atas X_2	63	0,05	0,200	Normal

Based on the table above, the Kolmogorov-Smirnov (Sig.) value of 0.200 is greater than α = 0.05; then all variables indicate that the sample distribution follows a normal distribution. In conclusion, the distribution of sustainable campus scores in the UI Green

Metric ranking of environmental understanding and environmental awareness scores is normally distributed.

3.2.2 Linearity test

The results of the linearity test for the sustainable campus variable in the UI Green Metric ranking for scores on environmental understanding and environmental concern using the test of linearity. The linearity test results are presented in table 3 as follows:

Table 3. Test Linearity With A Test Of Linearity

Group	N	α	*Test of linearity (Sig.)*	Conclusion
Y atas X_1	63	0,05	0,000	Linear
Y atas X_2	63	0,05	0,000	Linear

Based on the table above, the test of linearity (Sig.) value of 0.000 is smaller than $\alpha = 0.05$; then all variables indicate that the sample distribution is linear. In conclusion, the distribution of sustainable campus scores in the UI Green Metric ranking on environmental understanding and environmental awareness scores is linear.

3.2.3 Heteroscedasticity test

The results of the heteroscedasticity test for the sustainable campus variable in the UI Green Metric ranking for scores on environmental understanding and environmental concern using a scatterplot of the heteroscedasticity test. The linearity test results are presented in figure 2 as follows:

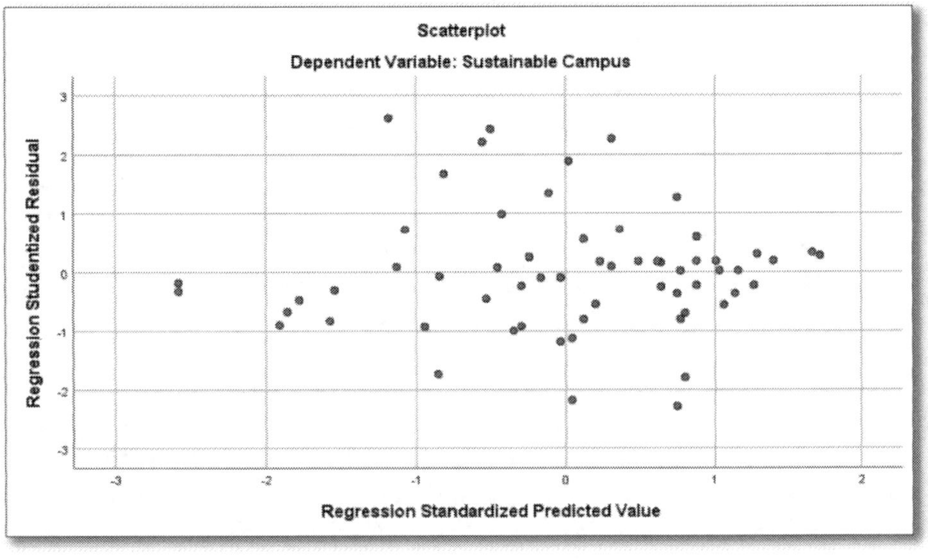

Fig. 2. Scatterplot of heteroscedasticity test

Based on the picture above, there is no particular pattern because the points are spread irregularly above and below the 0 axis on the Y axis. So it can be concluded that there are no symptoms of heteroscedasticity.

3.2.4 Multicollinearity test

Multicollinearity test results for the sustainable campus variable in the UI Green Metric ranking on environmental understanding and environmental awareness scores using tolerance and VIF. The results of the multicollinearity test are presented in Table 4 as follows:

Table 4. Multicollinearity Test Results

Group	N	*Tolerance*	VIF	Conclusion
Y atas X_1	63	0,326	3,064	Multicollinearity does not occur
Y atas X_2	63	0,326	3,064	Multicollinearity does not occur

From the table above, it can be seen that the Tolerance value is close to one for all variables and the VIF value is around one, thus the sustainable campus variable in the UI Green Metric ranking for environmental understanding and environmental concern scores shows that there are no cases of multicollinearity in the model.

3.3 Hypothesis testing and discussion of research results

Hypothesis testing in this research was carried out to draw conclusions that were supported by empirical data. This hypothesis testing was carried out using correlation and multiple or simultaneous regression analysis. The influence of environmental understanding and environmental concern together on sustainable campuses in the UI Green Metric ranking was analyzed using multiple or simultaneous regression. The summary of the results of multiple or simultaneous regression analysis is as follows.

Table 5. Joint Environmental Understanding and Environmental Concern Test Results for Sustainable Campuses in the UI GreenMetric Ranking

Sample (n)	*Constant* (B)	Environmental understanding (X_1)	Environmental awareness (X_2)	Correlation coefficient (*R*)	*R squar e*	F_{Count}	F_{table} (0,05)
63	11,420	2,070	2,767	0,722	0,521	32,69 5	3,15

The results of multiple or simultaneous regression analysis show the equation $\hat{Y}$ = 11.420 + 2.070X1 + 2.767X2. In the regression equation $\hat{Y}$ = 11.420 + 2.070 11.420. The correlation coefficient for environmental understanding and environmental concern together with a sustainable campus in the UI Green Metric ranking is 0.722.

The F test results for F_{count} are greater than F_{table}, namely 32.695 > 3.15 at $\alpha = 0.05$. This means that the correlation coefficient between environmental understanding and environmental concern together with a sustainable campus in the UI Green Metric ranking is significant. The conclusion is that there is a significant positive influence between environmental understanding and environmental concern together with a sustainable campus in the UI Green Metric ranking. This means, that the more environmental understanding and concern students increase, the more sustainable campuses will increase in the UI Green Metric ranking. The calculation of multiple or simultaneous regression test results can be summarized in Table 6 as follows:

Table 6. Multiple or Simultaneous Linear Regression Test Results

Variable	Beta (β)	T_{count}	Significance (*Sig.*)
X_1 to Y	0,324	2,070	0,043
X_2 to Y	0,433	2,767	0,008

Based on table 6 above, the results of multiple or simultaneous regression tests show that the highest influence on sustainable campuses in the UI Green Metric ranking, the first rank is an environmental concern with a beta value (β) = 0.433; tcount = 2.767; and sig. = 0.008. The second rank is environmental understanding with a beta value (β) = 0.324; t_{count} = 2.070; and sig. = 0.043. The coefficient of determination of the influence of environmental understanding and environmental concern together with a sustainable campus in the UI Green Metric ranking is obtained from the R square value of 0.521. So 52.1% of the variance in students' environmental understanding and concern can be explained by the sustainable campus variable in the UI Green Metric ranking.

The results of testing the three hypotheses proposed in this research concluded that all research hypotheses proposed in chapter II were accepted. This means that a sustainable campus in the UI Green Metric ranking of 52.1% can be clarified by students' environmental understanding and concern. Based on the description above, the independent variable that has the strongest influence and the greatest contribution to the dependent variable is students' environmental awareness. The research results obtained from testing the first, second and third statistical hypotheses do not contradict this research hypothesis. By testing each of the research hypotheses, the theoretical framework developed based on the theoretical description as outlined in Chapter II can be supported by empirical evidence through this research data. The conclusions that can be drawn from each hypothesis test are in accordance with the concepts used to explain the influence of the independent variable on the dependent variable. The conclusions drawn in this research mean that it can become a treasure that enriches scientific knowledge, especially about sustainable campuses in the UI Green Metric ranking. This is in accordance with the results of research conducted by Dewi, Kusumoarto, & Rejoni [5] said that a sustainable campus emerged from academics' concern and awareness of the importance of campuses contributing to maintaining the integrity of the current environment for future generations, starting in the campus environment. Marsudi et al. [24] said that UI Green Metric is one of the indicators used by universities in Indonesia and internationally as part of a strategy to increase sustainability.

Research by Muslih, Priyanto, & Muslim [11] shows that a sustainable campus is closely related to a university's high commitment to creating environmentally friendly campus infrastructure. Apart from that, the concept of integration in research, education and community service in environmental management. Thus, the role of campuses in environmental conservation efforts is very substantial. Research by Atici et al. [6] states that a sustainable campus is defined as a concept that prioritizes the practice of sustainable environmental protection, management and preservation efforts in educational institutions. Gunawan, Prayogo, & Mardiono [4] stated that sustainable campus development is a strategic step because the campus environment has a very large mass, so the frequency of use of clean water and energy in the campus environment is also large. Apart from that, waste and waste management that is not environmentally friendly will have an impact on environmental pollution. Pelcher et al., [7] said that the criteria for creating a sustainable campus include the layout and condition of campus facilities and infrastructure, energy utilization and anticipation of global warming, integrated waste management, economical use of water, use and creation of environmentally friendly transportation facilities, and environmentally friendly education. In Indonesia itself, this concept has begun to be widely implemented on various campuses. Phramesti and Yuliastuti [8] stated that a sustainable campus is a concept of campus management that considers ecological sustainability consisting of material, air, water, land and energy components, by creating an environmentally sound lifestyle through developing knowledge, community, health and institutions for society. which is part of the campus environment.

4. Conclusion

The findings show that there are partial and simultaneous influences on campus sustainability. A sustainable campus in the UI Green Metric ranking can be improved by paying attention to factors that can support environmental sustainability, namely by increasing environmental understanding and awareness in the academic community. There are several efforts to improve these factors as stated in the Implications as follows: 1) Policy implications; through efforts to increase students' environmental understanding and awareness in creating curiosity, activity, creativity, and ability to change attitudes, intentions and behavior related to sustainable campus efforts in the UI Green Metric ranking to become a campus policy; 2) Theoretical implications; can be done with or through anything that is lacking or not yet included in sustainable campus measurements in the UI Green Metric ranking or elements such as considerations for increasing students' environmental understanding and concern; and 3) Research implications; can be done by involving more variables using more complex methods and analysis such as experiments, ex post facto, path analysis, factor analysis and SEM (Structural Equation Modeling).

Based on the findings, conclusions and implications, the suggestions in this research are as follows: 1) For the Budi Luhur University campus, it is necessary to consider increasing students' environmental understanding and awareness as factors that influence a sustainable campus in the UI Green Metric ranking in every decision and planning campus activities; 2) For students in an effort to improve the sustainability of the campus in the UI Green Metric ranking, environmental action and training activities are needed that can increase knowledge, understanding and concern about the environment in everyday life; 3) For companies and private parties, it is hoped that there will be cooperation between companies and universities in which there are many experts and students so that they can bridge the relationship between lecturers and students and companies to carry out research, including in matters of funding and research permits for companies; and 4) For further research, you can examine sustainable campuses in the UI Green Metric ranking which is not only connected to students' environmental understanding and concern but there are many other variables that influence it, such as environmental action, ability, trust, justice and ethics.

Acknowledgments

Thank you to the Director of Research and Community Service for providing the opportunity and funding for this research and all those who have assisted in the completion of this research.

References

1. F. Wardhono, Sustainable city development (UCLG, Jakarta, 2012)
2. United Cities and Local Goverments, Sustainable Development Goals (UCLG, Jakarta, 2018)
3. Badan Pusat Statistik, " Indicators of Indonesia's 2018 Sustainable Development Goals (TPB) (BPS, Jakarta, 2018)
4. T. E. Gunawan, D. N. Prayogo, L. Mardiono, *Eco-sustainable Campus Initiatives: A Web Content Analysis*, in Proceedings of the 3rd International Conference on Technology and Operation Management (2012)
5. N. R. Dewi, A. Kusumoarto, R. Rejoni, Lakar J. Arsit. **6**, 1 (2023)
6. K. B. Atici, G. Yasayacak, Y. Yildiz, A. Ulucan, J. Clean. Prod. **125289**, 1 (2021)
7. J. Pelcher, S. McCullough, B. P. Trendafilova, Int. J. Sustain. High. Educ. **22**, 1 (2021)
8. R. Phramesti, N. Yuliastuti, J. Tek. PWK **2**, 1 (2013)
9. A. A. Utami, N. W. P. Susatyo, Oper. Excell. J. **10**, 1 (2018)
10. D. Moh. Muslih, Green Campus Series, in Green Campus Series, 1st ed. (Scientist Publishing, Pekalongan, 2022)
11. UI GreenMetric Team World University Rankings, *Guideline of UI GreenMetric World University Ranking,* in Guideline of UI GreenMetric World University Ranking, Depok Jawa Barat: UI GreenMetric (2017)
12. R. P. Buana, M. Wimala, R. Evelina, J. Online Inst. Teknol. Nas. **2**, 1 (2018)
13. UI GreenMetric Ranking Team, *UI GreenMetric Guideline,* in UI GreenMetric Guideline, Depok, West Java (2019)
14. UI GreenMetric Team World University Rankings, UI GreenMetric World University Rankings: Background of The Ranking, Depok Jawa Barat (2014)
15. I. P. Riyadi, B. A. Prayitno, P. Karyanto, *Standardization of Ecological Content for Sebelas Maret University Green Campus Students to Increase Ecological Literacy*, in *Proceeding Biology Education Conference* (2017)
16. A. Supriyantono, W. Usino, A. Sholicin, Preparing a Strategic Plan: Efforts Towards Budi Luhur Emas University 2029," in Sharing for the Country: Detachment's Work and Thoughts for Improving the Quality of Indonesian Higher Education (Deepblish Publisher, Jakarta, 2022)

17. A. Amran, I. Jasin, M. Perkasa, et al., Journal of Physics: Conference Series **1521**, 4 (2020)
18. A. Sudijono, Introduction to Educational Evaluation (Rajawali Pers, Jakarta, 2017)
19. S. Arikunto, Research Procedures A Practical Approach (Rineka Cipta, Jakarta, 2011)
20. Presiden Republik Indonesia, Law of the Republic of Indonesia Number 32 of 2009 concerning Environmental Protection and Management, Jakarta (2009)
21. Z. Iskandar, Environmental psychology: methods and applications (Refika Aditama, Bandung, 2013)
22. M. P. Pooraskari, Hossein, M., Mohseni and E. Shahriary, J. Curr. Res. Sci. **3**, 2 (2015)
23. G. K. Dagher, O. Itani, and A. N. Kassar, Contemp. Manag. Res. **11**, 2 (2015)
24. Marsudi, et al., J. Agribisnis Perikan **14**, 1 (2021)

Trend and research of Lego and Minecraft as learning media to realize 4th SDGs

Khoirun Nisa'[1], *Nadi* Suprapto[1*], *Hasan Nuurul* Hidaayatullaah[1], and *Husni* Mubarok[2]

[1]Department of Physics, Universitas Negeri Surabaya, Surabaya, Indonesia
[2]Graduate Institute of Digital Learning and Education, National Taiwan University of Science and Technology, Taipei, Taiwan

Abstract. The SDGs emphasize quality education. Lego and Minecraft can stimulate students' 21st-century skills. This research aims to identify trends and contributions of Lego and Minecraft in education. The study utilized bibliometric analysis to evaluate publications. Data was collected from Scopus and analyzed using VosViewer software. Over the past decade, research on Lego and Minecraft in education has been more stable, with a peak in 2017 due to its popularity. Minecraft research increased from 2013 to 2019 but declined in 2020. Lego and Minecraft education research was published in conference papers with Springer as publisher. The United States and Germany are the most productive countries with top ten sources, with 90% focusing on computer science. The top eleven subjects significantly impact education by developing systems, online courses, and digital media. Lego and Minecraft research uses quan-qual-mix studies to explore their application in education. They strongly relate to STEM education, digital learning, 21st-century skills, and game-based learning. Combining Lego and Minecraft with technologies can increase computational thinking, creative-critical thinking, collaboration, and engineering skills. Future studies must employ one or more keywords, and work with Google Scholar and Web of Science data for in-depth analysis.

1 Introduction

The United Nations set 17 SDGs and 169 targets that emphasize the importance of quality education in achieving these goals [1]. There needs to be more quality education, particularly in numeracy, literacy, mathematical science, and technology. So, the government must increase the quality of curricula, teacher training, and learning environments [2]. Implementing digital learning media in education can improve quality education and create a promising framework for sustainable development education. Hsiao [3] states that implementing digital learning media—robot education—in schools can reach the fourth goal of the SDGs. According to the fourth SDG, Quality Education, the right to quality education directly impacts the world's human resources and future

* Corresponding author: nadisuprapto@unesa.ac.id

human prosperity. In contrast, persistent inequality and inadequate education can keep children trapped in poverty and immobility for generations without a comparable and quality education [4].

Lego has developed educational tools for many instructional activities that stimulate play and creativity in students [5]. It helps students develop their cognitive-creative skills and improve their language and literacy skills [6]. Lego is used by governments worldwide, including Peru, Singapore, and the United States [7]. In addition, another type of digital learning media is game education, one of the most popular video games [8]. Minecraft is advantageous for gaming, simulation, and education because it fosters creativity and responsiveness [9]. Over the past ten years, Minecraft has progressively become a part of game education in humanities and sciences research [10-12].

This research elaborated on the trends and research of Lego and Minecraft in education to identify and compare its research, which can help researchers in future studies.

- Regarding research characteristics and features, what was the location, type of documents, sources, and authors of Lego and Minecraft education research?
- In terms of application, what were the dominant disciplines, and how are the contributions and differences between Lego and Minecraft education research?

2 Methods

Bibliometric is a statistical technique used to assess and quantify publications in a specific research field, reducing objectivity and potential bias [13, 14]. It helps researchers understand structure, patterns, and trends in research activities in various disciplines [15, 16].

2.1 Resources

The data were collected from the Scopus database with broader coverage in social science than WoS [17].

2.2 Data collection

The data were collected on August 28, 2023. The three steps of bibliometric analysis include keyword selection, refining, export data, and analysis data. In this research, authors selected two keywords, "Lego" and "Minecraft". In refining steps, authors limited the data by scope and publication year. The timespan covered the range from January 2013 to December 2022. As a result, 597 documents relevant to Lego education research and 136 documents to Minecraft education research. Each data is downloaded in .ris and .csv format.

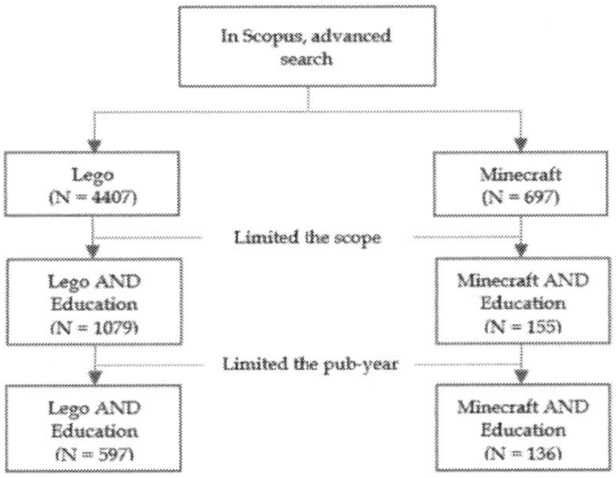

Fig. 1. Research map

2.3 Data analysis

The data was analyzed through VosViewer software to get the relationship among authors, documents, and keywords [18, 19]. The year of publication, country, affiliation, language, and other information were analyzed from the .csv file using Microsoft Excel. Microsoft Excel results were used to solve RQ 1 and VosViewer to solve RQ2 [20].

3 Results and discussion

3.1 Research characteristics and features

3.1.1 Year-wise distribution

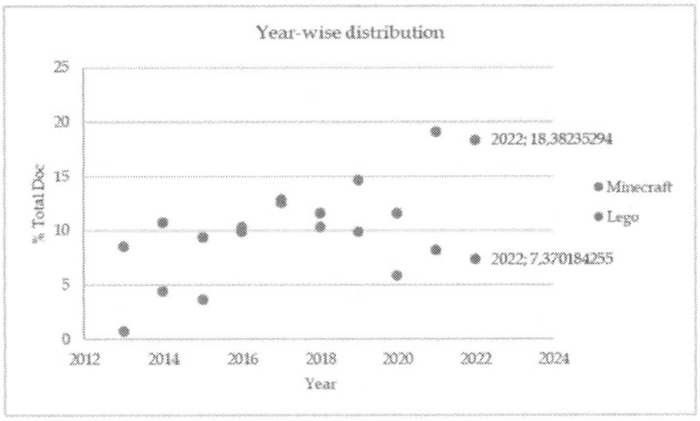

Fig. 2. The year-wise distribution of Lego-Minecraft education research

The Figure 2 shows research trends in Lego and Minecraft education over the last decade. There is more research on Lego than on Minecraft. The Lego research tends to be stable from 2013 to 2016, with the highest peak in 2017 (77 documents). Lego has been very popular since 1932 and was the most influential brand in the world in 2017 [21]. From 2018 to 2020, the publication percentage is the same as from 2013 to 2016.

In addition, Minecraft research increased from 2013 to 2019, declined in 2020, and increased from 2021 to 2022. The productive year was 2021, with 26 documents. The Lego research trend has declined in the last two years because of the Covid-19 pandemic. There was a change in the learning system that made students unable to meet to build Lego together. However, the pandemic helped Minecraft become successful and popular. Microsoft made the Minecraft education edition as an effort to assist students who are not able to return to school.

3.1.2 The most productive affiliation and region

Table 1 shows four public universities in Spain that participated in Lego Education. On the other hand, Tufts University gives the United States the first place of publications in Lego Education. Tufts University has Tufts CEEO with many exciting services and programs for educators. In addition, Aalborg University also participated in Lego and Minecraft Education and gave Denmark the third productive region. Denmark was a Lego block-maker in 1932. Minecraft was made in Sweden with two public universities producing five documents.

Table 1. Top five productive affiliation

Affiliation	Region	Total Doc	
		Lego	Minecraft
Tufts University	United States	12	-
Aalborg University	Denmark	6	8
Chalmers University of Technology	Sweden	6	-
Univerzita Komenskeho v Bratlslave	Slovakia		
Universidad del Pals Vasco	Spain		
ITMO University	Rusia	5	-
Arizona State University	United States		
Goteborgs Universitet	Sweden		
Universidad de Salamanca	Spain		
Universidad Politecnica de Madrid	Spain		
Universidad Nacional de Edcacion a Distancia	Spain		
Central Connecticut State University	United States		
Universitetet I Oslo	Norway	-	4
Waseda University	Japan		

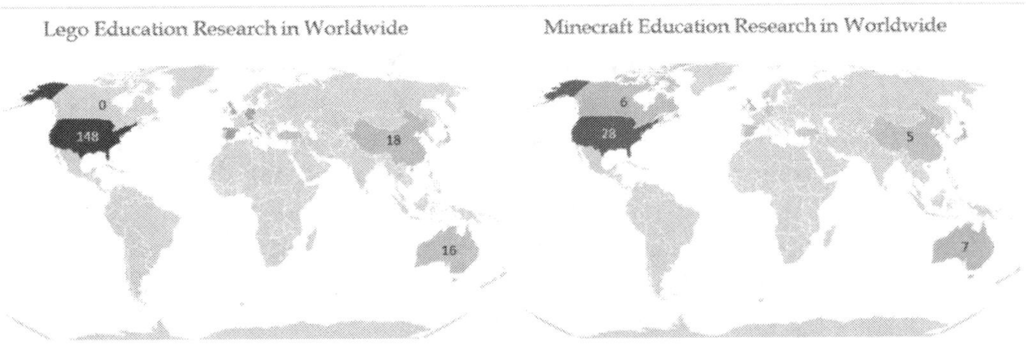

Fig. 3. The most productive region in Lego education research (left) and Minecraft education research (right)

Based on Figure 3, eight regions participated in Lego and Minecraft education research. The United States has a dark color (green and blue), which means the United States is the most productive region and has published 148 documents in Lego Education research and 28 research documents in Minecraft education. Italy and Greece have the last rank in Minecraft education. Canada and Norway in Lego Education.

3.1.3 Type of Research Documents

Half the document types in Lego and Minecraft education research are conference papers. Many international conferences have been held over the last decades (Table 2). The article takes second place in the research document on Minecraft education (200 documents; 33.5%) and Lego education (39 documents; 28.6%). The article was published in 39 journals not included in the top ten sources. The Education and Information Technologies journal was the most published article. Four types get less than 3%, such as review, note, book, and editorial. There is Lego education research published in an editorial. Minecraft has three types that have 0,7%, such as notes (SA Pharmaceutical Journal), books (The Music Technology Cookbook: Ready-made recipes for the classroom), and editorials (Interactive Learning Environments).

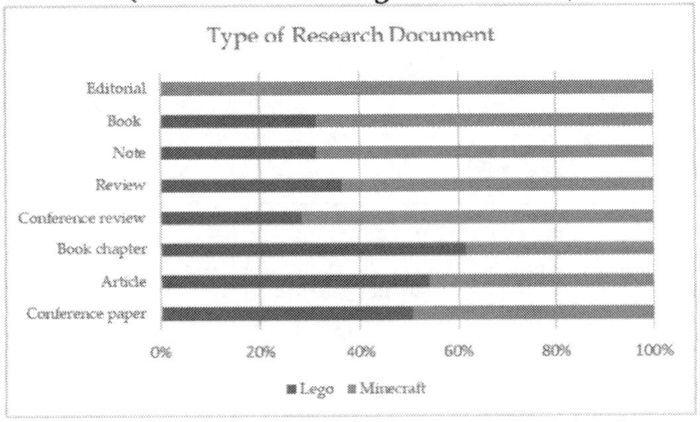

Fig. 4. Document types

According to Table 2, the top ten sources were published in Springer (3 sources), the association for Computing Machinery (2 sources), and another publisher. Springer has three sources with the highest H-index: Lecture notes in computer science (446), Communications in computer and information science (62), and Advances in intelligent systems and computing (58). The second highest H-index is the ACM international conference proceeding series published by the association for Computing Machinery. Advances in intelligent systems and computing and ASEE annual conference and exposition conference proceedings is the most productive source that published 32 Lego education research documents and Proceedings of the European Conference on game-based learning in Minecraft education research. According to Figure 3, the United States and Germany are the most productive countries with the top ten sources. 90% of sources are concrete in computer science scope.

Table 2. The top ten productive source

Source	Publisher	Subject Area	Country	H-Index	Total Doc	
					Lego	**Minecraft**
Advances in intelligent systems and computing (book)	Springer	Computer Science and Engineering	Germany	58	32	-
ASEE annual conference and exposition conference proceedings	American Society for Engineering Education	Computer Science and Engineering	United States	26	32	3
ACM international conference proceeding series	Association for Computing Machinery	Computer Science	United States	137	24	3
Proceedings of the European conference on game-based learning	Dechema e. V.	Computer Science, Engineering, and Social Sciences	Germany	16	-	16
Lecture notes in computer science	Springer	Computer Science and Mathematics	Germany	446	22	4
Ceur workshop proceedings	CEUR-WS	Computer Science	United States	62	8	7
IEEE global engineering education conference educon	IEE Computer Society	Decision Sciences, Engineering, and Social Sciences	United States	30	20	3
Proceedings-frontiers in education conference	Institute of Electrical and Electronics Engineers Inc.	Computer Science and Social Sciences	United States	45	9	3

Source	Publisher	Subject Area	Country	H-Index	Total Doc	
					Lego	**Minecraft**
Communications in computer and information science	Springer	Computer Science and Mathematics	Germany	62	9	-
Annual conference on innovation and technology in computer science education	Association for Computer Machinery	MBA and Social Sciences	United States	32	8	-

3.1.4 The top productive authors

Table 3 describes the top ten authors, the number of papers, the author's affiliation, and region. 80% of Lego-education researchers have the same total documents. Salto from Waseda University, Japan, and Hanghoi from Aalborg University, Denmark, are the researchers who have published many papers about Minecraft. Related to Figure 3, US researchers were dominant in Lego research. Although, Japan's researchers were dominant in Minecraft education.

Table 3. Top ten productive author

Lego Education				**Minecraft Education**			
Name	**Affiliation**	**Region**	**Total**	**Name**	**Affiliation**	**Region**	**Total**
Veselovska, M.	Comenius University	Slovakia	5	Salto, D.	Waseda University	Japan	5
Tobarra, L.	Spanish University for Distance Education	Spain	5	Hanghoj, T.	Aalborg University	Denmark	5
Ros, S.	Universidad Nacional de Educación a Distancia	Spain	5	Washlzaki, H.	Waseda University	Japan	4
Rogers, C.	Tufts University	United States	5	Fukazawa, Y.	Waseda University	Japan	4
Ludi, S.	University of North Texas	United States	5	Yi, S.	University of Illinois	United States	3
Kurkovsky, S.	Central Connecticut State University	United States	5	Morch, A. I.	University of Oslo	Norway	3
Castro, M.	Spanish University for Distance Education	Spain	5	Lane, H. C.	University of Illinois	United States	3
Albo-Canals, J.	Lighthouse-DIG	United States	5	Yamazaki, T.	Astrax Lab	Japan	2

Lego Education				Minecraft Education			
Name	**Affiliation**	**Region**	**Total**	**Name**	**Affiliation**	**Region**	**Total**
Pastor, R.	Spanish University for Distance Education	Spain	4	Winter, V.	University of Nebraska	United States	2
Mayerova, K.	Comenius University	Slovakia	4	White, J.	Kyoto University	Japan	2

3.2 Application of Lego and Minecraft in education research

3.2.1 Subject area

Eleven top subjects in Lego-Minecraft research and six were included in the top source's scope. It is related to Table 2; computer science, social science, engineering, mathematics, MBA, and decision science were dominant in this research. Computer science was dominated by Lego and Minecraft education. Computer science significantly impacts education by developing systems, online courses, and digital media. Research has often influenced disciplines like artificial intelligence, science, STEAM, and engineering. Computer science is considered closer to mathematics than many scientific disciplines. Computational social science uses computer science theories to contribute to social sciences, while social computing explores the intersection of social behavior and computational systems.

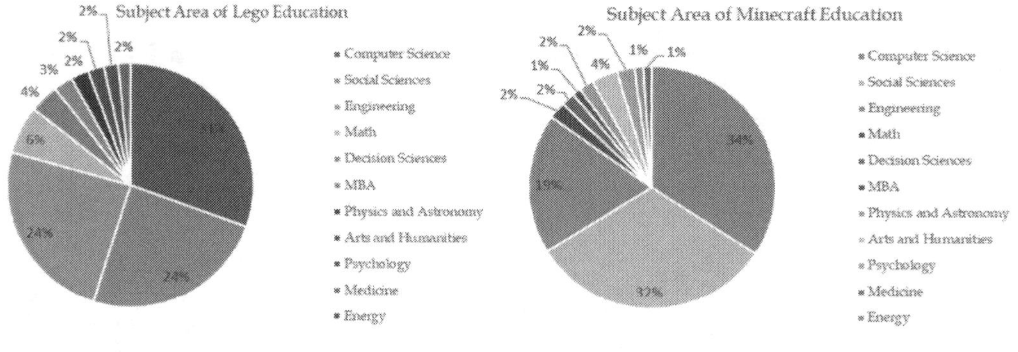

Fig. 5. Top subject area

3.2.2 Application Lego and Minecraft

The keywords in Lego and Minecraft research have similar forms, research methods, participants, and scope. According to Figures 6 and 7, Lego and Minecraft research used questionnaires, and interview studies to explore how Lego and Minecraft applied in education research: influence, participation, perspective, and awareness of students during learning [22-23]. Lego and Minecraft strongly relate to STEM education, digital learning, 21st-century skills, and game-based learning. Lego is related to robot education, and Minecraft is related to virtual game/world. Lego and Minecraft usually combine with

STEM education to increase computational thinking, creative-critical thinking, collaboration, and engineering skills. The realization of 4th SDGs also closely links to STEM education. It also strives to create and offer responses to global problems. Technology like Minecraft or Lego integrating robot education, AI, and VR is taking hold and is influencing all facets of our daily lives. Technology prompts a shift in STEM education, focusing on sustainable teaching that incorporates environmental understanding, morals, ethics, practical life skills, digital components, problem-solving, critical thinking, creativity, teamwork, and communication.

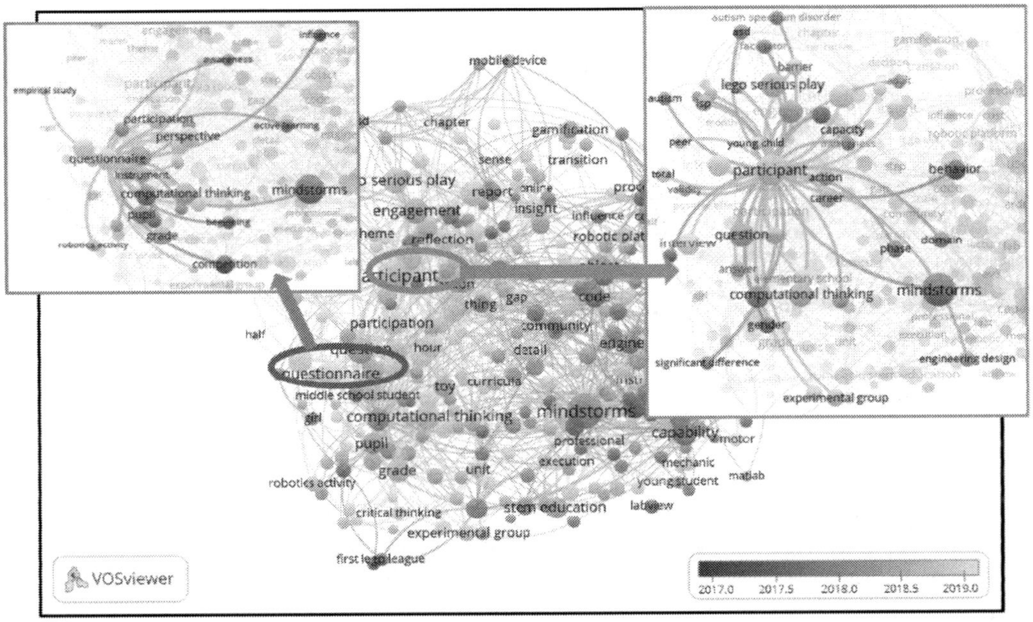

Fig. 6. Visualisation of Lego Education

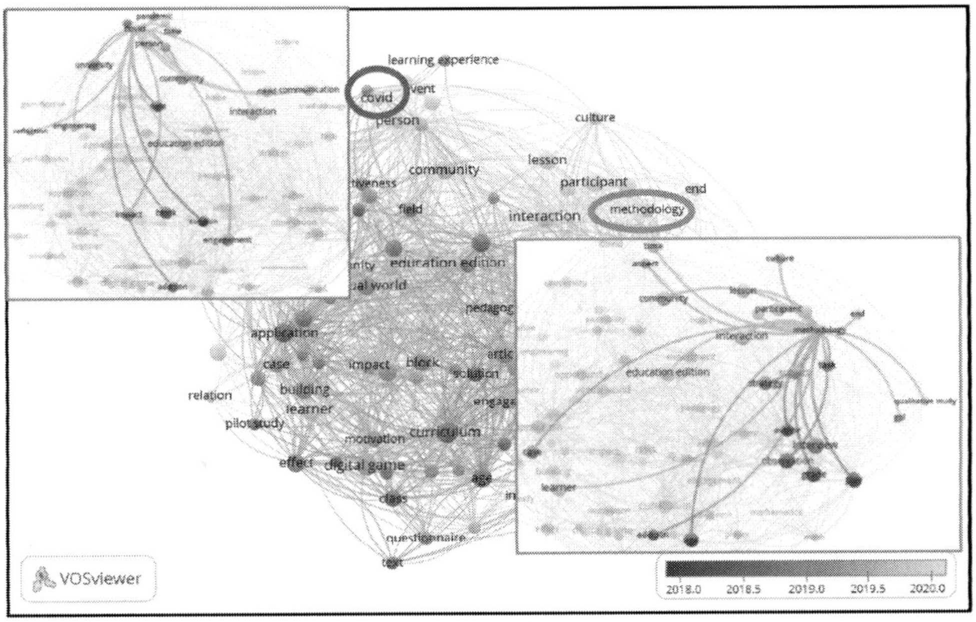

Fig. 7. Visualisation of Minecraft Education

Lego and Minecraft usually have other devices to develop quality learning activities like AR, VR, and system software. Lego and Minecraft have the difference in their year publication. In Figure 6, authors do not find the Covid or pandemic keyword, but this keyword was found in Figure 7 and linked to "solution". It is related to Figure 2; there are significant publications on Minecraft education during the pandemic. Microsoft made the education edition as an effort to assist the students who are not able to return to school. In Minecraft Education, there are SDGs challenges in several themes (Government and Politics; Climate and Environment; Service Learning) so students can discover SDGs while creating a virtual world. The literature review about Lego and Minecraft education is in Table 4.

Table 4. The literature review about Lego-Minecraft Education

Category	Minecraft	Lego
Technology/ device	- GIS [24] - Python and JavaScript [26] - 3D printers [28] - Kodu [30] - AI and Metaverse [32]	- AR [25] - Arduino and Microcontrollers [27] - Robot [29] - VR [31] - 3D Printed [33] - Matlab [34]
Skills	Collaborative, problem-solving, and creative-critical thinking [30]	Problem-solving, creative-critical thinking, collaboration, communication [35, 36]
How to Use	Using block by block to build digital world	Using groups of bricks in a variety of colors and sizes to build prototype.
Education Major	STEM	STEAM and medicine
Learning activities	- Collaborative learning [37] - Group-based learning [39] - Game-based learning [41] - PjBL [43] - PBL - Simulation [46]	- Game-based learning [38] - Robot education [40] - PjBL - PBL [42] - Demonstration [44] - Workshop [45] - Blended learning [47]
Product	- Minecraft extension [48] - Minecraft Earth and MineCraft-Urban Planning [24] - Minecraft Education [51]	- More To Math kit - STEAM Park and My XL World [49] - Coding Express - StoryTales [50] - BricQ Motion and Spike Prime Set [6] - Lego Exoskeleton [52]
Participant	- Primary students [53] - Secondary students [55] - Pre-service teacher [37] - College students [59] - Teacher [61]	- Pre-K/Kindergarten [54] - Primary [56] - Secondary [57] - College [58] - Pre-Service teacher [60]

Category	Minecraft	Lego
		- Teacher [62]
Methods	- Qualitative (Observation) - Quantitative (Questionnaire, Interview) - Descriptive (Case study) - Explorative - Experiment (Program Development)	- Experimental design (System development, Quasi-experiment) - Mix methods (Participatory research) - Quantitative (Questionnaire/Survey, meta-analysis) - Descriptive (Phenomenology, Case study, systematic review - Qualitative (Action research, Interview, Literature studies)

Many types of Lego (Table 5) are used in robotic education. Students learn to build the robots, use code, and use sensors to move them [63]. Lego and Minecraft are related to Bruner's theory. Bruner's theory of learning uses the discovery method to construct knowledge and link new facts with previous information, which helps learners acquire knowledge from smaller components, building a more substantial base [64].

4 Conclusion

This research explores the role of Lego and Minecraft in education, focusing on their trends, applications, and contributions. The study found that Lego and Minecraft had a stable trend in education over the past decade, with a peak in 2017 due to their popularity. Minecraft research increased from 2013 to 2019 but declined in 2020. The research primarily consists of conference papers, with the top ten sources being Springer and the association for Computing Machinery. Lego and Minecraft research focuses on eleven top subjects, such as computer science, social science, engineering, mathematics, MBA, and decision science, and their impact on education through systems, online courses, and digital media. They strongly relate to STEM education, digital learning, 21st-century skills, digital technologies, and game-based learning. This research only used metadata from Scopus, so there are several articles not indexed. Future research can combine metadata from Web of Science, Science Direct, and Scopus. This study offers insights for further researchers and educators to advance virtual-game learning, promoting cooperation between universities and other affiliations for increased publications and reference information.

Acknowledgements

The acknowledgements, thanks to Universitas Negeri Surabaya and the National Taiwan University of Science and Technology to supporting this research.

References

1. T. P. L. Nguyen, T. H. Nguyen, T. K. Tran, Sustainability (Switzerland), **12**, 21 (2020)
2. E. Ota, R. Murakami-Suzuki, Sustainability, **14,** 2988 (2022)

3. J.-M. Hsiao. *Cross-domain Sharing of Robots in the Community Caring and Practice of University Social Responsibility*. Proceedings of International Conference on Artificial Life and Robotics, **28**, (2023)
4. X. Wang, G. W. Young, A. Plechatá, C. Mc Guckin, G. Makransky. Computers & Education, **201**, 104815 (2023)
5. M. Moore, D. O'Sullivan, International Journal of Mentoring and Coaching in Education, **12**, 3 (2023)
6. P.-N. Chou, R.-C. Shih, Lecture Notes in Computer Science, **13449** LNCS (2022)
7. I. Bello, S. Kazibwe, Problemy Polityki Spolecznej, **58**, 3 (2022).
8. Z. C. Zhang, H. B. Li, E. W. M. Lee, Y. Ma, W. K. Zhang, M. Shi, Physica A: Statistical Mechanics and Its Application, **624**, 15 (2023).
9. K. Wilbert, P. Lesley, K. Maaike, Physics Education, **58**, 2 (2023)
10. T. -H., Nguyen, T. -L., Nguyen, T., -B. Nguyen, *Design and evaluate a stem-oriented education teaching plan by exploiting the strength of minecraft education game-based platform*, AIP Conference Proceedings, **2685**, 030014 (2023)
11. I. Maria, C. Anne-Marie, Research in Science and Technological Education, **41** (2023)
12. K. Kalofolia, K. Siountri, *Inclusion of the Minecraft digital game in the theatre education course: theoritical approaches and an interactive experiment,* International Archives of the Photogrammetry, Remote Sensing and Spatial Information Sciences - ISPRS Archives, **48** (2023)
13. K. Fellnhofer. Educational Research Review, **27** (2019).
14. R. Khodabandelou, N. Ale Ebrahim, A. Amoozegar, G. Mehran, Iranian Journal of Comparative Education, **2**, 1 (2019)
15. L. Zhang, J. Ling, M. Lin, Energy Reports, **8**, (2022).
16. A. Gatto, C. Drago, M. Ruggeri, Environmental Science and Pollution Research, **30**, 15 (2023)
17. S. M. Jamali, N. Ale Ebrahim, F. Jamali, International Journal Technology and Design Education, **33**, (2023)
18. D. Guleria, G. Kaur, Library Hi Tech, **39**, 4 (2021).
19. O. T. Oladinrin, M. Arif, M. Q. Rana, L. Gyoh, Construction Innovation, **23**, 3 (2023)
20. B. K. Prahani, K. Nisa', B. Jatmiko, N. Suprapto, T. Amelia, E. Candrawati, International Journal of Online and Biomedical Engineering, **18**, 6 (2022)
21. B. Çavaş, L. Özge Güney, E. Karagöz, P. Çavaş, Science Education International, **31**, 1 (2020)
22. Y. N., Lee, M. Zhu, Computers in the Schools, **39**, 3 (2022).
23. M. Opmeer, E. Dias, B. De Vogel, L Tangerman, H. J. Scholten. *Minecraft in support of teaching sustainable spatial planning in secondary education lessons learned from the Marker Wadden-project*. In Proceedings of the 10th International Conference on Computer Supported Education (CSEDU 2018) (2018).
24. R. Adanali, Review of International Geographical Education Online, **11**, 1 (2021)
25. Z. Shaghaghian, H. Burte, D. Song, W. Yan, W. *Learning Geometric Transformations for Parametric Design: An Augmented Reality (AR)-Powered Approach,* Communications in Computer and Information Science, vol **1465**. Springer, Singapore (2022).

26. K.-T. Förster, M. König, R. Wattenhofer. *A concept for an introduction to parallelization in Java: Multithreading with programmable robots in Minecraft.* SIGITE '16: Proceedings of the 17th Annual Conference on Information Technology Education (2016)
27. O. Hrybiuk, O. Vedyshcheva, P. Lukavyi, A. Ivaniuk, N. Kulish, *Engineering in educational institutions: Standards for arduino robots as an opportunity to occupy an important niche in educational robotics in the context of manufacturing 4.0.* ICTERI Workshops (2020)
28. B. Love, C. Corritore, V. Winter, D. Faimon, *Creating an environment in which elementary educators can teach coding,* IDC '16: Proceedings of the 15th International Conference on Interaction Design and Children, (2016)
29. J. Gao, J. Gong, G. Zhou, H. Guo, T. Qi, *Learning with Yourself: A Tangible Twin Robot System to Promote STEM Education,* 2022 IEEE/RSJ International Conference on Intelligent Robots and Systems (IROS), Kyoto, Japan (2022)
30. H.-J. So, M. Gaydos, M. Seo, Y. Jung, H. Lee. *Learning with Minecraft and Kodu: Examining complex problem-solving strategies.* Workshop Proceedings of the 25th International Conference on Computers in Education. New Zealand: Asia-Pacific Society for Computers in Education (2017).
31. O. O. Doma, S. M. Şener, Interaction Design and Architecture Journal, **52** (2022)
32. J. M. Wei-KocsisSabounchi, B. Yang, T. Zhang. *Cybersecurity Education in the Age of Artificial Intelligence: A Novel Proactive and Collaborative Learning Paradigm.* IEEE Frontiers in Education Conference (FIE), Uppsala, Sweden (2022)
33. S. M. S. Ardakani, J. Ellis. *Developing a 3D-Printed Statics Modeling Kit.* ASEE 2022 Annual Conference, Minnesota, June (2022).
34. K. Pribilova, D. Gabriska. *Use of Lego Mindstorms EV3 MATLAB/Simulink with a focus on technical education.* 19th International Conference on Emerging eLearning Technologies and Applications (ICETA), Košice, Slovakia (2021).
35. H. Güneş, Z. Genç, Malaysian Online Journal of Educational Technology, **9**, 4 (2021).
36. S. Chu, R. Renolds, N. Tavares, M. Notari, S. Lee, *21st century skills development through inquiry-based learning.* Springer Science (2017).
37. D. Sun, F. Ouyang, Y. Li, H. Chen, Journal of Educational Computing Research, **59**, 4 (2021)
38. A. Körei, S. Szilágyi, *Parametric Graph Project - Using LEGO Gears for Drawing Curves, Advanced Research in Technologies, Information, Innovation and Sustainability, ARTIIS 2022.* Communications in Computer and Information Science, 1675. Springer, Cham. (2022)
39. S. Nebel, S. Schneider, G. D. Rey, Computers in Human Behavior, **55** (Part A), (2016).
40. X. Cao, Z. Li, R. Zhang, International Journal of Emerging Technologies in Learning (iJET), **16**, 5 (2021)
41. J. White, *Gender differences in GBL: The Japanese higher education context,* in European Conference on Games Based Learning, (2020)
42. S. Siouli, I. Dratsiou, P. E. Antoniou, P. D. Bamidis, Interaction Design and Architecture Journal, **42,** (2019)
43. S. N.Musa, N. A.Zulkafly, *Collaborative virtual environment to enhance learning experience,* In International Conference on Education and E-Learning, (2017)
44. T. Heinze, Natural Hazards, **104**, (2020)

45. J. Anthoney, R. Stead, K. Turney, Knowledge Management & E-Learning: An International Journal. **9**, 3 (2017).
46. E. G. Fedorenko, N. V. Kaidan, V. Velychko, Ye, V. N. Soloviev, *Gamification when studying logical operators on the Minecraft EDU platform*, 4th International Workshop on Augmented Reality in Education, May 11, 2021, Kryvyi Rih, Ukraine (2021).
47. I. Marulcu, M. Barnett, Res Sci Educ. **43** (2013)
48. D. Saito, H. Washizaki, Y. Fukazawa, *Analysis of the learning effects between text-based and visual-based beginner programming environments,* in IEEE 8th International Conference on Engineering Education (ICEED), Kuala Lumpur, Malaysia, (2017)
49. B. Autukevičienė, G. Stonkuvienė, Pedagogika, **148**, 4 (2022).
50. M. T. Billen, N. A. Ward, J. D. De-Hart, R. R. Moran, S. Yang, *Story visualizer by LEGO education: Using digital storytelling to integrate STEM and literacy,* In Connecting Disciplinary Literacy and Digital Storytelling in K-12 Education. IGI Global Publishing (2021)
51. V. Masiulionytė-Dagienė, T. Jevsikova, *Assessing Computational Thinking: The Relation of Different Assessment Instruments and Learning Tools,* In Informatics in Schools. A Step Beyond Digital Education, ISSEP 2022. Lecture Notes in Computer Science, 13488. Springer, Cham. (2022)
52. A.V. Prituja, H. Ren, *Lego exoskeleton: An educational tool to design rehabilitation device*, in IEEE International Conference on Real-time Computing and Robotics (RCAR), Okinawa, Japan, (2018).
53. P. Näykki, J. Laru, E. Vuopala, P. Siklander, S. Järvelä, Front. Educ. **4**, (2019).
54. W. Sung, J. Ahn, J. B. Black, Journal of Research in Childhood Education, **36**, 4, (2022)
55. A.I. Morch, S. Eie, L. Mifsud, *Tradeoffs in combining domain-specific and generic skills' practice in minecraft in social studies in teacher education*, in International Workshop on Cultures of Participation in the Digital Age, Italy, (2018)
56. T. Papazoglou, C. Karagiannidis, S. Mavropoulou, *Educational robotics can foster social inclusion and social status of children with autism,* in International Conference on Advanced Learning Technologies (ICALT), Tartu, Estonia, (2021)
57. O. Kadeeva, A. Belov, V. Syritsyna, S. Maksimenko, V. Sinko, *Methodological aspects of the implementation of additional education in robotics*, in E3S Web of Conference, **210**, 22040 (2020)
58. S. W. H., Lee, P. S. Saw, American journal of pharmaceutical education, **85**, 8 (2021)
59. S. Supraja, S. Tan, F. S. Lim, Ng B. Koon, S. Y. Ho, A. W. H. Khong, *Freshmen Orientation Program Using Minecraft: Designed by Students for Students during the Covid-19 Pandemic*, in IEEE Frontiers in Education Conference (FIE), Uppsala, Sweden, (2022)
60. S. Casler-Failing, Research in Learning Technology, **29** (2021)
61. L. López López, L. de Wildt, N. Moodie, British Journal of Sociology of Education, **40**, 8 (2019)
62. L. Fortunati, A. M. Manganelli, G. Ferrin, Int J Technol Des. Educ. **32**, (2022).
63. J. Leonard, A. Buss, R. Gamboa, et al., J Sci Educ Technol, **25**, (2016).
64. B. Altakhayneh, European Journal of Educational Research, **9**, 2 (2020)

Sustainable campuses: Building self-efficacy among academic, staff, and students of Green Office Model

Nuphanudin Nuphanudin[1*], *Helda Kusuma* Wardani[2], *Asep Iwa* Sumantri[3], *Erny* Roesminingsih[1], *Pramudya* Cahyandaru[4], and *Umi* Salamah[3]

[1]Manajemen Pendidikan, Fakultas Ilmu Pendidikan, State University of Surabaya, 60213, Indonesia
[2]Pendidikan Guru Sekolah Dasar, Fakultas Ilmu Pendidikan, State University of Surabaya, 60213, Indonesia
[3]Akademi TNI Angkatan Laut, Surabaya, 60178, Indonesia
[4]Universitas Sarjanawiyata Tamansiswa, Yogyakarta, 55167, Indonesia

Abstract. The goal of this study is to draw attention to the critical value of self-efficacy strategies, particularly the Green Office (GO) Model, which aims to improve collaboration among all stakeholders in higher education communities and serves as an example of what a campus sustainability office might entail. A two-step technique guides the research that is presented. To better understand the forces and obstacles preventing the institutionalization of GOs, first the features and similarities of currently existing GOs were examined. Based on this, a peer-to-peer digital training program was created and put to the test with the goal of giving students, employees, and academics the knowledge and abilities they need to start new sustainability offices and enhance the operations of those that already exist. The lack of adaptation of foreign experiences to different university environments, the absence of educational programs, and the scope of public relations were found to be the three key barriers to the spread of the GO model. The results demonstrate that GO offers a reproducible methodology that may be scaled internationally and modified for use in various university situations. The availability and capacity to obtain information from the actors at the sampled colleges outside of the immediate GO community have placed restrictions on this research.

1 Introduction

The potential of student design for the reform of universities toward sustainability is significant because it affects how future generations will approach the societal difficulties that lie ahead [1]. Higher education is essential to sustainable development [2, 3]. It emphasizes how these challenges go beyond education for creating future career profiles and claims that universities are required to impart not only the skills needed to advance successfully in a globalized world but also a positive attitude towards environmental issues and cultural diversity in their students, faculty, and staff [4].

* Corresponding author: nuphanudinnuphanudin@unesa.ac.id

The success of the teaching and learning processes is a duty shared by both professors and students as part of university education for sustainable development. In addition, higher education for sustainable development benefits university participants while also having positive effects off campus [5]. It is crucial to inspire other university actors to collaborate for a better partnership on campus. This is illustrated by numerous examples in the field of service learning, a teaching and learning strategy developed in the United States that combines subject-specific study with social engagement and provides students with solutions to unsustainable developments in ecological and social life areas [6, 7].

Students are not to be seen solely as addressees but rather as contributors to effective educational activities, according to the principle of involvement in the knowledge of (education for) sustainable development [8, 9]. Students may, for instance, take on the following increasingly responsible roles as part of higher education change processes: 1) as evaluators of their own learning processes and university experience; 2) as participants in decision-making processes; 3) as partners, co-creators, and experts; and 4) as change operators [10].

As the interpersonal relationship between teachers is still frequently dominated by a traditional understanding of education, in which a hierarchical understanding of roles dominates, numerous studies have shown that in practice there is a strong participation gap between the four levels, from selective cooperation to genuine participation [11]. Accordingly, the amount of participation decreases steadily from the first to the fourth stage [12]. Student councils are even less common [13], and the role of sustainability organs in institutional governance typically declines to a sustainability office or advisory committee.

It is crucial to consider what levers within higher education may be used to give more space to student involvement because the majority of these projects are simply a few examples of the higher education landscape [14]. It is beneficial to take a systematic look at various roles that students can perform in universities and partnerships for such a strategic examination of student design potential. This framework is covered in a recent publication from the UK's Higher Education Academy (2014) in four overlapping areas: curriculum design, pedagogical planning, and consultancy, subject-based research and inquiry, and scholarship of teaching and learning (both within and across disciplines) (see Figure 1).

In light of this, student demands for cutting-edge educational approaches to sustainable development can only be seen as consistent [15–20]. especially when these demands are combined with those for rigorous effectiveness and efficiency tests. But in addition to the demands put forth by the student body, innovative concepts have also been effectively implemented at a number of colleges [21–24].

However, despite being hopeful and effective, all of these instances only make up a small portion of the higher education scene [25]. But what worries me the most is that these numerous individual success stories continue to be fragmented [26]. In other words, there is no strategic approach that would methodically assist motivated students in launching their own sustainability-oriented model by arming them with the essential information and useful tools [27]. Last but not least, the preservation of information is essential in light of student cohort turnover rates that typically last only a few years at universities [28].

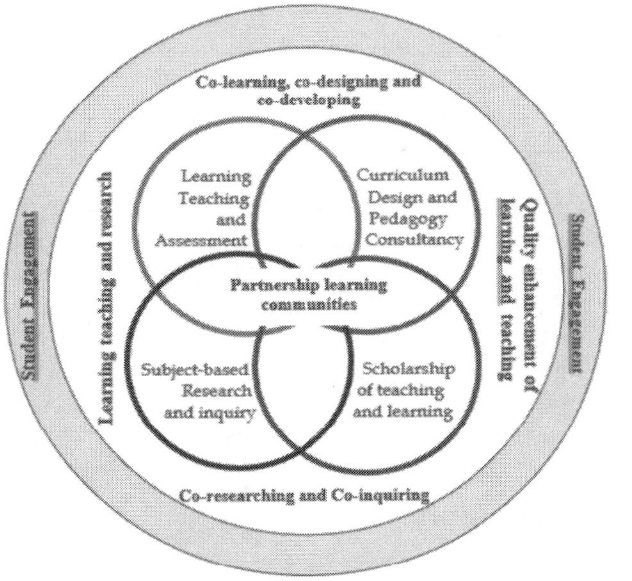

Fig. 1. Framework enabling students to be partners in learning and teaching in higher education

The Green Office (GO) Model, which is outlined in the following, offers a solution to these problems: It attempts to deliver general, practical information materials for a quick start (such as for the identification of strategic leverage points) while also providing essential adaptability to specific institutional framework conditions.

From an institutional perspective, it would be desirable to invite all stakeholder groups to contribute their specific expertise to achieve a joint vision of a sustainable university. However, the compartmentalisation of universities often makes it difficult for people to collaborate across departmental boundaries. This is a problem, as sustainability efforts remain constrained to small islands of activity. Platforms need to be established to create bridges across those islands and offer more opportunities for people to contribute.

A GO can be seen as such a platform, as it informs, connects and supports students and staff to act on sustainability [29]. At the same time, a GO develops its own ideas to better embed sustainability in education, research or operations. The first GO was established at Maastricht University in 2010. Since then, the model has been replicated by 35 universities across Europe, and won many awards, among them the UNESCO-Japan Prize on Education for Sustainable Development. The model is open-source to tailor-make it to individual institutional needs.

2 Methods

At the beginning of a two-year project, a closer analysis of 13 existing GOs revealed that the average GO consists of five student employees, who work 10 hours a week, and one staff member. Some GOs also engage students as volunteers. The average funding for a GO is $30,000 per year to pay for project expenditures and salaries. It is made visible through permanent office space and an online presence.

Additionally, according to the research, there are three different kinds of GOs. Universities without official sustainability personnel or working groups are most suited for the student-led sustainability office. Then, with staff interaction at the university, the GO is established as a department of the institution run entirely by students. This adaption offers excellent student empowerment and grassroots engagement while being an economical solution to institutionalize sustainability initiatives and advance them. However, it should only be a temporary fix because the students do not have the institutional backing they require to achieve completely while having all the duties of a full-fledged sustainability department. Institutions with a professional sustainability coordinator or team but limited student involvement in the university's sustainability initiatives would benefit from having a student engagement unit. The GO was developed to encourage student engagement and staff-student collaboration. Students and one to three staff members make up the GO, which collaborates with the current sustainability team. As a result, staff can interact with more students, and students can contribute to the university's sustainability initiatives. The sustainability team can frequently put it up without a protracted and difficult lobbying effort. The roles of the GO as contrasted to the staff members run the risk of being confused or misinterpreted by both parties or outside parties. A student or employee who manages the GO and regularly participates in the sustainability team can combat this.

3 Results and Discussion

Students and faculty who want to start a GO, however, must address the crucial issues of knowledge transfer and capacity growth. Workshops, Skype calls, and downloadable resources were used to facilitate the transfer of prior knowledge. It constrained the scope of influence. The notion of expanding the GO approach by establishing an online course to draw in and support more students and university personnel arose.

Where there are numerous ongoing sustainability projects among students and staff that are poorly connected and undetectable, a central sustainability platform is a smart option. In a single GO, students and staff work together as a cohesive team. Through a single department, this strategy can effectively improve sustainability across the university by encouraging close collaboration between students and faculty. This method runs the risk of having staff members dominate the team and weakening student leadership.

Through financing, office space, and a mandate, the university offers institutional support for GOs. Students and employees who want to start a GO must enlist this institutional backing. Students and staff have similar obstacles to overcome even if every institution has its own unique processes. Recognize the GO Model and the organization's current sustainability initiatives. Organize a group of allies to advocate for their GO. Make compelling reasons for why the university ought to support a green office. develop a GO that is customized to their university using the GO model. Determine potential financial sources, create a funding application, and seek help.

Students and faculty who want to start a GO, however, must address the crucial issues of knowledge transfer and capacity growth. Workshops, Skype calls, and downloadable resources were used to facilitate the transfer of prior knowledge. It constrained the scope of influence. The notion of expanding the GO approach by establishing an online course to draw in and support more students and university personnel arose.

The failure to adapt the global experience surrounding the GO Model to different university environments, the absence of educational offers, and the narrow reach of public relations are the three key barriers to the spread of the GO model. In response, a two-year initiative supported by the Indonesia Environmental Foundation aims to promote this self-efficacy paradigm throughout Indonesia and beyond. It is being carried out by rootAbility and the Indonesia University of Education.

The idea that creating a GO is always a hands-on process serves as the foundation for the learning philosophy and approach. Therefore, lengthy lectures or theoretical talks are not very helpful. Previous research demonstrated that the most effective method involved the autonomous production of specific tasks and questions, debate in small groups, and personalized feedback.

The following components make up the didactical layout: Online guides: Six online tutorials are completed by the participants. These seminars allow for plenty of discussion and questions and are interactive. Videos: In advance of each online tutorial, participants are emailed links to quick videos (between 5 and 7 minutes long) that introduce them to the topic. Prior to each online tutorial, participants must complete a job in addition to the videos. It takes 30 to 60 minutes, depending on how diligently they work. Individual feedback: On more significant preparation activities, such as the initial GO design, individual feedback is provided. Intermittent online instructional questions can be submitted to the teachers or posted in the project group on platform n, a social media site run by the German sustainability network netzwerk n. Website and e-books: There is a website dedicated to the GO Model (www.GreenOfficeMovement.org) as well as two e-books, one of which serves as an introduction to the GO Model and the other is devoted to the findings of 23 case studies examination.

The core of the learning support consisted of online tutorials. The participants engaged in six online tutorials in both English and German, including an introduction to the GO Model, building the GO Initiative, analyzing current sustainability initiatives, designing the goals and structure of the GO, identifying funding needs and options, and funding application.

An overview of the contact information for student organizations and university staff at 152 Indonesian universities and colleges with a focus on sustainability was produced to promote the Indonesia online course. Keyword searches on individual university websites were conducted for this purpose, and Hoch-N partner network participating universities' contact information was gathered. At least one person who received email notification of the start of the online course could be located for 79 percent of the universities. Several hundred emails were sent to various stakeholder groups to advertise the online Indonesia course. Additionally, 20 university groups that registered on the platform n and had a focus on sustainability were approached directly.

The first cycle of the course was taught in English and Indonesia between February and July 2022. 82 applications were received in total, and 73 of them were chosen to participate. 34 people participated in at least four of the six online courses. The 60-minute lessons were scheduled every two weeks. Zoom was employed for the implementation (a conference software). A Facebook group was utilized as an online forum for the online English language course, and platform-n from netzwerk-n was used for the online Indonesian language course, where the preparation assignments for each tutorial and the results were uploaded.

Participants evaluated the first cycle of the course through pre- and post-surveys as well as participant interviews. In order to determine the elements that enabled participants to effectively complete the course, interviews based on the Success Case Methodology were conducted. Participants were asked to rate the complexity of the main steps in the process of creating a GO in the pre- and post-surveys. The difficulty of the steps was rated as being 17 percent less challenging in the post-surveys.

On a scale from 1 to 10, participants were also asked how likely they were to suggest the course to others. Respondents gave an 8.2 on a scale of 10, with 10 being the average response. The participants identified two areas for improvement: better workload communication and paying more attention to the participants' various language proficiency levels (in the English course). Simplified texts and a workload estimate in the registration form will address these difficulties. The interaction between participants and the group as well as the individual feedback on more substantial assignments were both highly valued by the participants. Along with the participants' enthusiastic responses, the training has produced distinct, observable results. Eight new GOs have been founded since the project's inception, and 25 universities have started new GO-creation initiatives.

The online course's subsequent cycle will run from October 2022 to January 2023. To increase interaction and promote more independent study, the course materials will be relocated to an interactive platform created with Articulate Storyline. There will only be four calls for the group tutorials. After taking into account participant input, each participant or participating team will receive a regular individual call. The course will remain accessible after the project is finished for self-study on www.GreenOfficeMovement.org, with the option to ask questions and schedule support calls.

4 Conclusion

There are many different strategies to achieve sustainability, according to research on the topic of sustainability in higher education. While the creation of a strong institutional culture of sustainability can be considered as a crucial success element for the improvement of universities, just little attention is paid to these issues. Their findings are supported by those of Stephens and Graham (2010), whose investigation of Transition Management provided pertinent insights that enabled them to create an empirical study agenda on sustainability in higher education.

When considering students' roles and appropriate levers towards sustainability-oriented transformation, the following questions are suggested as the central discussion points against this backdrop of developments within higher education institutions.

What adds to the Green Office Model and how does it vary from existing sustainability governance approaches? The Green Office Model (GO) is a reproducible, international sustainability governance approach that can be scaled internationally and modified to suit various university situations. It offers a structured method for enhancing student leadership in university sustainability initiatives, collaborating with other student-driven initiatives or staff-led teams. However, undemocratic nations may hinder sustainable future development due to governance regimes, limiting students' voices.

The first round of the online course's results indicate that it is a scalable and efficient means to disseminate information about the GO Model establishment process to students and employees all around Europe—and perhaps beyond. Providing remote support to participants individually or traveling to each site and offering workshops and guidance at their institutions were two alternatives to the online course. The online course has several advantages over these options, including peer learning, a decrease in travel emissions, and the creation of learning resources that will be available after the second round of learning support is finished. In order to encourage other institutions to pursue the route towards sustainability, transferability and scalability are critical strategic and operational aspects. But the authors are adamant that the GO Model's peer-to-peer nature and adaptability make it a promising tool for reorienting higher education institutions toward sustainability.

References

1. L. V. Trevisan, J. H. P. P. Eustachio, B. G. Dias, W. L. Filho, E. Á. Pedrozo, Environ. Dev. Sustain. **25** (2023)
2. T. F. A. C. Sigahi et al., Ergonomics, **66**, 9 (2022)
3. A. C. F. Caldana, J. H. P. P. Eustachio, B. Lespinasse Sampaio, M. L. Gianotto, A. C. Talarico, dan A. C. da S. Batalhão, Int. J. Sustain. High. Educ. **24**, 2 (2023)
4. E. AbuKhousa, M. S. El-Tahawy, Y. Atif, Futur. Internet **15**, 2 (2023)
5. L. Janssens, T. Kuppens, I. Mulà, E. Staniskiene, A. B. Zimmermann, Int. J. Sustain. High. Educ. **23**, 8 (2022)

6. M. Yemini, L. Engel, A. Ben Simon, Educ. Rev. **75** (2023)
7. S. Chansaengsee, P. Niramitchainont, A. Leksansern, P. Longpradit, *The Blueprint of Transdisciplinary Concepts of Education for Sustainable Development (ESD) in Thai Tertiary Education*, in Fostering excellent next generation, 1st ed., Chiba University, Chiba (2022)
8. N. Gericke, T. Torbjörnsson, J. Environ. Educ. **53**, 4 (2022)
9. C. Scalabrino, A. Navarrete Salvador, J. M. Oliva Martínez, Environ. Educ. Res. **28**, 5 (2022)
10. K. E. Matthews, M. Dollinger, High. Educ. **85**, 3 (2023)
11. J. H. Nieminen, Teaching in Higher Education **27** (2022)
12. R. S. P. Singh et al., Clin. Pharmacol. Ther. **112**, 1 (2022)
13. S. Kraus, D. K. Kanbach, P. M. Krysta, M. M. Steinhoff, N. Tomini, Int. J. Entrep. Behav. Res. **28**, 9 (2022)
14. X. Kang, W. Zhang, Interact. Learn. Environ. **31**, 2 (2023)
15. F. Zamora-Polo, J. Sánchez-Martín, M. Corrales-Serrano, L. Espejo-Antúnez, Sustain. **11**, 13 (2019)
16. M. V. Eitzel et al., Dev. Eng. **3** (2018)
17. W. M. Purcell, H. Henriksen, J. D. Spengler, Int. J. Sustain. High. Educ. **20**, 8 (2019)
18. I. Bello, M. D. K. Bin Shariffuddin, M. F. Othman, N. Osman, Innov. Eur. J. Soc. Sci. Res. **36**, 3 (2023)
19. M. T. Pasara, D. Mhlanga, Emerald Open Res. **4** (2022)
20. B. B. Gupta, A. Gaurav, P. K. Panigrahi, Int. Entrep. Manag. J. **19**, 2 (2023)
21. D. Edem Adzovie, A. B. Jibril, Cogent Educ. **9**, 1 (2022)
22. M. A. Mohamed Hashim, I. Tlemsani, R. Duncan Matthews, Educ. Inf. Technol. **27**, 7 (2022)
23. I. A. Akour, R. S. Al-Maroof, R. Alfaisal, S. A. Salloum, Comput. Educ. Artif. Intell. **3** (2022)
24. M. Sofi-Karim, A. O. Bali, K. Rached, Educ. Inf. Technol. **28**, 1 (2023)
25. T. C. Young, K. Malone, Teach. High. Educ. **28**, 5 (2023)
26. M. Eby, Work and Occupations, **50** (2023)
27. S. Simon, I. Vieira, M. Jecu, Front. Educ. 8 (2023)
28. M. F. Larsen, J. Valant, Journal of Research on Educational Effectiveness **16** (2023)
29. T. Strasser, J. De Kraker, R. Kemp, Glob. Sustain. **5** (2022)

The pathway of digital learning environments in advancing Sustainable Development Goals (SDGs): A Bibliometric analysis covering three decades of research

Hanandita Veda Saphira[1], *Binar Kurnia* Prahani[1*], *Eko* Hariyono[1], *Firmanul Catur* Wibowo[2], *Muhammad Abd Hadi* Bunyamin[3], and *Sugeng* Wahono[4]

[1]Faculty of Mathematics and Natural Science, State University of Surabaya, 60231 Surabaya, Indonesia
[2]Faculty of Mathematics and Natural Science, State University of Jakarta, 13220 Jakarta, Indonesia
[3]Faculty of Social Sciences and Humanities, Universiti Teknologi Malaysia, 81310 UTM Skudai, Johor, Malaysia
[4]Akademik Angkatan Laut, 60178 Surabaya, Indonesia

Abstract. A digital learning environment can play a pivotal role in advancing the objectives of Goal 4 and, in doing so, can contribute to several other SDGs. This study purposed to investigate the impact of digital learning on sustainable growth in institutions by emphasizing the educational procedures that were implemented in extended years (three decades) throughout the global and broader countries. Furthermore, this research also analyzes the challenges and opportunities for digital learning in educational institutions, primarily to support the SDGs This research employs a descriptive bibliometric analysis, utilizing Scopus as the primary data source. The research findings indicate a consistent upward trend in publication rates each year, with articles being the most prevalent document type, followed by conference papers. Potential areas for future research include the integration of e-learning practices into various educational levels, from universities to high schools and early education. Additionally, there are opportunities to investigate the development of innovative learning techniques and the emergence of computer-assisted learning as an educational system. Notably, the most influential studies fall within the Q1 category. The implications of this research extend to providing a comprehensive overview of trends in DLE publications from 1993 to 2022 and their potential contributions to the SDGs. Furthermore, future researchers can build upon these findings to develop and enhance digital learning environments in educational contexts, ultimately contributing to the attainment of SDG goals.

[*] Corresponding author: binarprahani@unesa.ac.id

1 Introduction

The UN 2030 Sustainable Development Goal (SDGs) Goal 4 (Quality education) specifically focuses on education, while the concept of a digital learning environment can intersect with several of the targets and indicators associated with Goal 4. Digital technology can play a pivotal role in advancing the objectives of Goal 4 and, in doing so, can contribute to several other SDGs [1–3].

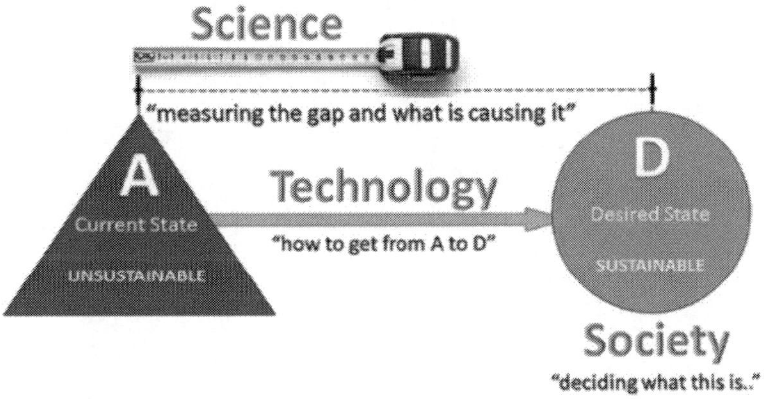

Fig. 1. Educational framework for sustainability transformation (Picture Source : [4]).

Fig. 1. shows that digital learning environments (DLE) support the goal's aims by enhancing access to education, improving the quality of education, promoting inclusivity, and offering lifelong learning opportunities, among other benefits [5]. DLE also supports economic growth, reduces inequalities, and fosters innovation and collaboration, making them a valuable tool for achieving a sustainable future [6]. Previous studies state that the digital transition [7], including e-learning [8], also highlights the importance of critically and practically engaging with what "transformational" education means in the context of the SDGs. Furthermore, previous studies [9,10] on the possibilities and necessities demonstrate an essential and beneficial connection between technological transformation achievement and global colleges and universities, including the opportunities for e-learning regarding SDGs, creative thinking, and organizational preparedness PRISMA statement. However, previous research [9] is still limited to a decade of metadata on DLE analysis and for specific countries. Hence, the purpose of this research is to investigate the impact of digital learning on sustainable growth in institutions by emphasizing the educational procedures that were implemented in extended years (three decades) through the global and broader countries. Furthermore, this research also analyzes the challenges and opportunities for digital learning in educational institutions, primarily to support the SDGs. This study, on the other hand, helps us comprehend the present condition of learning and instruction in universities. It is also crucial to comprehend how to improve the studied situation based on the recommendations offered in this study.

2 Methods

This research is a descriptive analysis using bibliometrics. The primary source database used is Scopus [11–13]. However, Scopus has become a leading resource with a publication loading 70% higher than other database sources [14–16]. Bibliometrics is divided into four stages: (1) creating a research design, (2) gathering information using criteria, (3) data analysis, and (4) representing and displaying data [17–20]. The process of this research is to determine the use of DLE, as in Fig. 2.

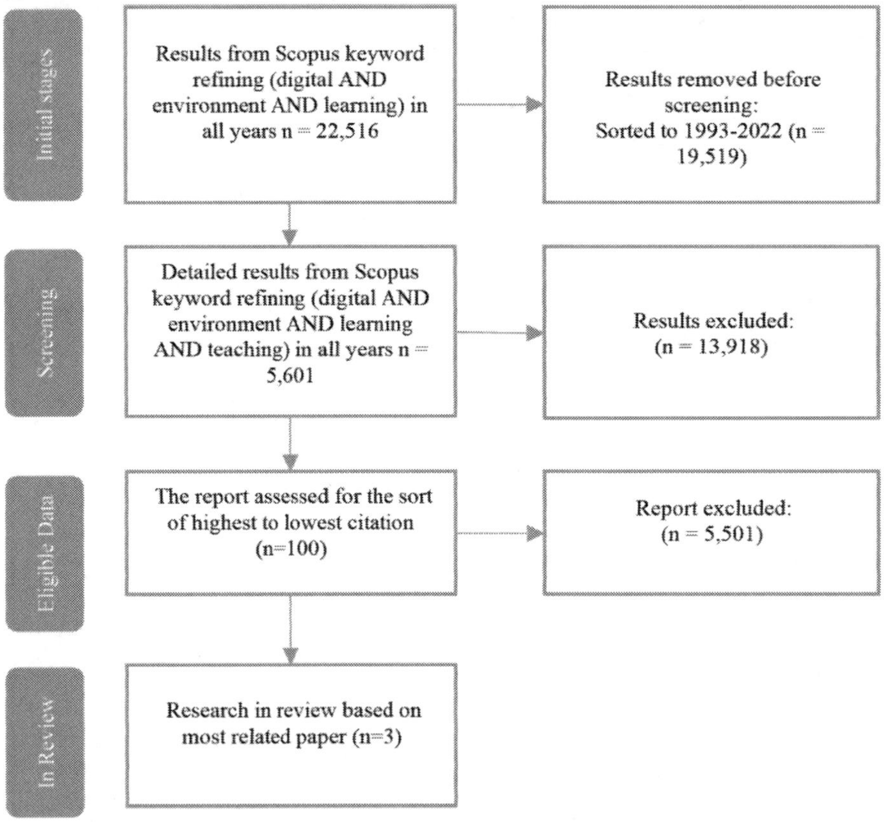

Fig. 2. Research Flowchart.

Gathering information on the metadata is conducted in October 2023. The results are chosen from the highest citation amount. Then, the metadata is saved as a .ris and .csv document types. Furthermore, the metadata are imported to assisted tools to relieve interpretation data analyzed. VOSViewer maps visualization to determine the link between published articles [18]–[21]. Then, selected articles are analyzed and reviewed to find the findings and recommendations to strengthen novelty besides mapping visualizations.

3 Results and discussion

3.1 DLE trendline during three decades

Based on the metadata gathered, the trend research of DLE during three decades is known. It is depicted in **Fig. 3.**

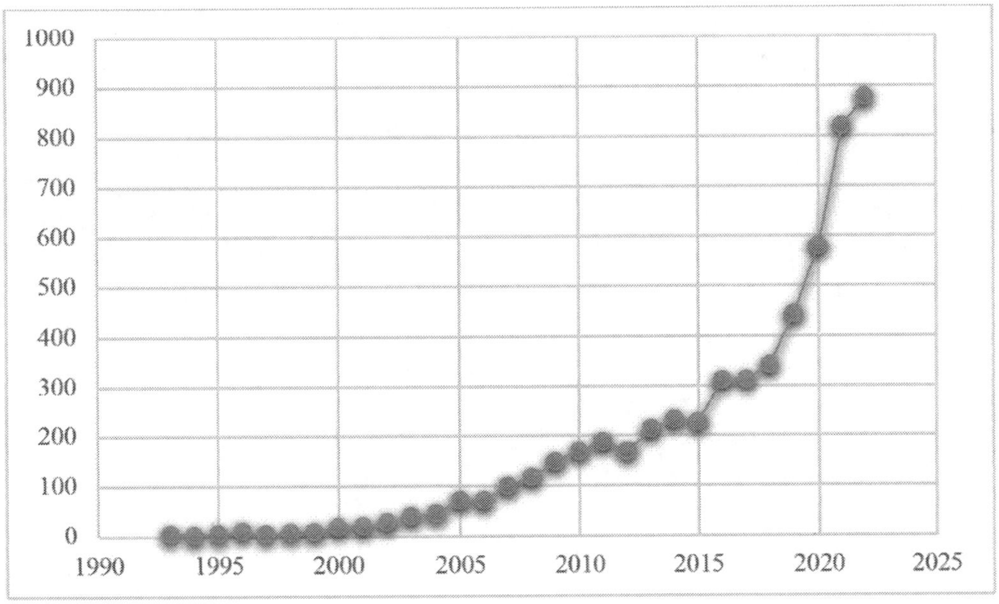

Fig. 3. Publication Trendline of DLE.

According to the research findings, as depicted in **Fig. 3**, there is an evident and stimulating annual growth trend in publication rates. This surge can be attributed to the growing interest among researchers in standardizing broadband cellular networks, particularly DLE. DLE is currently under consideration and is rapidly gaining prominence in the realms of research and education [25–27]. This shift led to a new set of DLE services characterized by diverse and technologically advanced design demands.

The pattern of DLE publications encompasses a wide array of available materials. According to the collected data, various document types were issued between 1993 and 2022, providing valuable insight for prospective researchers. **Fig. 4** illustrates the assortment of published documents related to DLE.

Fig. 4. Document Type of DLE Publications.

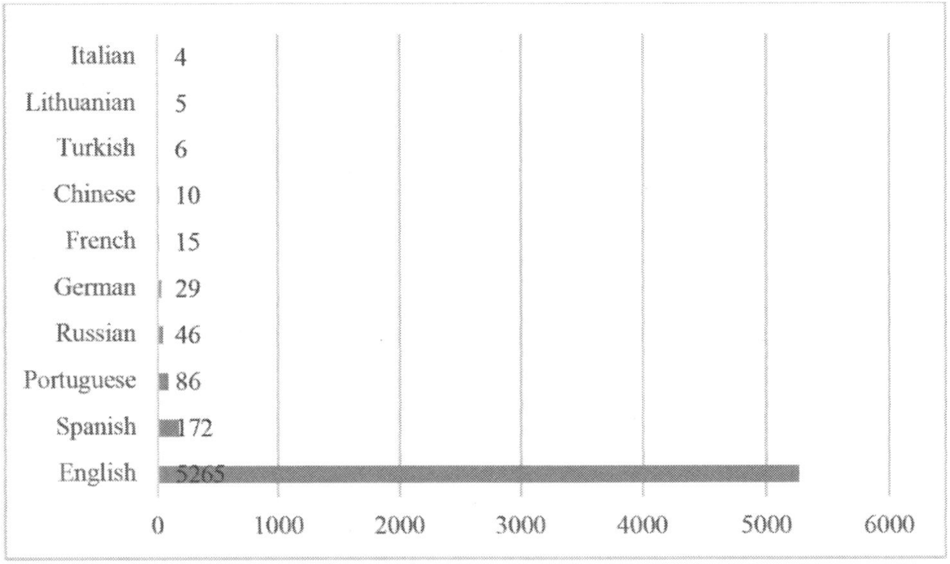

Fig. 5. Top Publications Language of DLE.

Fig. 3 reveals that articles are the most frequently employed publication format, accounting for 2,429 documents. Conference papers are the second most prevalent type with a count of 2,261. Furthermore, the predominant language for DLE publication research is English, as demonstrated by the data presented in **Fig. 5**. the information collected over the span of three decades indicates that English is the most commonly utilized language for disseminating DLE research in educational contexts [31,32].

3.2 Novelties of the DLE publications during three decades

Fig. 6. delves into the mapping visualization of metadata, which aims to explore the connections between relatively fewer keywords, thereby discerning the research's novelty from the mapping outcomes.

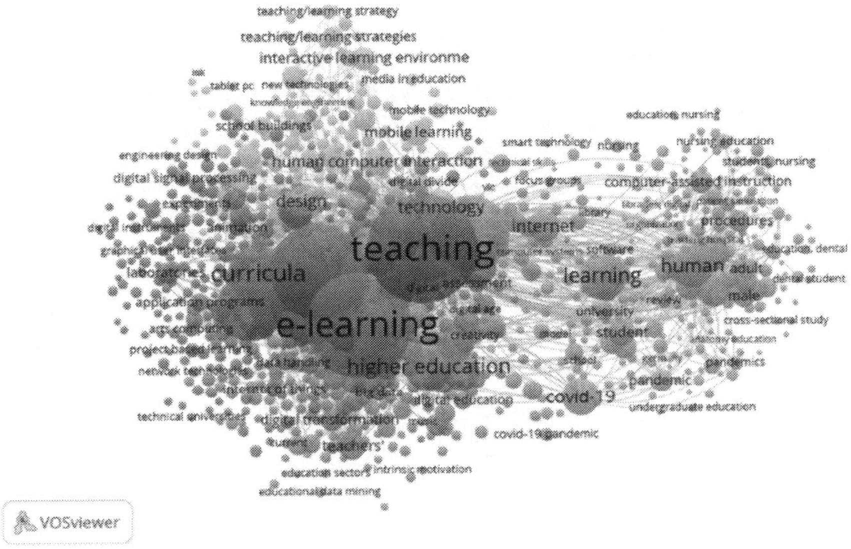

Fig. 6. Mapping Visualisation of DLE Publications.

Based on **Fig. 6.** it is known that six clusters, namely 1) Cluster 1 (n=308) with red node, consist of e-learning systems, platforms, teaching, teaching systems, virtual reality, virtual laboratory, university education, and so on. 2) Cluster 2 (n=284) with green node consists of COVID-19, digital technology, education, health education, high school, online systems, postgraduate students, school students, etc. 3) Cluster 3 (n=240) with blue node consists of collaborative learning, DLE, digital teaching, gamification, ICT, LMS, MOOCs, online environments, virtual education, and so on. 4) Cluster 4 (n=158) with the yellow node consists of active learning, practical learning, learning achievement, learning outcomes, mobile learning, web-based learning, and so on. 5) Cluster 5 (n=7) with purple node consists of data mining, educational data mining, primary education, learning analytics, etc. 6) Cluster 6 (n=3) with cyan node consists of PBL (problem based-learning), research question, and secondary school. Hence, from the DLE during three decades, the novelties are found in **Fig. 7** and **Fig. 8**.

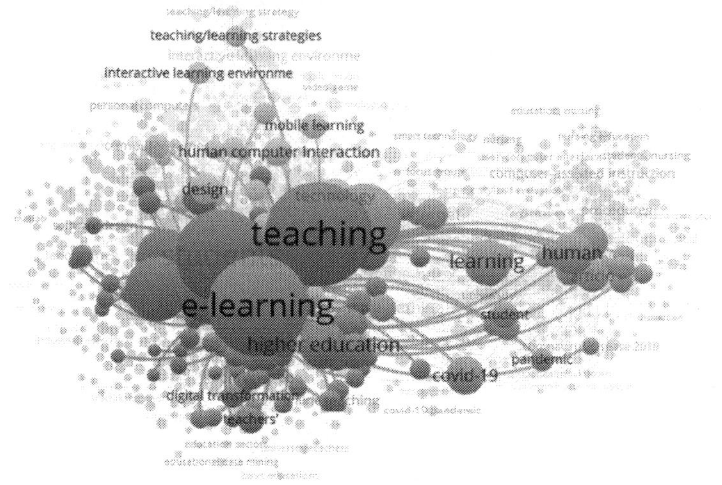

Fig. 7. The Top Occurance Keywords.

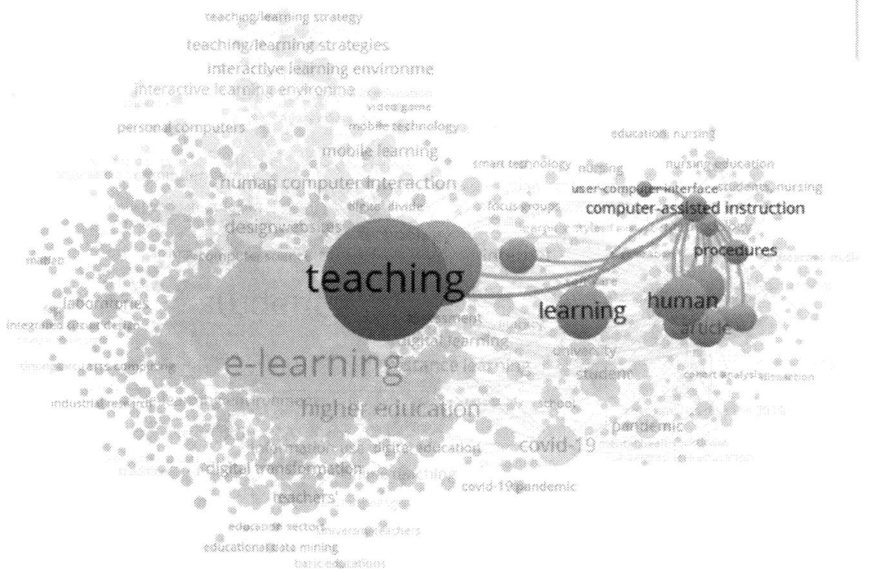

Fig. 8. The Few Occurance Keywords.

In order to identify the originality of prior research, an examination of metadata keywords through mapping is conducted [33]. Therefore, within the realm of DLE publications spanning three decades, the novel aspects are depicted in **Fig. 7** and **8**. As shown in **Fig. 7**, there is a noticeable research interest in e-learning and teaching, as evidenced by its substantial total link strength. Consequently, forthcoming research endeavors could focus on enhancing e-learning methods, such as integrating e-learning activities into universities, high schools, and early education. Alternatively, researchers could explore the application of e-learning across all educational institutions and evaluate its impact. On the other hand, **Fig. 8**, with its more limited set of keywords, highlights research opportunities related to developing learning techniques and the emergence of computer-assisted learning as an instructional system. These keywords can be utilized to uncover novel approaches and recommend actions such as implementation, development, enhancement, installation, or further research.

3.3 Review of DLE's most cited papers from 1993 to 2022

Table 1 provides an evaluation of the three most frequently cited publications considered to be highly influential in the field of DLE from 1993 to 2022. Each of these articles underwent analysis based on their citations, Scimago Journal and Country Rank (SJR) data retrieved from www.scimagojr.com as well as CiteScore data accessed on www.scopus.com as of October 2023. This analysis includes a review of the findings and recommendations presented in each publication [34–37].

Table 1. Review of Most Cited Papers on DLE Publication from 1993 to 2022.

Author(s)	Citation	SJR	CiteScore	Findings & Recommendations
Papastergiou, (2009) [38]	1,111	Q1 (3.68)	23.8	The findings indicate that educational computer games can serve as practical and motivating learning environments in high school computer science, regardless of the student's gender.
Rapanta et al. (2020) [39]	888	Q1 (2.45)	16.5	The creation of learning tasks possessing specific attributes, the integration of three presence types (social, cognitive, and facilitatory), and the necessity to adjust assessments to meet the evolving learning demands.
Robin & Robim (2008) [40]	608	Q1 (1.25)	4.6	In recent years, digital storytelling has gained prominence as an effective educational tool that captivates both educators and their students. Even so, up until recently, there has been a need for more emphasis on the development of a theoretical framework

Author(s)	Citation	SJR	CiteScore	Findings & Recommendations
				that could enhance the efficacy of technology as a classroom tool.

Furthermore, based on **Table 1**, these publications become fundamental for future research, so they have outstanding citations and impact DLE development. As of October 2023, a list of the most highly cited publications in Quartile 1 (Q1) ranked journals, with CiteScores ranging from 23.8 to 4.6, is provided. This indicates that these publications hold significant influence and unquestionable credibility, mainly due to the reputable nature of the publishers. The analysis of the SJR indicator assigns varying scores to citations based on the significance of the source journal from which the citation originates. Consequently, citations from influential journals hold more significant value, and the journals receiving such citations tend to gain increased recognition and prominence.

4 Conclusion and remarks

Based on the findings and subsequent discussion, it is evident that there is a consistent upward trajectory in publication trends on an annual basis. Articles represent the most prevalent document type, with conference papers following as the second most common form of publication. Opportunities for future research include the exploration of novel approaches through the integration of e-learning into educational institutions, ranging from universities to high schools and early education. On the other hand, a more limited set of keywords points to prospects for research in the realm of developing innovative learning techniques and the establishment of computer-assisted learning as an instructional framework. Notably, the most influential studies belong to the Q1 category.

The research's broader implication is to provide a comprehensive understanding of the trends in DLE publications from 1993 to 2022 and their potential contributions to the Sustainable Development Goals (SDGs). Moreover, future researchers can leverage this knowledge to refine and advance research in the field of educational DLE, ultimately contributing to the attainment of SDG objectives. This research framework enables subsequent researchers to establish a document profile that serves as a focal point for further exploration. By identifying the most pertinent topics within Scopus' DLE, researchers can effectively narrow down the emerging trends in this research domain.

References

1. S. Clark, M. Maclachlan, K. Marshall, N. Morahan, C. Carroll, K. Hand, N. Boyle, K. O'Sullivan, Sustain. **14** (2022)
2. M. Akintolu and C. Uleanya, Univers. J. Educ. Res. **9**, 713 (2021)
3. D. Rad, A. Redeş, A. Roman, S. Ignat, R. Lile, E. Demeter, A. Egerău, T. Dughi, E. Balaş, R. Maier, C. Kiss, H. Torkos, G. Rad, Front. Psychol. **13** (2022)

4. N. Voulvoulis, M. A. Burgman, Crit. Rev. Environ. Sci. Technol. **49**, 1079 (2019)
5. M. Laufer, A. Leiser, B. Deacon, P. Perrin de Brichambaut, B. Fecher, C. Kobsda, and F. Hesse, Int. J. Educ. Technol. High. Educ. **18** (2021)
6. A. Amran, I. Jasin, M. Perkasa, et al., Journal of Physics: Conference Series **1521**, 4 (2020)
7. J. Holmes, O. R. Moraes, L. Rickards, W. Steele, M. Hotker, A. Richardson, Int. J. Sustain. High. Educ. **23**, 503 (2022)
8. S. Ghanem, Int. Sustain. Resil. Conf. Technol. Innov. Build. **12** (2020)
9. N. S. Alotaibi, Sustain. **14** (2022)
10. V. Vázquez-Verdera, J. Domingo, E. Dura, D. Gabaldón-Estevan, E. López-Baeza, S. M. López, F. Meco-Tébar, S. Rueda, J. J. Serrano-Lara, I. Signes-Soler, M. L. V. de Á. Pascual, and E. Martínez-García, Sustain. **13**, 1 (2021)
11. Y. J. Utama, B. Setiyono, M. Tauviqirrahman, and H. Susanto, E3S Web Conf. **125**, 1 (2019)
12. Z. Hakimova, E3S Web Conf. **413**, 1 (2023)
13. H. Purnaweni, A. Roziqin, S. P. Hadi, E3S Web Conf. **317** (2021)
14. R. Pranckutė, Publications **9**, (2021)
15. F. J. Agbo, S. S. Oyelere, J. Suhonen, M. Tukiainen, Smart Learn. Environ. **8**, 1 (2021)
16. M. Shareefa, V. Moosa, Eur. J. Educ. Res. **9**, 331 (2020)
17. P. Song and X. Wang, Asia Pacific Educ. Rev. **21**, 473 (2020)
18. Z. Liu, J. Moon, B. Kim, and C. P. Dai, Educ. Technol. Res. Dev. **68**, 1931 (2020)
19. C. T. Ha, T. T. P. Thao, N. T. Trung, L. T. T. Huong, N. Van Dinh, T. Trung, Eurasia J. Math. Sci. Technol. Educ. **16** (2020)
20. H. L. T. Thu, T. Tran, T. T. T. Phuong, T. L. T. Tuyet, H. Le Huy, T. V. Thi, Educ. Sci. **11**, (2021)
21. D. F. Al Husaeni, A. B. D. Nandiyanto, R. Maryanti, Indones. J. Teach. Sci. **3**, 1 (2023)
22. S. M. A. H. Gillani, A. B. A. Senin, J. Bode, Muniba, S. M. A. H. Gillani, Int. J. Interact. Mob. Technol. **16**, 48 (2022)
23. Y. Hou and Z. Yu, Educ. Sci. **13**, 125 (2023)
24. H. Soegoto, E. S. Soegoto, S. Luckyardi, A. A. Rafdhi, Indones. J. Sci. Technol. **7**, 89 (2022)
25. I. Rodr and N. Shirvanizadeh, Int. J. Environ. Res. Public Health **18**, (2021)
26. X. Cheng, Y. Wu, S. Member, G. Min, A. Y. Zomaya, IEEE J. Sel. AREAS Commun. (2020)
27. V. Seters, Comput. Educ. Artif. Intell. **1**, 1 (2020)
28. R. Davidson, Y. Liu, Lang. Cult. Curric. **33**, 32 (2020)
29. S. N. Sari, D. Aminatun, J. English Lang. Teach. Learn. **2**, 16 (2021)
30. R. Rabiman, P. Sudira, H. Sofyan, M. Nurtanto, Int. J. Interact. Mob. Technol. **15**, 130 (2021)
31. S. Hussain, Alford Counc. Int. English Lit. J. **2**, 6 (2019)
32. V. Ramírez-Castañeda, PLoS One **15**, 1 (2020)
33. S. H. P. W. Gamage, J. R. Ayres, M. B. Behrend, Int. J. STEM Educ. **9**, (2022)
34. M. Torres-Samuel, C. L. Vásquez, A. Viloria, N. Varela, L. Hernández-Fernandez, R. Portillo-Medina, Lect. Notes Comput. Sci. (Including Subser. Lect. Notes Artif. Intell. Lect. Notes Bioinformatics) **10943 LNCS**, 188 (2018)

35. L. Sun, Learn. Publ. **32**, 207 (2019)
36. I. Ianoş, A. I. Petrişor, Publications **8**, (2020)
37. S. Kasper, Int. J. Psychiatry Clin. Pract. **25**, 1 (2021)
38. M. Papastergiou, Comput. Educ. **52**, 1 (2009)
39. C. Rapanta, L. Botturi, P. Goodyear, and L. Guàrdia, Postdigital Sci. Educ. **2**, 923 (2020)
40. B. R. Robin, B. R. Robin, Theory Pract. **37** (2009)

Trends in research publication topics related to eco enzymes with bibliometric analysis

Winarsih[1*], *Achmad Wachidul* Kohar[2], *Yanisworo Wijaya* Ratih[3], *Ahmad Zainul* Aziz[4], and *Wahyu Adi* Nugroho[5]

[1,4,5]State University of Surabaya, Surabaya, Indonesia
[2]EDP Mathematical Sciences, University of Michigan, Michigan, United States
[3]Pembangunan Nasional Veteran University, Yogyakarta, Indonesia

Abstract. Ecoenzyme is a potential sustainable waste management solution that could make waste management more beneficial. This study seeks to analyze the evolution of ecoenzyme-related research publication topic trends over the past decade, then visualize them and identify potential ecoenzyme-related research topics for future study. Using Publish or Perish software with Google Scholar database sources and VOSviewer software, this research method employs a literature review and bibliometric analysis. The investigation and search produced a collection of published document metadata regarding ecoenzymes that included up to 666 documents from 2013 to 2023. During that time, ecoenzyme-related research publications covered a variety of topics, including composition, training, and product development. N. Ginting and T. Wikaningrum are the authors who have contributed the most to the publication of the ecoenzyme topic, with a total of 12 publication documents. Further research can be conducted on the development of ecoenzymes, such as the investigation of the efficacy and effectiveness of using ecoenzymes and the investigation of other ecoenzyme research involving various other scientific disciplines. This study has limitations due to its reliance on data obtained through Google Scholar and its lack of a rigorous screening process. Therefore, it is recommended that future research take additional measures to filter more collected data.

1 Introduction

The technique of organic waste fermentation has garnered significant interest in recent times owing to its capacity for ecoenzyme production. Ecoenzymes refer to enzymes that are derived from the process of fermenting organic waste. These enzymes find utility in a range of applications, such as wastewater treatment, agriculture, and food processing

* Corresponding author: winarsih@unesa.ac.id

[1]. The use of ecoenzymes derived from the fermentation of organic waste is in accordance with the Sustainable Development Goals (SDGs) established by the United Nations, specifically SDGs [2], which seek to promote sustainable consumption and production patterns [3, 4, 5]. Organic wastes, such as citrus fruit peels, vegetable trash, and food waste, have been utilized in the production of ecoenzymes [6, 7, 1]

Microorganism identification from fermented biowaste is critical for the synthesis of ecoenzymes. Microbial entities, namely bacteria and fungi, exert a substantial influence on the process of fermentation and have the capacity to generate a diverse array of enzymes, encompassing cellulases, amylases, and proteases [8]. The utilization of effective microorganisms (EMs) for the purpose of generating ecoenzymes from food waste has been the subject of investigation, and the findings indicate that the application of EMs can augment the production of ecoenzymes [7]. The utilization of food waste for the generation of ecoenzymes can also facilitate the mitigation of food waste and the advancement of a circular economy [6].

Due to the ability of ecoenzymes to degrade organic pollutants, they have been utilized in wastewater remediation. The utilization of ecoenzymes in the process of wastewater treatment is in accordance with Sustainable Development Goal 6 (SDG 6), which seeks to guarantee the accessibility and sustainable administration of water and sanitation for all [1]. Ecoenzymes made from vegetable and fruit waste have been employed in wastewater treatment, and the results have shown that they can successfully lower the chemical oxygen demand (COD) and biological oxygen demand (BOD) of wastewater [7]. Ecoenzymes can help treat wastewater while reducing the need for chemical treatments, which could have an adverse effect on the environment.

The utilization of ecoenzymes in specific applications has the potential to diminish dependence on chemical treatments and foster the adoption of sustainable practices. The generation of ecoenzymes through the process of fermenting organic waste has the potential to contribute to the mitigation of greenhouse gas emissions by decreasing the quantity of organic waste that is disposed of in landfills [6]. The objective of this paper is to perform a bibliometric analysis of ecoenzymes derived from the fermentation of organic waste and their application, with a particular focus on their relevance to the Sustainable Development Goals (SDGs). This paper aims to conduct a comprehensive examination of the existing body of research pertaining to the production, characterisation, and applications of ecoenzymes derived from the process of organic waste fermentation. Furthermore, it seeks to analyze the potential role of these ecoenzymes in fostering sustainable development.

2 Methods

This research employs a systematic review methodology coupled with bibliometric analysis. The integration of these two approaches will provide a comprehensive analysis of the progression of scientific research both statistically, by employing bibliometrics, and qualitatively, through conducting systematic reviews on specific subjects and content [9]. Bibliometric analysis refers to a quantitative research approach that utilizes the Publish or Perish software in conjunction with the Google Scholar source database. The VOSViewer software was utilized to visualize the publishing trend mapping, with the aim of assessing the extent to which publications have contributed to the advancement of knowledge across diverse literary sources. Bibliometric analysis enables the identification of novelty and trends in research [10]. The present study was conducted using an online search on September 13, 2023. A comprehensive web search was undertaken by researchers, wherein the terms 'Ecoenzyme', 'Eco-enzyme', and 'Eco enzyme' were entered in the title area, covering the period from 2013 to 2023.

3 Result and discussion

The findings of the interpretation analysis conducted using VOSViewer, utilizing publication data extracted from the Google Scholar database, yielded three distinct types of bibliometric mapping visualization. These include network visualization (figure 1a), overlay visualization (figure 1c), and density visualization (figure 1b). The occurrence of each term will be visually represented by a colored circle accompanied by a connecting line, which serves to illustrate the relationship between one keyword and another. There is a positive correlation between the size of a circle representing a term and the frequency of its occurrence in the abstract and title of the gathered publishing data [11].

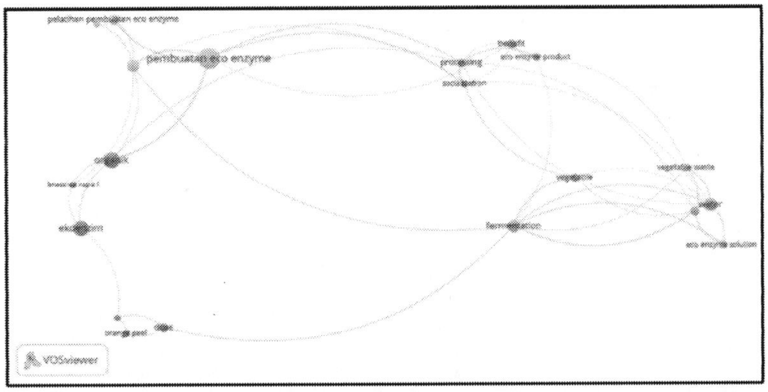

(a)

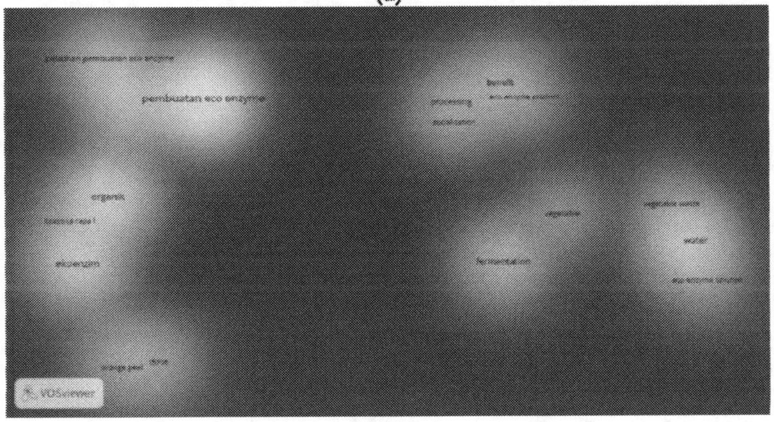

(b)

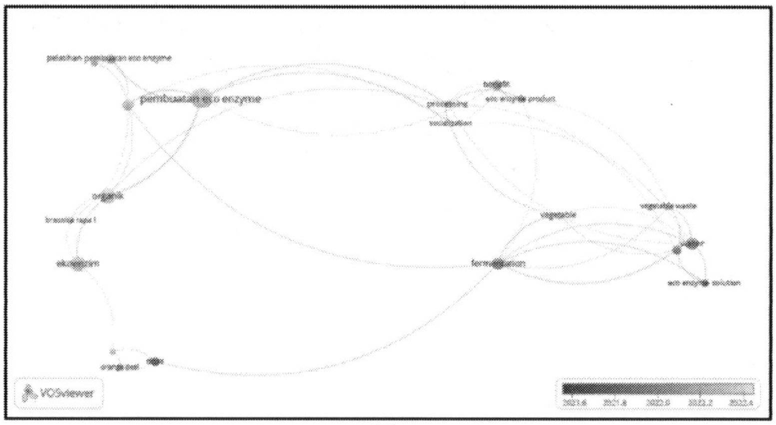

(c)

Fig. 1. Visualization results of bibliometric mapping **(a)** network, **(b)** density, and **(c)** overlay based on the collection of publication data for the last 10 years on the topic of ecoenzymes from the Google Scholar database.

The ecoenzyme topic network visualization reveals that throughout the past decade, publications on ecoenzymes may be categorized into four distinct clusters (Figure 1a). Clusters are classified according to the overarching theme of the frequency of commonly occurring keywords. The initial cluster exhibits a red color and comprises specific keywords, including organic, ecoenzyme, dosage, orange peel, Allium ascalonicum L., and Brassica rapa L (Figure 2a). The cluster of articles identified by these six keywords mostly focuses on the discourse surrounding the production of ecoenzymes and the evaluation of ecoenzyme treatment methods. This includes investigations into the optimal dose levels and the utilization of organic materials in the ecoenzyme manufacturing process. In their study, [12] demonstrated that the ecoenzyme derived from orange peel waste (Citrus sp.) cage cleaning solution possesses inhibitory properties against the growth of Escherichia coli and Staphylococcus aureus. The findings demonstrate that ecoenzymes derived from orange peel waste exhibit a bactericidal effect that is fivefold greater than that of detergent when applied to a standardized test area. Additionally, other studies demonstrate that the usage of ecoenzymes from orange, chicory, and pineapple peels can decrease the dry weight of shallot bulbs (Allium ascalonicum L.) by 18.41, 37.78, and 26.29%, respectively [13]. These two papers represent a subset of the several forms of publications that pertain to similar issues within this cluster, which focuses on development trends seen over the past decade.

As shown in figure 2b, the second cluster is colored green. The cluster encompasses the following keywords: fermentation, vegetable, vegetable waste, water, sugar, and ecoenzyme solution. The keywords present in this cluster suggest that the publications within this cluster primarily focus on the examination of processes related to ecoenzymes. These processes encompass various aspects, including the composition of ecoenzymes and the fermentation process involved in the production of ecoenzyme solutions. The research conducted by [14] demonstrates alterations in both color and volume observed during the organoleptic evaluation of the ecoenzyme solution under investigation. The observed variation might be attributed to disparities in water content within the organic waste utilized, as well as the duration of the fermentation process involved in the production of ecoenzymes, culminating in the formation of the final ecoenzyme product. The findings presented in this publication represent a specific type of scholarly output that pertains to a similar subject matter within the second cluster category throughout the past decade

(a)

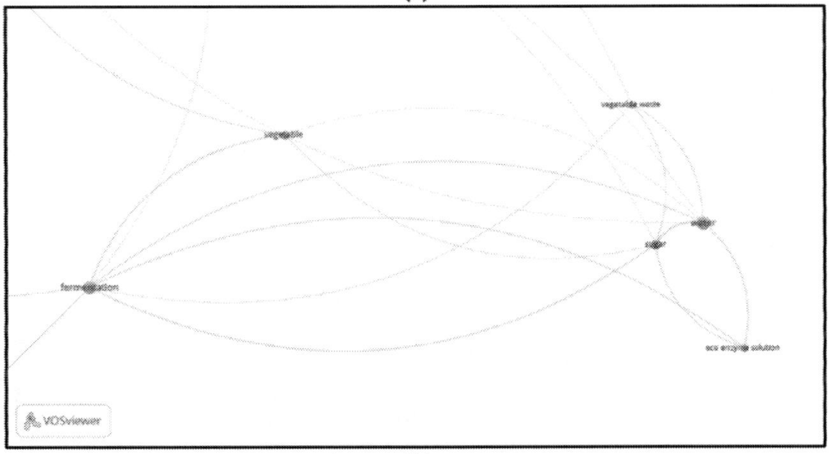

(b)

https://doi.org/10.1051/e3sconf/202345001006

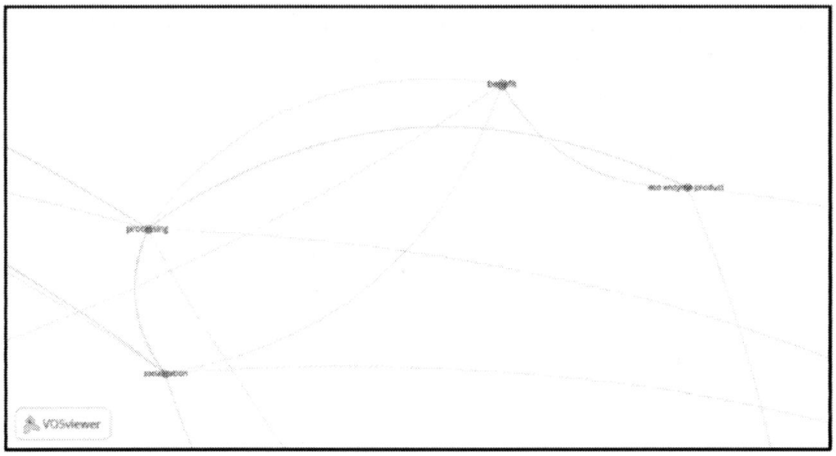

(c)

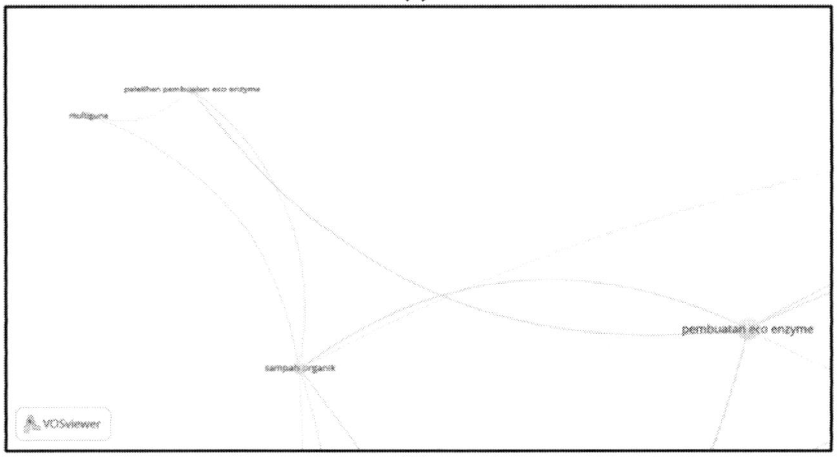

(d)

Fig. 2. Keyword members in **(a)** cluster 1, **(b)** cluster 2, **(c)** cluster 3, and **(d)** cluster 4 ecoenzyme publication topics

The classification of the next cluster is represented by the blue color in figure 2c. The cluster under consideration encompasses the keywords processing, benefit, socialization, and eco enzyme product. These keywords serve as indicators for the categorization of publications within this cluster, which predominantly focus on the analysis of the advantages and applications of ecoenzyme goods. In a recent study conducted by Wikaningrum and [15], it was observed that the utilization of eco enzymes at varying concentrations (0%, 2%, 4%, and 6%) over a period of 10 hours resulted in a corresponding decrease in nitrite levels by 0.3%, 20%, 29%, and 35% correspondingly. The application of greater concentrations, specifically 10%, 15%, and 20%, resulted in nitrite removal efficiencies of 35.7%, 36.7%, 46.7%, and 49.4% correspondingly after a 7-hour exposure period. The findings of these articles suggest that there exists a correlation between the topics covered in publications belonging to the third cluster category within the past decade.

The last cluster depicted in figure 2d is denoted by the color yellow. The members of this fourth cluster contain key terms such as the production of ecoenzymes, training in ecoenzyme production, organic waste management, and multipurpose applications.

Publications within this cluster mostly focus on the examination of endeavors aimed at fostering the development of ecoenzymes within the community, often through means like as training initiatives. [16] elucidated in their publication that conducting training and socialization programs on the utilization of vegetable and fruit waste for ecoenzyme production will enhance the community's capacity to repurpose trash into economically valuable products. In addition, it was emphasized that a crucial factor for the activity's success lies in fostering community knowledge regarding the need to adhere to health norms and comprehending the potential of the village. This understanding enables the conversion of trash into ecoenzyme products that possess versatility and offer numerous advantages. The subject matter under consideration has emerged as a prominent area of focus within the realm of scholarly publications over the past decade, falling within the purview of the fourth cluster type.

Figure 1b illustrates a density representation that showcases the hierarchical arrangement of themes mentioned, as determined by the frequency of keywords present. The intensity of the color of the keyword signifies the frequency of its occurrence or discussion within the article. According to the density mapping visualization, it is evident that the subject matter extensively addressed in ecoenzyme literature over the past ten years pertains to the production of ecoenzymes, particularly those that are closely associated with instruction and familiarization in ecoenzyme manufacturing (Figure 1a). The paper of [2] has demonstrated. Training activities were conducted to teach the production of ecoenzymes, which were subsequently applied with success in Karang Joang Village. The innovative approach to waste management and processing implemented by the individuals in question serves as a cost-saving alternative by offering educational programs and training on the production of disinfection liquid. In contrast, the overlay graphic presents a depiction of year-on-year patterns throughout the past decade pertaining to publications on ecoenzymes, utilizing frequently employed keywords. According to the data presented in Figure 1c, it can be observed that the term "making ecoenzymes" is frequently mentioned in scholarly papers throughout the approximate timeframe of 2022. The data presented in Figure 3 indicates a notable surge in the number of publications related to ecoenzymes in the year 2022, as compared to preceding years within the past decade. Specifically, there were 277 publications on this topic out of a total of 666 publications. The individuals who have made regular contributions to the authorship of ecoenzyme papers throughout the past decade include N. Ginting and T. Wikaningrum.

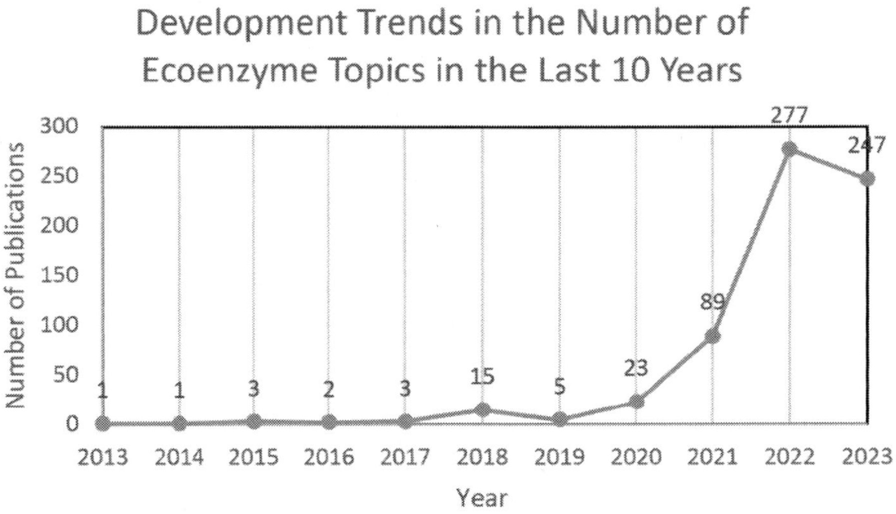

Fig. 3. Trend development in the number of ecoenzyme publications in the last decade

The analysis of keyword mapping for ecoenzyme publications over the past ten years reveals the recurrence of some subjects that serve as updates to the ongoing discourse on ecoenzymes. The scarcity of discussions pertaining to this topic within the field of ecoenzyme studies in past years highlights the need for further exploration and the acquisition of updated knowledge in this area. In addition, it is conceivable that the discourse on ecoenzymes will encompass novel areas of study, including investigations into the usefulness and effectiveness of ecoenzyme utilization, as well as explorations of ecoenzyme research within other scientific disciplines. The bioinformatics approach is a scientific methodology that can be employed for the advancement of ecoenzymes. The field of bioinformatics has experienced significant growth and has had a transformative impact on scientific advancements. The application of computational tools and methodologies is employed to evaluate and understand biological data. The application of bioinformatics has been observed across diverse domains, including genomics, proteomics, and metabolomics, wherein it facilitates the identification and analysis of biological molecules and their relationships. The application of bioinformatics has facilitated the advancement of novel pharmaceuticals, immunizations, and diagnostic instruments [17, 18]. Bibliometrics has an important contribution in research [19-23]. The novelty and importance of further research can be obtained from bibliometric analysis [24-28]. Furthermore, this technique has been employed for investigating the evolutionary processes of biological systems and elucidating the intricate characteristics and functionalities of proteins, such as the enzyme found in ecoenzymes.

4 Conclusion

Based on bibliometric evaluation of the growth of ecoenzyme trend publications. There remain numerous avenues for further advancement in the field of ecoenzyme research, including investigations into the efficacy and effectiveness of ecoenzyme utilization, as well as the exploration of interdisciplinary approaches involving bioinformatics and other scientific disciplines. This can serve as a foundation for future advancements in ecoenzyme research

References

1. N. Benny, et al., J Agric Food Res. **13** (2023)
2. R. G. Harahap, et al., Jurnal Sinar Sang Surya **5** (2021)
3. M. Bilal, H. M. N. Iqbal, Food Res. Int. **123** (2019)
4. M. Lucarini, et al., Molecules **26** (2021)
5. Y. Song, et al., Int. J. Environ. Res. Public Health **18** (2021)
6. G. Dinesh, Nat. Environ. Pollut. Technol **22** (2023)
7. L. C. Wen, R. L. Z. Ling, S. Teo, Appl. Microbiol. Biotechnol **2** (2021)
8. R. L. Z. Ling, et al., Environ. Res. **218** (2023)
9. S. Pizzi, A. Caputo, A. Corvino, J. Clean. Prod. **276** (2020)
10. N. Suprapto, et al., Libr. Philos. Pract. **5788** (2021)
11. I. Hamidah, S. Sriyono, M. N. Hudha, Indones. J. Sci. Technol. **5** (2020)
12. A. Mahdia, et al., Jurnal Ilmu Produksi Dan Teknologi Hasil Peternakan **10** (2022)
13. Y. Hasanah, J. Ginting, A. S. Syahputra, Asian J. Plant Sci. **21** (2022)
14. D. Larasati, P. Astuti, A. T. Maharani. *Uji Organoleptik Produk Eco-Enzyme Dari Limbah Kulit Buah (Studi Kasus Di Kota Semarang)*, in Seminar Nasional Edusainstek **4** (2020)
15. T. Wikaningrum, P. L. Anggraina, IOP Conference Series: Earth and Environmental Science **1065** 012023 (2022)
16. S. P. A. Alkadri, K. D. Asmara, Jurnal Buletin Al-Ribaath, **17** (2020)
17. A. Bayat. *Clinical review Science, medicine, and the future Bioinformatics.* in BMJ-British Medical Journal-International Edition **324** (2002)
18. J. Wang, Z. Li, J. Zhang, BMC Bioinform. **23** (2022)
19. H. N. Hidaayatullaah, N. Suprapto, E. Hariyono, B. K. Prahani, D. Wulandari, J. Phys. Conf. Ser. **2110**, 012026 (2021)
20. B. K. Prahani, J. Alfin, A. Z. Fuad, B. Jatmiko, E. Hariyono, N. Suprapto, Int. J. Emerg. Technol. Learn. **17**, 17 (2022)
21. Nurhasan, B. K. Prahani, N. Suprapto, M. A. A. Ardha, Int. J. Instruction, **15**, 3 (2022)
22. B. Jatmiko, B. K. Prahani, N. Suprapto, S. Admoko, U. A. Deta, N. A. Lestari, M. N. R. Jauhariyah, M. Yantidewi, D. Muliyati, J. Phys. Conf. Ser. **2110**, 012018 (2021)
23. B. K. Prahani, M. Z. B. Amiruddin, N. Suprapto, U. A. Deta, T. -H. Cheng, Int. J. Educ. Method, **8**, 3 (2022)
24. B. K. Prahani, I. A. Rizki, B. Jatmiko, N. Suprapto, T. Amelia, Int. J. Emerg. Technol. Learn. **17**, 8 (2022)
25. N. Suprapto, A. Kholiq, B. K. Prahani, U. A. Deta, J. Phys. Conf. Ser. **012017** (2021)

26. B. Jatmiko, T. Sunarti, B. K. Prahani, E. Hariyono, Dwikoranto, F. C. Wibowo, S. Mahtari, Misbah, M. Asy'ari, J. Phys. Conf. Ser. **2110**, 012020 (2021)
27. B. K. Prahani, M. Z. B. Amiruddin, B. Jatmiko, N. Suprapto, T. Amelia, Int. J. Interactive Mobile Technologies, **16**, 8 (2022)
28. U. A. Deta, I. Laeliyah, N. A. Lestari, M. Yantidewi, M. N. R. Jauhariyah, B. K. Prahani, AIP Conf. Proc. **2595**, 020013 (2023)

Strengthening the ecosystem for preventing and handling sexual violence on campus to support achievements SDGs gender equality

Mutimmatul Faidah[1*], *Dina* Rakhmawati[2], *Iman* Pasu[3], *Diana* Rahmasari[4], and *Rojil Nugroho* Bayu Aji[5]

[1]Cosmetology Education, Universitas Negeri Surabaya, Surabaya, Indonesia.
[2]Communication Science, Universitas Negeri Surabaya, Surabaya, Indonesia.
[3]Pancasila and Civic Education, Universitas Negeri Surabaya, Surabaya, Indonesia.
[4]Psycology, Universitas Negeri Surabaya, Surabaya, Indonesia.
[5]History Education, Universitas Negeri Surabaya, Surabaya, Indonesia.

Abstract. This research aims to describe (1) the ecosystem for preventing and handling sexual violence (PHSV) on campus and (2) the effectiveness of implementing an ecosystem for PHVS on campus. This qualitative research with interviews, FGDs, and document studies. The subjects were University Leaders, PHSV Task Force, Anti-Sexual Violence Ambassadors, lecturers, students, and educational staff at Surabaya State University. The data analysis technique uses the Miles and Haberman pattern. The research results show that the ecosystem for preventing and handling sexual violence (PHSV) on campus includes (a) campus community, (b) physical elements, (c) PHSV responsive curriculum, (d) regulation, and (e) interactions between campus community. Strengthening the academic atmosphere through the design of a hidden curriculum in the form of non-violent interactions and the integration of anti-sexual violence in learning. Prevention strategies effectively create a positive, friendly, safe, equal, and participatory campus culture fighting sexual violence. Case-handling strategies have a positive impact from various sides. Victims dare to speak up and receive psychological, social, and legal assistance. The perpetrator will receive sanctions according to existing regulations, thereby ensuring that there is no recurrence of the case and increasing the commitment of the campus community not to commit acts of sexual violence.

1 Introduction

Sexual violence is still a global problem that significantly impacts women [1], including in Indonesia. Sexual violence can be defined as an attack that contains sexual elements, whether sexual intercourse occurs or not, regardless of the relationship between the victim and the perpetrator [2]. During the 2015-2021 period, the National Commission

*Corresponding author: mutimmatulfaidah@unesa.ac.id

on Violence Against Women noted 67 cases of violence against women in educational environments with details: sexual violence at 87.91 percent, psychological and discrimination at 8.8 percent, and physical violence at 1.1 percent. Universities rank first for sexual violence in educational environments, with 35 reported cases. This number is like an iceberg because only a handful of cases of sexual violence have been revealed. To create an academic environment that is supportive, inclusive, and free from sexual violence, appropriate regulations and synergy from various parties are needed [3].

The Indonesian Government responded to this issue by issuing Minister of Education, Culture, Research and Technology Regulation No. 30 of 2021 concerning the Prevention and Management of Sexual Violence in Higher Education Environments. This regulation is fresh air for the entire higher education academic community. This rule is a guideline in formulating policies and taking steps to prevent and handle sexual violence. It also provides legal certainty for university leaders to take firm action against perpetrators of sexual violence. Surabaya State University is committed to realizing a campus with zero sexual violence to ensure all campus community safety, comfort, and benefit and achieve the Sustainable Development Goals, especially gender equality (SDG 5).

Research related to sexual violence on campus has been carried out using various approaches. O'Connor [4] researched Predictors of campus sexual violence perpetration. Research that reviewed 28 articles found that the general predictors of sexual violence on campus were personality factors, attitudes, and negative life experiences as the main predictors. In contrast, other factors, such as sexual behavior and past experiences, received less attention.

Martin-Storey [5] researched sexual violence on campus for sexual minorities and transgender people. These findings highlight the high levels of vulnerability to sexual violence among gender minority students. The report also indicates the context in which the violence occurred and suggests specific strategies for prevention. Kettrey [6] examines the implementation of the US Elimination of Sexual Violence on Campuses (SaVE) Act, which mandates that all higher education institutions offer prevention and awareness-building programs addressing sexual violence to new students. This research recommends that programming efforts not only focus on individuals but also adopt an ecological perspective that targets individuals, social relationships, community, and societal factors.

Senn et al. [7] researched the effectiveness of self-defense and resistance programs against sexual assault for women. Programs with evidence-based content formats and providing practice opportunities are effective in reducing levels of sexual victimization for women. This program can also increase confidence, knowledge in self-defense, willingness to use self-protection strategies, and even reduce attitudes of self-blame. This program can increase women's assertiveness in sexual situations and reduce fear of violence in the future. This research recommends a self-defense program for women.

In contrast to the research above, which focuses more on prevention programs, this research views sexual violence as an action involving various elements. The novelty of this research is a systemic approach to photographing the prevention and handling of sexual violence. This research examines (1) elements in the ecosystem for preventing and handling sexual violence on campus and (2) the effectiveness of implementing the ecosystem for preventing and handling sexual violence at Surabaya State University. This issue is essential to study as a reference in ending sexual violence on campus through a systemic approach.

2 Methods

This research is descriptive qualitative research, where qualitative research is research that aims to interpret phenomena by involving various methods [8]. Data collection techniques using interviews and focus group discussion (FGD). For data completeness, relevant document studies were also carried out in the form of journals or supporting information from mass media or the Internet.

The subjects of this research were University Leaders, the PHSV task force, Anti-Sexual Violence Ambassadors, lecturers, students and educational staff at Surabaya State University. The location of this research is at the State University of Surabaya, Ketintang and Lidah Wetan. The research was carried out in March – October 2023. The data analysis technique used was the Miles and Haberman pattern through data condensation, data presentation and drawing conclusions.

3 Result and discussion

3.1 Sexual violence prevention and management ecosystem

Regulation of the Minister of Education and Culture of the Republic of Indonesia No. 30 of 2021 explains all acts that constitute sexual violence, namely: (1) Speech that discriminates or insults the victim's physical appearance, body condition, and/or gender identity; (2) Intentionally exposing his genitals without the victim's consent; (3) Delivering remarks containing sexual advances, jokes, and/or whistling to the victim; (4) Staring at the victim with sexual overtones and causing discomfort; (5) Sending messages, jokes, images, photos, audio, and/or videos with sexual nuances to the victim even though the victim has prohibited it; (6) Taking, recording, and distributing photos and audio and/or visual recordings of the victim that have sexual nuances without the victim's consent; (7) Uploading photos of the victim's body and/or personal information that have sexual nuances without the victim's consent; (8) Disseminating information regarding the victim's body and personality that has a sexual nuance without the victim's consent; (9) Peeping or deliberately looking at the victim who is carrying out activities privately and/or in a private space; (10) Persuading, promising, offering something, or threatening the victim to carry out sexual transactions or activities that the victim disagrees with; (11) Giving punishment or sanctions that have a sexual nuance; (12) Touching, rubbing, touching, holding, hugging, kissing, and/or rubbing body parts on the victim's body without the victim's consent; (13) Opening the victim's clothes without the victim's consent; (14) Forcing the victim to carry out sexual transactions or activities; (15) Practicing a community culture of Students, Educators and Education Personnel with nuances of Sexual Violence; (16) Attempted rape, but penetration did not occur; (17) Committing rape includes penetration with objects or body parts other than the genitals; (18) Forcing or deceiving the victim to have an abortion; (19) Forcing or deceiving the victim to become pregnant; (20) Intentionally allowing sexual violence to occur; (21) Committing other acts of sexual violence

In higher Education, the potential and opportunity for sexual violence can occur between (1) lecturers and students, (2) lecturers with educational staff, (3) lecturers and campus community, (4) students with students, (5) students with Education; (6) students and campus community; (7) educational staff with educational staff; (8) staff with campus community; and (9) lecturers, students and staff with external parties. Violence can occur during interactions in learning, research, community service, student activities, actual work lectures, internships, and outside the lecture context. Violence can occur verbally, non-verbally, physically, and digitally.

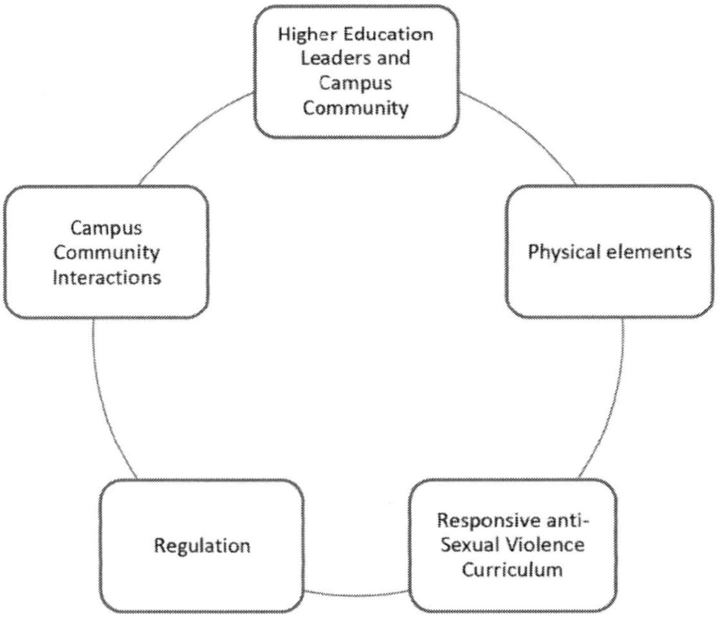

Fig. 1. PHSV Ecosystem Elements on Campus

Prevention and treatment efforts involve various elements, namely campus community, physical elements, PHSV responsive curriculum, regulations, and interactions between campus community.

a. Campus community, including higher education leaders, lecturers, students, education staff, and parties involved in the *Tri Dharma* of Higher Education. A massive and systematic anti-sexual violence education and outreach program is needed to target these components.
b. Physical elements include buildings, parks, fields, places of worship, laboratories, libraries, and other facilities. Sexual violence can occur anywhere and at any time. This requires the establishment of campus facilities that are conducive to preventing sexual violence. The instruments needed are (1) bright lighting; (2) the existence of an SOP for space use that is responsive to anti-sexual violence issues; (3) friendly for people with disabilities; (4) there is an operational time limit for the use of the

facility; (5) control and monitoring continuity from the security team; (6) ease of room access; and (7) the existence of an alarm that can be used as a sign of danger.

c. A curriculum that is responsive to anti-violence issues is developed through three patterns, namely (1) special courses that discuss gender and sexual violence, such as courses on Gender and Media, Gender in Law, Sociology of Gender, and others; 2) integrating the issue of sexual violence in relevant courses in one discussion chapter, such as courses in Islamic Religious Education, Pancasila Education, Citizenship Education, Family Welfare Science, and others; (3) anti-sexual violence is a value that is built into learning interactions; and (4) participation in anti-sexual violence education and training which is mandatory for all lecturers, students and educational staff.

d. Regulations, Regulation no. 30 of 2021, which is the legal umbrella for preventing and handling sexual violence on campus, needs to be operationalized through (1) the chancellor's regulation discussing PHSV; (2) the PHSV guidebook, which is a reference for the PHSV implementation; (3) granting authority to the PHSV Task Force and guaranteeing protection to the team; (4) Leadership commitment to enforcing rules without discrimination; and (5) supporting the program in the form of adequate funding.

e. Campus community interaction. The formation of non-violent habits is the core of interactions in learning or outside of learning, both directly and indirectly. To develop humanistic, respectful and mutually respectful interactions, systematic habituation is required. Strengthening this habit can be outlined in interaction guidelines on campus.

The ecosystem concept in this research is in line with Regehr [9] that Universities are large, complex organizations. The population and physical environment are similar to those of a small town or large city. Like a small town, a university has a community and workers, libraries, public spaces, office buildings, roads, parking lots and more. On campus interact diverse individuals with mental health and behavioural challenges similar to society at large. Universities have an obligation to safeguard and protect their citizens, ensuring health, security and mental health.

The importance of tertiary institutions' commitment is in line with the article [10], schools and universities must provide more intense support to survivors of sexual violence and dismantle the structures that perpetuate and create conditions for harassment. Sara Ahmed photographed the existence of a culture of sexual harassment and violence in the educational environment. This research provides recommendations for the importance of understanding gender and sexual violence in the context of education to accelerate social transformation.

In the digital era, violence has shifted to the mode of cyber-based sexual violence or what is usually called Online Gender-Based Violence (GBV). GBV is a form of verbal violence that occurs due to the development of information and communication technology. GBV is carried out through the media online with the intention of harassing the victim based on gender [11]. There are at least nine forms of GBV according to the Commissioner of the National Commission on Violence Against Women, Sitohang V [12], namely; (1) Cyber harassment is a threat to the victim to be afraid of using technology and social media; (2) Cyber hacking is the most common cybercrime act experienced by many people; (3) Impersonation is taking another person's identity to humiliate, insult the victim, or create false documents; (4) Cyber recruitment manipulates the victim, leading them into a detrimental and dangerous situation; (5) Cyber stalking is an act of misuse of technology to stalk other people, either by direct observation or investigating the victim's traces; (6) Malicious distribution is the digital distribution of content to damage the reputation of victims or organizations defending women's rights; (7) Revenge porn is distributing pornographic videos or photos of victims aimed at revenge; (8) Sexting is sending pornographic images or videos to the victim; (9) Morphing, namely changing an image or video to damage the person's reputation in the photo/video.

In the context of Surabaya State University, reports of sexual violence until September 2023, 80% were dominated by gender-based violence in the form of revenge porn, Cyber recruitment, Malicious distribution, Sexting, and Morphing. Mendes et al. [13], in research on Sexual violence in contemporary educational contexts, found that increasing time spent interacting digitally increases the risk of sexual violence in the form of online sexual harassment and image-based sexual harassment in some teenagers. Filice [14] examines the rise of online dating websites and applications that offer the convenience of establishing new relationships. However, this platform has great potential as a catalyst for sexual violence, whether inside or outside the network. This research reveals an increase in the prevalence of sexual violence through online dating.

There are variations in the relationship between perpetrators and victims based on the reports received. Most violent activities are carried out in romantic relationships, friendship relationships, and as a form of fraud. Sexual violence in romantic relationships is in the spotlight of Miller [15]. He revealed that there was violence in dating or harassment in the interactions between men and women who were romantically linked. This violence commonly occurs since middle school adolescence and peaks in early adulthood. Victims of dating violence experience a variety of negative impacts on physical and mental health, including unwanted pregnancies and sexually transmitted infections. This research recommends universal Education and applicable sexual violence prevention guidelines. Referring to this phenomenon, being active in maximizing digital literacy education and understanding forms of gender-based violence is a necessity.

3.2 Implementation of an ecosystem for preventing and handling sexual violence

Prevention of Sexual Violence at Surabaya State University is carried out through various strategies:

a. Learning by requiring new students, lecturers, and education staff to participate in violence prevention lessons through the Learning Management System. Hopefully, this strategy will provide the foundation for forming insight, understanding, and attitudes toward anti-sexual violence. In this context, it is necessary to develop Semester Lecture Plans, learning media, and textbooks that discuss sexual violence and integrate anti-violence values in lectures. The pattern that has been developed is the integration of anti-sexual violence in Islamic religious education courses. This theme becomes a chapter discussed and integrated into the RPS and textbooks.

b. Strengthening campus governance with a pattern of (1) policies that support PHSV in Higher Education; (2) forming a Task Force to oversee PHSV; (3) compiling PHSV guidelines; (4) limiting meetings between campus community outside operational hours or outside campus areas; (5) availability of easily accessible sexual violence reporting services; (6) providing Education and training for students, lecturers, education staff and campus community regarding PHSV efforts; (7) provide reasonable accommodation for persons with disabilities for PHSV; and (8) cooperation with related agencies for the protection of victims and others.
c. Strengthening the community culture of students, educators, and educational staff can take the form of communication, information, and Education regarding the prevention and handling of sexual violence in various activities.

Prevention programs are effective in reducing cases of sexual violence committed by lecturers and education staff. In 2021 – 2022, there were four reported cases of lecturers committing acts of violence, and sanctions were determined. 2023, there will be a decrease; there is one case report involving an educational worker. On the other hand, student cases experienced a spike, reaching 23 cases until September 2023. This spike in cases is understood as (1) the courage of student victims to report and dare to speak up as a product of the massive Education and outreach program carried out; (2) increasing understanding regarding sexual violence among lecturers and education personnel so that they do not become perpetrators or victims; and (3) new patterns of digital violence have not been addressed appropriately among students.

This finding aligns with research by Rominski et al. [16], which states that sexual violence prevention education programs for students can change attitudes and behavior about sexual violence. Acceptance of rape myths is common, and educational programs can reduce acceptance of rape myths. Fayaz [17] highlights poor interactions between lecturers and students, students in the same class and seniors, and friendships. Fayaz recommended creating healthy relationships at various levels, increasing emotional intelligence, and introducing seniority awards among students. Efforts to end campus violence must be supported and institutionalized.

The effectiveness of preventing sexual violence is researched [1]. The implementation of Global Consent for students in Vietnam has been proven to reduce the occurrence of sexually violent behavior. Through legal Education, the dangers of sexual violence and the importance of empathizing with victims can increase student pro-social behaviour. Efforts to prevent sexual violence should ideally be carried out from an early age across levels of Education, in line with the findings of Anderson et al. [18], who studied the cycle of violence. Violence and poly victimization experienced in childhood are associated with violent acts in adolescence and adulthood. The existence of this attachment necessitates intervention to break the cycle of violence. The results of this study underscore the need to screen for and identify violence in childhood to stop the cycle of violence.

Teachers and education personnel have the potential to become reliable Anti-Sexual Violence reporters. Experiences of childhood sexual abuse among college students are associated with their current depressive symptoms. Schools, families, and society need to be encouraged to provide healthy sexual behavior prevention interventions related to sexual violence to childhood students. In this context, the Indonesian Government has created regulations for the Prevention and Handling of Violence in Schools by promoting the Child-friendly School program.

Apart from preventing sexual violence, universities are also required to handle sexual violence by:

a. Assistance to victims or witnesses who are students, educators, education staff, and the community. The assistance provided is in the form of counseling. Counseling by an on-site psychologist adjusts the victim's comfort. Counselors apply pro-victim principles and fulfill victims' needs. Apart from counseling, health services, legal assistance, advocacy, and social and spiritual guidance are also provided. People with disabilities receive assistance by paying attention to the needs of victims. Accompanying victims or witnesses is, of course, carried out based on agreement. Then, when the victim or witness cannot consent, consent can be given by parents, guardians, or companions. This program is effective in restoring the victim's psychological condition.
b. Protection is provided to victims or witnesses who are students, educators, education staff, and campus community. Protection can be in the form of (1) guarantee of continuity in completing Education for students; (2) guarantee of continued employment as an educator and/or educational staff at the University concerned; (3) guarantee of protection from physical and non-physical threats from the perpetrator or other parties or recurrence of sexual violence in the form of facilitating reporting of physical and non-physical threats to law enforcement officials; (4) protection of identity confidentiality; (5) providing information regarding protection rights and facilities; (6) providing access to information on protection implementation; (7) protection from attitudes and behavior of law enforcement officers that demean and/or reinforce stigma against victims; (8) protection of victims and/or reporters from criminal prosecution; (9) civil lawsuits for reported incidents of sexual violence; (10) provision of safe houses; and/or (11) protection of security and freedom from threats relating to the testimony given.
c. Imposition of administrative sanctions when the perpetrator is proven to have committed sexual violence. The imposition of administrative sanctions is determined by a Higher Education Leadership Decree based on recommendations from the PHSV Task Force. Administrative sanctions are classified into three categories: light administrative, medium administrative, and heavy administrative. Light administrative sanctions in the form of a written warning or a written statement of apology published internally on campus or in the mass media. After completing light administrative sanctions, the perpetrator must participate in a counseling program at an institution appointed by the PHSV Task Force. The perpetrator bears funding for the counseling program. The report on the counseling program results is the basis for Higher Education Leaders to issue a certificate that the perpetrator has carried out the sanctions imposed. Administrative sanctions are in the form of temporary dismissal from office without obtaining office rights or reduction of rights as a student, which include: (1) postponement of attending lectures (suspension); (2) revocation of scholarship; or (3) reduction of other rights. Meanwhile, heavy administrative sanctions are in the form of permanent dismissal as a student or permanent dismissal from a position as an educator, education staff,

or campus member by the provisions of the laws and regulations of the relevant higher education institution. The imposition of sanctions is carried out proportionally and fairly by the recommendations of the PHSV Task Force. Higher Education Leaders can impose administrative sanctions that are more severe than the administrative sanctions recommended by the PHSV Task Force, taking into account (1) the victim is a person with a disability, (2) the impact of sexual violence experienced by the victim; and (3) the reported party or perpetrator is a member of the PHSV Task Force, head/head of a study program, or head of department. In 2023, four sanctions have been imposed on perpetrators of sexual violence in the severe category. Imposing sanctions is effective in minimizing cases of sexual violence committed by lecturers.

d. Victims recover through medical procedures, physical therapy, psychological therapy, and social and spiritual guidance. Victim recovery can involve doctors/other health workers, counselors, psychologists, community leaders, religious leaders, and other companions according to needs, including those of victims with disabilities. The victim's recovery period does not reduce the victim's rights in the learning process, employment rights, or other rights by statutory provisions. The characteristics of perpetrators are an interesting study as a basis for developing sexual violence prevention programs on campus [19]. There are four areas of consideration discussed: (1) the perpetrator's modus operandi or pattern of actions, (2) the perpetrator's risk factors, (3) the perpetrator's perception of norms, and (4) situational risk factors. Recommendations in the form of programs to prevent and handle sexual violence at individual and community levels as a support for creating a friendly and comfortable campus.

Referring to the description above, the formula for handling sexual violence at Surabaya State University boils down to three approaches: mentoring, protection, and providing sanctions. The principle emphasized in handling sexual violence is the pro-victim principle. This principle ensures that what the victim wants and needs becomes a priority. The imposition of sanctions for perpetrators of sexual violence is based on the University's code of ethics and civil servant disciplinary regulations [20]. If the victim wants the case to be processed legally, the PHSV Task Force will provide legal assistance for reporting the campus to the authorities. In this context, Indonesia has enacted the Criminal Act of Sexual Violence. Laws in Indonesia regarding sexual violence include prevention, fulfillment of victims' rights, recovery of victims, and regulation of handling during the legal process. Criminal sanctions will be imposed if proven wrong. This differs from handling on campus; the sanctions given are administrative for the perpetrator. The highest administrative sanction, namely being dismissed as a lecturer and education staff.

Results should be clear and concise. The discussion should explore the significance of the results of the work, not repeat them. A combined Results and Discussion section is often appropriate. Avoid extensive citations and discussion of published literature.

4 Conclusion

From this research, it can be concluded that the ecosystem for preventing and handling sexual violence on campus includes (a) biotic elements consisting of higher education leaders, students, lecturers, education staff, and other campus community, as mentioned in the *Tri Dharma* of Higher Education; (b) physical elements, in the form of adequate buildings or facilities; (c) PHSV responsive curriculum; (d) regulation; and (e) interactions between campus community. Implementing the PHSV ecosystem at Surabaya State University relies on PHSV responsive leadership policies in creating a positive, friendly, safe, equal, and participatory campus culture in fighting all forms of sexual violence. Apart from that, strengthening the academic atmosphere through a designed hidden curriculum in the form of non-violent interactions between campus communities and the integration of anti-sexual violence in learning.

Acknowledgments

Thank you to the Rector of Universitas Negeri Surabaya who has provided support and funding for this research

References

1. K. M. Yount, et al., Soc. sci. med. **313**, 115402 (2022)
2. M.T. Indrayana, Malay Health Journal **1,** 1 (2017)
3. K. Imania, A. Santoso, Educ. Res. **33**, 2 (2023)
4. J. O'Connor, et al., Aggress Violent Behav. **58** (2021)
5. A. Martin-Storey, et al., J Adolesc Health **62**, 6 (2018)
6. H. H. Kettrey, et al., J Adolesc Health **72**, 6 (2023)
7. C. Senn, J. Hollander, C. Gidycz, What Works? Critical Components of Effective Sexual Violence Interventions for Women on College and University Campuses, (Academic Press, 2018)
8. A. Anggito, J. Setiawan, Qualitative Research Methodology (CV Trace, Sukabumi, 2018)
9. C. Regehr, et al., Int J Law Psychiatry **4** (2017)
10. S. Ahmed, Complaint! (Duke University Press, Durham, 2021)
11. R. A. Musyaffa, S. Effendi, Communicology: Scientific Journal of Communication Science **19**, 2 (2022)
12. A. L. Hoffmann, New Media Soc. **23**, 12 (2021)
13. K. Mendes, T. Horeck, J. Ringrose, Gend Educ. **34**, 2 (2022)
14. E. Filice, et al., Aggress Violent Behav. **67** (2022)
15. E. Miller, Pediatr Clin North Am. **64**, 2 (2017)
16. S. Rominski, E. Darteh, M. Munro-Kramer, Lancet Glob. Health **5**, 1 (2017)
17. F. Fayaz, et al., SSHO **7**, 1 (2023)
18. K. Anderson, et al., SSM - Population Health **18**, 101103 (2022)

19. J. Zatkin, et al., Perpetrator characteristics: Implications for programming, policy, and sanctioning on college campuses (Engaging Boys and Men in Sexual Assault Prevention, Academic Press, 2022)
20. Z. Turhan, Aggress Violent Behav. **50** (2020)

Analysis of the pre-service physics teacher's ability to develop SDGs-oriented multimedia

Abd Kholiq[1*], *Imam* Sucahyo[1], *Mita* Anggaryani[1], *Muhammad* Satriawan[1], and Muhammad Habibbulloh[1]

[1]Physics Education Study Program, Universitas Negeri Surabaya, Surabaya, Indonesia

Abstract. The teacher's ability to develop multimedia is one of the supporting factors in communicating physics learning. Besides that, another requirement for a teacher is to introduce the concept of SDGs to students to prepare generations to achieve sustainable development goals for a better future. Therefore, this study aimed to analyze the abilities of pre-service physics teachers in developing SDGs-oriented multimedia. This study is descriptive research on the pre-service physics teacher who joined multimedia lectures in the Physics Education study program at one of the state universities in Surabaya, Indonesia. Data was collected through interviews and using product assessment sheets, then analyzed descriptively. Based on the results of data analysis, of the six indicators of the ability to develop multimedia, only two indicators were in the high category, namely the ability to edit videos and Accurate presentation of SDG content, and one indicator in a low category was ability to present animations, while the others three in the medium category. Therefore, it can be concluded that pre-service physics teachers' ability to develop SDGs-oriented multimedia is in the medium category.

1 Introduction

In the era of globalization and rapid technological developments, education is one of the critical aspects that must be continuously updated to face the challenges of the times. Education is tasked with forming the next generation who are highly competitive, have a deep understanding of global issues, and can contribute to achieving sustainable development goals [1-3]. One of the fields of education that has an essential role in forming a quality generation is physics education. Physics education strategically shapes students' analytical, critical, and creative thinking patterns and helps them understand the natural phenomena around them. In addition, multimedia technology in the learning process has become increasingly relevant, especially to meet the demands of a curriculum oriented toward Sustainable Development Goals (SDGs).

The SDGs are a global framework established by the United Nations (UN) to address various global challenges, including poverty, inequality, and environmental damage [4]. In education, the SDGs aim to improve the quality and accessibility of education for all, ensure

* Corresponding author: kholiq@unesa.ac.id

inclusive and equitable education, and encourage innovation and continuous learning. To meet the demands of the SDGs, physics teachers have a vital role in developing innovative and relevant learning methods and media [5,6]. Using SDGs-based multimedia in physics learning is expected to help increase students' understanding of the relationship between physics and global issues and motivate them to actively participate in achieving sustainable development goals.

However, special abilities are needed for pre-service physics teachers to develop effective SDGs-based multimedia. The ability to integrate physics concepts with SDG issues, identify student needs and characteristics, and master multimedia technology is a critical factor in the development process [7]. Therefore, research on analyzing the abilities of pre-service physics teachers in developing SDGs-based multimedia is very relevant for exploration. In addition, based on the results of the analysis carried out by researchers from several previous research articles in developing learning multimedia, developers often need help with several obstacles, namely the low ability to produce exciting and high-quality multimedia content [8]. Another obstacle is the difficulty in creating engaging content due to a low understanding of the topics discussed and the need for more ability to package them in an exciting and easy-to-understand way [9,10]. From these various problems, it is necessary to carry out a more profound analysis regarding the developer's ability, in this case, pre-service physics teachers. From this problem, they can be used as a priority scale in teaching pre-service physics teachers regarding the development of multimedia physics learning, primarily in developing SDGs-based physics learning multimedia.

Research related to physics learning multimedia has been carried out by previous researchers, including the use of multimedia in physics learning [11,12], development of physics learning multimedia [13-15], review of literature related to learning multimedia [16-18], analysis of the effect of multimedia on physics learning outcomes to improving specific skills [19-21]. From a number of these studies, there has been no research related to the analysis of the ability of teachers or pre-service physics teachers to develop multimedia physics learning. Likewise, the content in the multimedia produced still needs to emphasize the contents of the SDGs. Therefore, this study aimed to evaluate pre-service physics teachers' ability to design SDGs-based multimedia in physics learning. The results of this research certainly provide valuable input for educational institutions in developing better physics teacher training and development programs. They can provide valuable insights for the development of SDGs-based education more broadly.

2 Method

The research methods used are descriptive quantitative and qualitative methods [22]. The data collected in a descriptive quantitative manner is used to provide an overview of the actual situation and answer questions related to the status of the research subjects. Meanwhile, qualitative data strengthens and complements quantitative data on the problem under study. The research sample is a study program of physics education students contracting multimedia lectures. Based on the data, students were selected by dividing the categories of high ability (S1), moderate ability (S2), and low ability (S3). The indicators of the ability to develop multimedia can include the following: (1) Graphic Design Skills, namely the ability to use graphic design software such as Adobe Photoshop, Adobe Illustrator, or CorelDRAW to create graphics, logos, and other visual elements in multimedia; (2) Video Editing Capability, namely the ability to edit videos using video editing software such as Adobe Premiere, Final Cut Pro, or Sony Vegas Pro, including cutting, merging, color adjustment, and special effects; (3) Animation skills include creating animations using animation software such as Adobe Animate, Blender, or Maya, including 2D and 3D animation, characters, effects, and object movement; (4) Understanding of Design Concepts, namely the ability to understand multimedia design principles, such as good layout, effective use of colors, proper typography, and visual consistency, as well as the topic of the content loaded; (5) Creativity and Innovation, namely the ability to produce creative and innovative ideas in developing attractive multimedia that attracts the audience's attention; and (6) Accurate presentation of Sustainability development goals (SDGs) content.

Table 1. The level scale of multimedia development ability

Level of multimedia developing ability	Score
High	$75<x\leq 100$
Moderate	$60<x\leq 75$
Low	$x\leq 60$

Quantitative data was collected through a product assessment rubric and product dissemination, while qualitative data was collected through observation. Furthermore, the collected data is analyzed using an inductive approach in which conclusions are drawn from minor case investigations in detail to provide a big picture. Furthermore, determining the ability level of clone physics teachers in developing SDGs content-based physics learning multimedia follows Table 1.

3 Results and discussion

Developing multimedia learning has many benefits and importance in modern education. Multimedia helps maintain student interest and makes the learning process more enjoyable. Learning materials presented in different ways help facilitate a more profound understanding. With multimedia learning, students can learn independently. Learners can choose the time and place to study according to their preferences. It allows learners to learn at their own pace, increasing the independence of their learning, thereby enabling each learner to learn in the most effective way for them. Therefore, teachers must be able to develop teaching materials in the form of multimedia in learning, likewise with physics teachers. Based on the data analysis results, pre-service physics teachers generally have abilities in the moderate category of developing multimedia learning physics with an average score of 69.95, as shown in Table 2.

Table 2. The level of PSPT ability in developing multimedia of each indicator

Indicators	Score	Level
Graphic design ability	67.20	Moderate
Video editing ability	69.80	Moderate
Ability to present animations	59.30	Low
Understanding of design concepts	77.70	High
Creativity and Innovation	69.40	Moderate
Accurate presentation of SDGs content	76.30	High
Average	69.95	Moderate

Based on Table 2, the ability of pre-service physics teachers to develop learning multimedia needs to be improved again. It can be seen from all indicators of the ability to develop multimedia physics learning; only two indicators are in the high category, namely the ability of pre-service physics teachers to understand multimedia design concepts with an average score of 77.70 and the accuracy of pre-service physics teachers in presenting SDGs content with a score of 76.30. And there is one indicator that is in the low category, namely the ability to present animation. In contrast, the other three indicators are in the medium category. The ability of pre-service physics teachers to present SDGs content can be seen in the content of the videos developed; in this case, the prospective teachers raise the theme of environmental pollution, as shown in Figure 1.

In Figure 1, it can be seen that prospective teachers present content related to the environment, which initially presented sources of environmental problems such as excessive use of air conditioners, use of vehicles, forest fires, and industrial disposal. This phenomenon is presented and narrated with good audio. Then, at the end of the video, the prospective physics teacher invites them to prevent global warming by planting trees, saving energy, etc.

Furthermore, if viewed from the participation of the number of teachers related to the ability to develop SDGs-based physics learning multimedia, it was found that only 36.51% of physics teacher candidates could develop physics learning multimedia in the high category, 54.76% in the medium category, and 8.73% in the low category. More details can be seen in Figure 2.

Fig. 1. An example of multimedia produced by PSPT.

Based on Figure 2, it can be seen that of the six indicators of the ability to develop SDGs-based and physical learning multimedia, there is only one indicator where the percentage of pre-service physics teachers is in the high category, namely the ability to present SDGs content in the developed multimedia. In the other five indicators, the percentage of the number of teachers is dominated by the medium category for each indicator. It is, of course, due to various factors. Based on the results of observations, the level of mastery and expertise in using graphic design software such as Adobe Photoshop, Illustrator, InDesign, CorelDRAW, or other software still needs to be improved. This results in an imbalance and disharmony in composing the visual elements of a design, including the arrangement of negative and positive space [23]. In addition, the quality of the video could be better with clear, not blurred or distorted images [24]. This is due to the need for more skills from pre-service physics teachers in editing videos. Edits should be done carefully to ensure visual consistency and quality [25,26]. If the video has a storyline, video editing skills should reflect a clear and logical storyline. Transitions between clips and scenes should be seamless and help build the narrative [27]. The use of transitions and visual effects should be relevant to the style and message of the video.

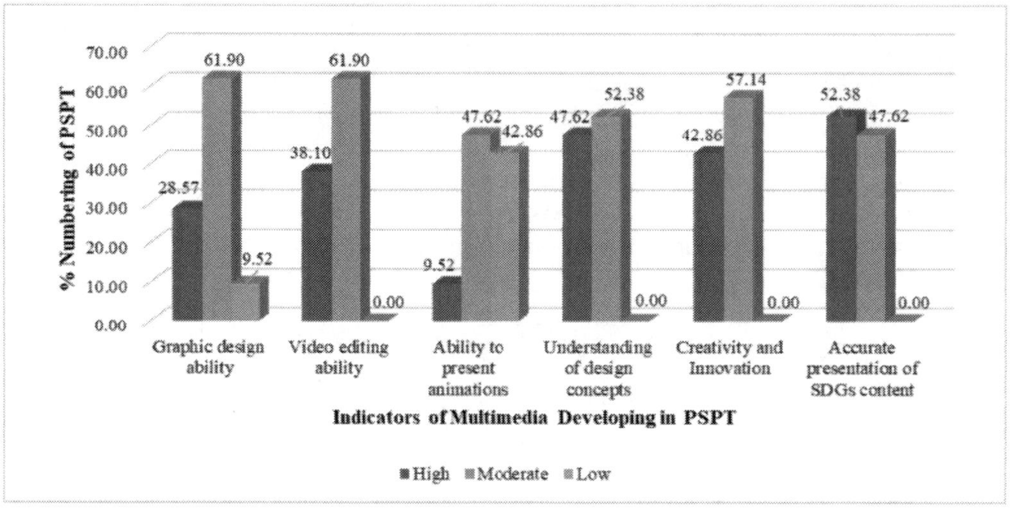

Fig. 2. Percentage of PSPT's level of each multimedia developing ability indicator.

Another factor that causes indicators of the ability to develop learning multimedia to be in a low category, especially in the aspect of presenting animation, is that the presentation of animation needs better visual quality with attractive character designs and backgrounds and inaccurate illustrations [28]. There is no visible use of keyframes and in between to ensure natural and authentic movement and an engaging storyline. Besides that, the creativity and innovation of pre-service physics teachers in developing multimedia physics learning still need to cover various aspects that can create attractive, interactive, and compelling content [29,30]. However, the creativity shown by pre-service physics teachers has revealed new, original content, not just copying material from other sources but also presenting information uniquely and excitingly. Using attractive, aesthetic designs and following the target audience will help increase student interest and involvement. Likewise, the content presented has presented lessons that lead to efforts to introduce the SDGs theme.

4 Conclusion

Based on the research and discussion, the ability of pre-service physics teachers to develop multimedia physics learning is still in the medium category. However, when viewed from all indicators, only two ability indicators are in the high category, namely (1) the ability of pre-service physics teachers to understand the concept of multimedia design and (2) the accuracy of pre-service physics teachers presenting SDGs content. In contrast, one indicator is in the low category, namely the ability of pre-service physics teachers to present animation in the developed multimedia. The limitations of this research are that the research only uses product data and observations in analyzing the abilities of pre-service physics teachers, and interview data may be needed to dig deeper into the constraints pre-service physics teachers face while developing multimedia. This research is beneficial as input for tertiary institutions that produce pre-service physics teachers to provide them with the ability to develop multimedia. Furthermore, for future research, it is necessary to develop teaching facilities or materials to improve the ability of pre-service physics teachers to develop learning multimedia.

Acknowledgment

The authors would like to thank Fakultas Matematika dan Ilmu Pengetahuan Alam, Universitas Negeri Surabaya, who founded this research.

References

1 R Bali-Swain, F. Yang-Wallentin, International Journal of Sustainable Development & World Ecology **27,** 2 (2020)

2 Y. Huan, T. Liang, H. Li, C. Zhang, Science of the Total Environment **752**, 1 (2021)

3 W. G. Santika, T. Urmee, Y. Simsek, P. A. Bahri, M. Anisuzzaman, Energy for Sustainable Development **59**, 1 (2020)

4 A. L. Salvia, Leal, W. Filho, L. L. Brandli, J. S. Griebeler, Journal of cleaner production **208**, 1 (2019)

5 J. Holloway, K. Mengersen, Remote Sensing **10**, 9 (2018)

6 R. Vinuesa, B. Sirmacek, Nature Machine Intelligence **3**, 11 (2021)

7 A. El Saddik, IEEE multimedia **25**, 2 (2018)

8 Y. Deldjoo, M. Schedl, P. Cremonesi, G. Pasi, ACM Computing Surveys CSUR **53**, 5 (2020)

9 W. Kurniawan, D. Darmaji, A. Astalini, D. A. Kurniawan, M. Hidayat, N. KurniawaN, L. Z. N. Farida, International Journal of Evaluation and Research in Education **8**, 4 (2019)

10 R. E. Mayer, Journal of computer assisted learning **33**, 5 (2017)

11 R. Kuba, S. Rahimi, G. Smith, V. Shute, C. P. Dai, Educational Technology Research and Development **69**, 1 (2021)

12 A. Henukh, H. Rosdianto, S. Oikawa, Jurnal Ilmu Pendidikan Fisika **5**, 1 (2020)

13 A. R. Hasanah, M. A. Salam, S. Mahtari, *Developing the interactive multimedia in physics learning*, In Journal of Physics: Conference Series 1**171**, 1 (2019)

14 D. Djamas, V. Tinedi, Research Anthology on Developing Critical Thinking Skills in Students **14**, 1 (2021)

15 M. Nasir, Applied Science and Technology **1**, 1 (2017)

16 M. Maruf, A. Setiawan, A. Suhandi, P. Siahaan, Jurnal Pendidikan Fisika **9**, 3 (2021)

17 M. D. Abdulrahaman, N. Faruk, A. A. Oloyede, et al., Heliyon **6**, 11 (2020)

18 R. Girwidz, L. J. Thoms, H. Pol, et al., International journal of science education **41**, 9 (2019)

19 Y. R. Liana, P. I. Nursuhud, Jurnal Ilmu Pendidikan Fisika **5**, 2 (2020)

20 G. Gunawan, A. Harjono, L. Herayanti, S. Husein, Journal for the Education of Gifted Young Scientists **7**, 4 (2019)

21 W. Kurniawan, D. Darmaji, A. Astalini, et al., International Journal of Evaluation and Research in Education **8**, 4 (2019)

22 N. Taguchi, System **75**, 1 (2018)

23 Y. Wei, X. Wang, L. Nie, X. He, T. S Chua, *Graph-refined convolutional network for multimedia recommendation with implicit feedback*, In Proceedings of the 28th ACM international conference on multimedia (2020)

24 D. Li, T. Jiang, W. Lin, M. Jiang, IEEE Transactions on Multimedia **21**, 5 (2018)

25 R. Liu, R. Yang, S. Li, Y. Shi, X. Jin, Multimedia Tools and Applications **79**, 1 (2020)

26 Z. Akhtar, T. H. Falk, IEEE access **5**, 1 (2017)

27 Y. Deng, *ProtoLife: Enhancing Mobile Multimedia Narratives Through Prototyping Techniques*, In International Conference on Human-Computer Interaction pp 46-53 Cham: Springer Nature Switzerland (2023)

28 C. Liu, P. Elms, Research in Learning Technology **27**, 2 (2019)

29 W. Wiana, *Interactive multimedia-based animation: a study of effectiveness on fashion design technology learning*, In Journal of Physics: Conference Series **953**, 1 (2018)

30 N. D. Rahayu, I. Yatri, *Animated video media based on adobe after effects AEF application: an empirical study for elementary school students*, In Journal of Physics: Conference Series **1783**, 1 (2021)

 https://doi.org/10.1051/e3sconf/202345001009

Bibliometric analysis of Mosquito population research peaks

Winarsih[1*], *Sri* Mulyati[2], *Wahyu* Adi Nugroho[3], *Raden Ahmad* Zainul Aziz[1], and *Husni* Mubarok[4]

[1]Biology, State University of Surabaya, Surabaya, Indonesia
[2]Veterinary Medicine, Airlangga University, Surabaya, Indonesia
[3]Mathematics, Surabaya State University, Surabaya, Indonesia
[4]Graduate Institute of Digital Learning and Education, National Taiwan University of Science and Technology, Taipei, Taiwan

Abstract. Research regarding the topic of mosquito population has been on the rise in the last 13 years, and because of that, it is an interesting topic to analyze the trends and growth of. This paper aims to analyze research papers and journals, then visualize research trends on the topic of mosquito population in the last 13 years and identify potential research topics related to mosquito population for future research. The method of this study uses a literature review and bibliometric analysis using Google Scholar, DBLP, and VOSviewer databases, in which 999 documents were collected. The results illustrate that mosquito population research has been on the rise gradually, reaching its peak in 2020 and showing a decline afterwards. The author with the highest contribution is Giovanni Belleni, and the most common type of document is a research paper. More research regarding mosquito populations can be used to analyze the transmission of diseases and the effects of climate change. This study is limited in its scope because most of the data was obtained through Google Scholar and DBLP without thorough filtering. Hence, it is recommended for future research to go through additional steps to filter more of the obtained data.

1. Introduction

Mosquitoes are a type of insect that consists of over 1000 species. Their habitat is scattered on almost every continent, with the exception of Antarctica, making them one of the most widespread insects in the world [1]. They are known to cause the most insect bites in human-populated areas compared to other insects [2], spreading diseases such as dengue, malaria, zika, yellow fever, and West Nile virus [3]. Because of this, mosquitoes pose a legitimate risk to human society, making them a large part of human day-to-day life.

One of the aspects that is being researched is that of the mosquito population. The spread and density of a mosquito population can highly affect the human population [4],

* Corresponding author: winarsih@unesa.ac.id

making its control a high priority. There are many factors in a mosquito population, such as temperature, humidity, and precipitation [5]. The behavior of mosquitoes can also change depending on the temperature, while humidity affects the longevity and blood-feeding behaviours of mosquitoes [6]. Because of this, human society will benefit from more research on how to control the population of mosquitoes, making coexistence a possibility in the future [7].

The benefits of research on mosquitoes don't stop at only designing human settlements; this also includes controlling the spread of diseases, improving quality of life, and balancing ecosystem biodiversity [8]. To figure out other possible research topics regarding mosquito populations, a bibliometric analysis is essential to discover results and create a research trend map [9]. Not only that, bibliometric analysis helps visualize data comprehensively [10], providing guidance for future research. Hence, it is important to conduct a bibliometric study on the topic of mosquito populations. The goal of this study is to do a literature review on mosquito population research papers from 2010–2023 with bibliometric analysis to conclude future research subjects using databases such as Google Scholar and DBLP. The research questions of this study are: (1) What is the trend of published articles on the topic of mosquito population during the last 13 years? (2) How is the visualization of mosquito population research trends during the last 13 years? (3) What is the most common type of document for mosquito population research in the last 13 years? (4) Who has been the top author in mosquito population research in the last 13 years? (5) How has the amount of research on mosquito populations changed in the last 13 years? (6) What are the opportunities for the publication of research on mosquito populations?

2. Methods

This research uses a review study with the bibliometric analysis method. Bibliometric analysis is a quantitative study using Google Scholar and the DBLP database. The analysis method was used to visualize the trend mapping of mosquito population research from various sources using statistical methods. The research was conducted through an online database search on August 29, 2023. An online search was done using the Publish or Perish application, entering 'Mosquito Population' in the keywords from 2010 to 2023.

Bibliometric studies use the Google Scholar database in Publish or Perish as the main data source, while DBLP data sources are used for document types and publication trends in the past years. By entering "Mosquito Population" in the application, 999 documents were gathered this way. 26 additional documents that were gathered from DBLP will also be used in this study.

The Google Scholar data will then be processed using VOSViewer to visualize the keywords and author mapping. Excel will also be used to process data from DBLP to visualize the fluctuation of the publication trend and show the most common type of document.

 https://doi.org/10.1051/e3sconf/202345001009

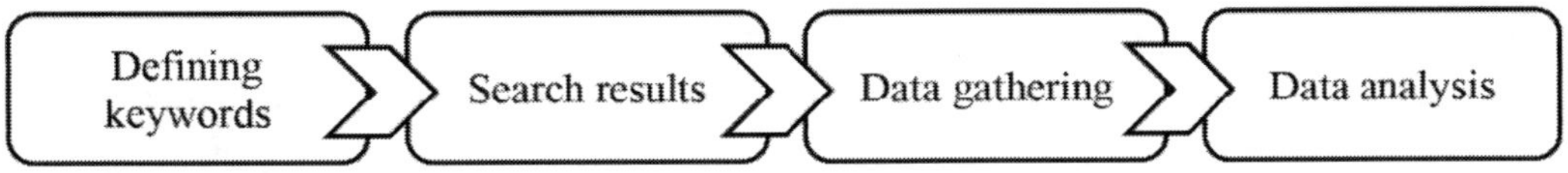

Fig. 1. Method Steps

3. Result and discussion

Bibliometrics has an important contribution in research [11-15]. The novelty and importance of further research can be obtained from bibliometric analysis [16-20]. Figure 2 represents the mosquito population research articles in the past 13 years. From the Google Scholar database, it can be seen that the development of research on mosquito populations occurred at a steady pace, between 105 and 115 documents per year. Before declining in 2016, with no sign of increasing in the future. Meanwhile, from the DBLP database, it can be seen that there is a gradual increase in research, reaching its peak in 2020 before declining again afterwards. Despite the decrease in research, there are still many potentials in mosquito population research that haven't been explored yet. especially in the current age where the earth's temperature rises each year.

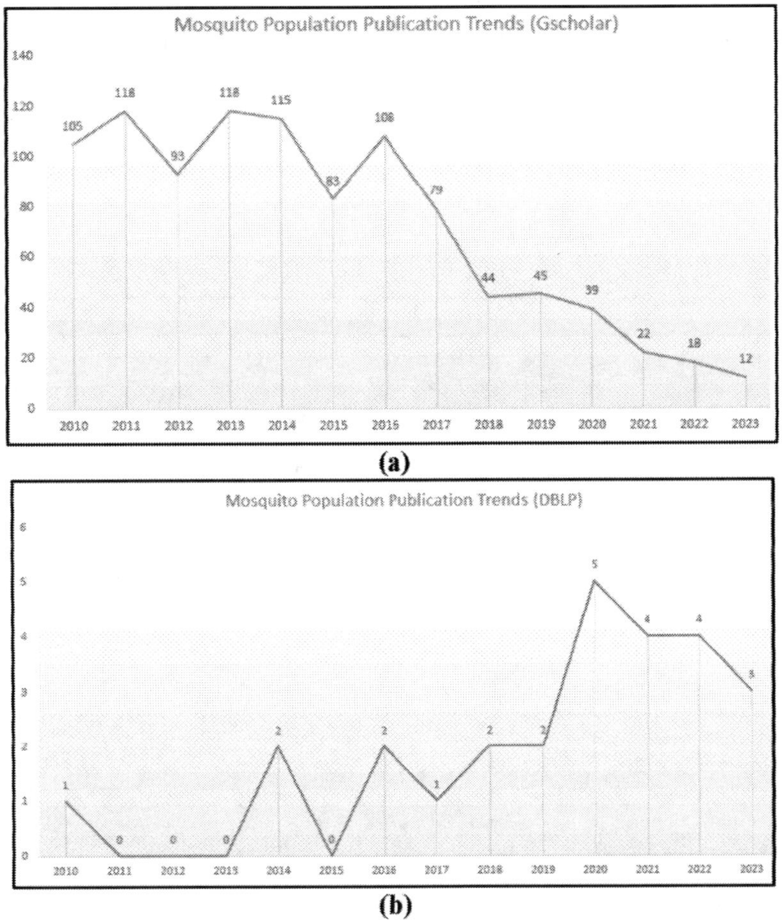

Fig. 2. Mosquito population publication trend in the last 13 years from two databases

From Figure 2, it can be seen that there are contradictions between the two databases, suggesting that the sample size of the DBLP database is too small for the study and cannot be reliably used to represent the publication trend of mosquito population research. The sample size of the Google Scholar database can be reliably used with a sample size of 999.

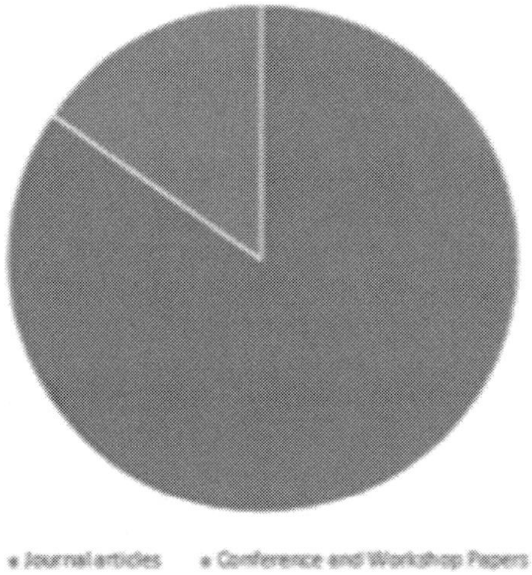

Fig. 3. Document types of mosquito population research

From Figure 3, from the 26 documents that were gathered from DBLP, it can be concluded that most documents consisted of journal articles with 22 documents, and only 4 documents were conference and workshop papers.

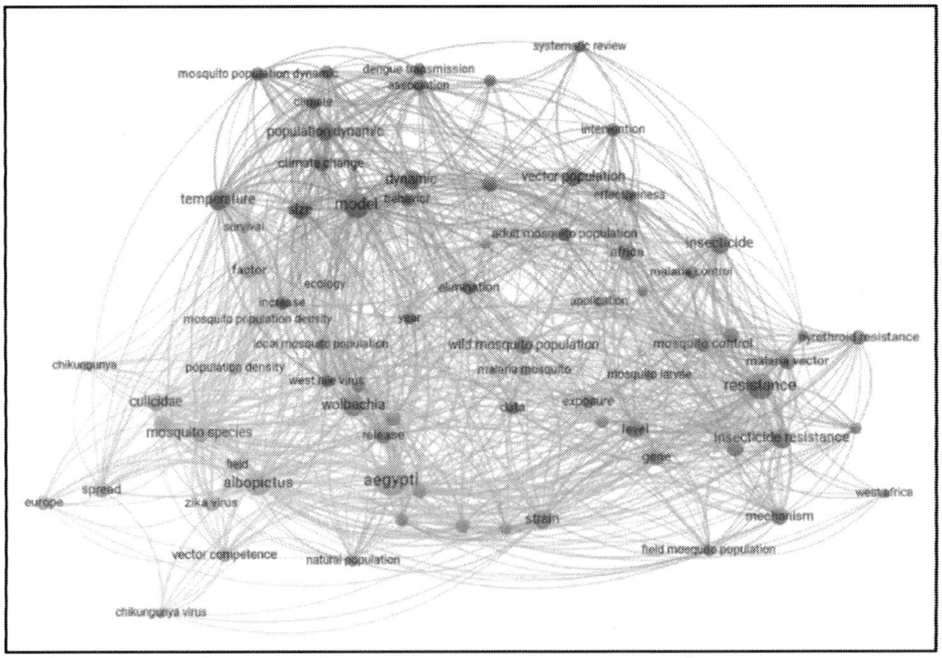

Fig. 4. Keyword visualization mapping across mosquito population research in the past 13 years

Figure 4 shows the keyword visualization mapping in mosquito population subjects between the years 2013 and 2023. There are four main clusters in the visual mapping, which consists of 77 items. The first cluster in red (n = 23) focuses on the dynamics and model of mosquito populations. And then, the second cluster in green (n = 19) is composed of topics on natural and wild mosquito populations. Next, the third cluster in blue (n = 18) focuses on the mosquito insecticide resistance level. Lastly, the fourth cluster in yellow (n = 17) focuses on the density of the mosquito population. The number of occurrences and the total link strength of an item suggest researchers' interest in the topic. The keyword that has the highest occurrence is aegypti, with 108 occurrences in total, followed by resistance, model, albopictus, and wolbachia. This suggests that research on the subject of mosquito populations mainly focuses on the population model and resistance of certain species of mosquitoes. By looking at the relations between large and small keywords, most of the research has gone to the Aegypti mosquito. While research on the topic of climate change affecting mosquito populations and density is still understudied, this shows the potential for future research.

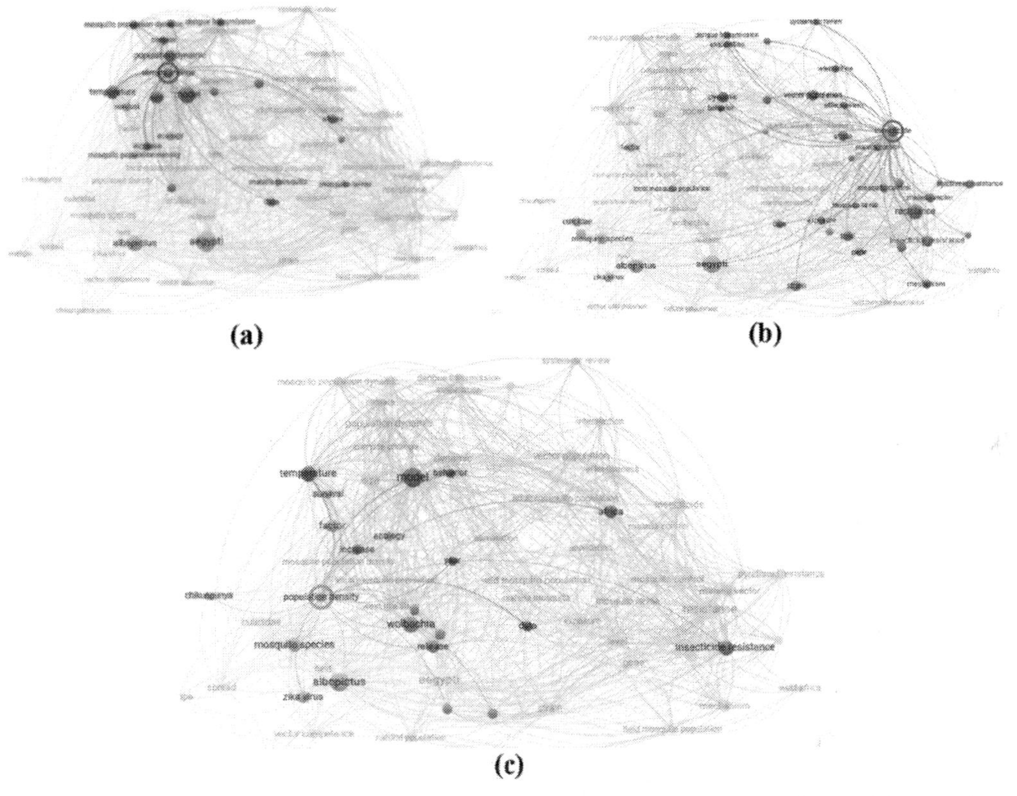

Fig. 5. Specific keywords mapping with topics **(a)** climate change, **(b)** insecticide, and **(c)** population density

Figure 5 gives an example of specific keywords such as population density, insecticide, and climate change. The figures above give examples of specific keywords such as population density, insecticide, and climate change. The trends in mosquito population research with climate change can be promising to research by showing how climate change affects mosquito populations. Further research on the population density of mosquito populations can also be beneficial to visualize the changes in mosquito populations over time. Figure (b) also shows the high interconnectivity of the insecticide topic to other topics, making it a good topic candidate for future research on tackling mosquito populations.

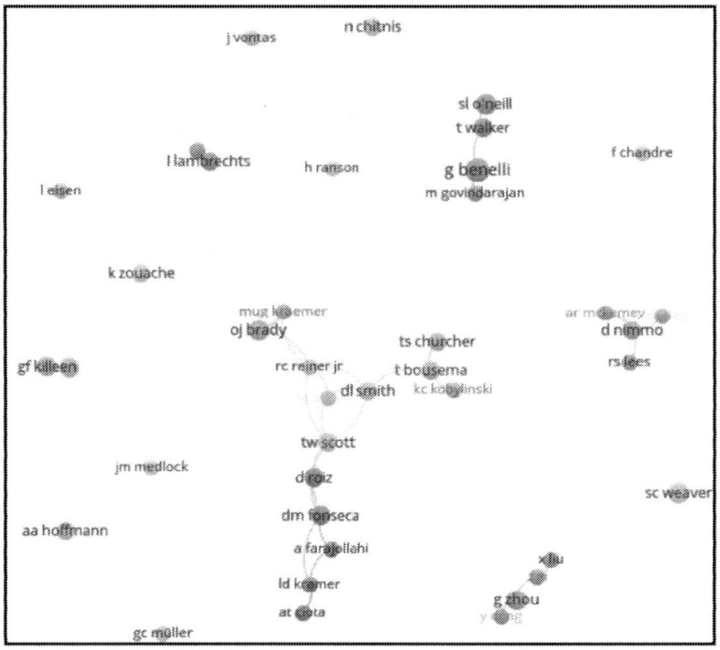

Fig. 6. Authorship mapping of mosquito population research in the last 13 years

In Figure 6, the authors who have worked on mosquito population research are visualized. The map is composed of 40 items that are grouped into 19 clusters. It can be seen that the interconnectivity between authors is low, with the highest link between authors being 5. The author with the most documents is G. Benelli with 11 documents, followed by O.J. Brady with 8 documents. This shows that in the last 13 years, there haven't been many authors who have done research on mosquitoes, making the field open for opportunities for other researchers.

4. Conclusion

Based on the conducted bibliometric analysis, these discoveries were gathered: The trend of mosquito population topics was steady from 2010 to 2016. But afterward, it decreased rapidly without increasing, making it a challenge for future researchers. At the same time, this also creates opportunities for new researchers. Based on research mapping, it can be gathered that the research trend on mosquito population still focuses on certain species of mosquitoes and the model of population, not other topics such as the effect of climate change on the mosquito population and the population density of mosquitoes. Based on authorship mapping, it can be concluded that there are not many authors who have researched mosquito population topics, making it hard for other researchers to analyze another researcher's work.

It can be concluded that the research on mosquito populations is still understudied and mainly focuses on only certain topics, making the topic ripe for further research. The researchers recommend using better methods of data filtering, making sure that there is no contradiction between databases and the data that is used to be accurate. This research can be utilized as a reference for further research on mosquito populations. It is also recommended to use other databases than DBLP and Google Scholar.

Acknowledgments

Thank you to my students who told me about the high mosquito population on campus. Based on the students' complaints, it is necessary to carry out a bibliometric study of mosquitoes.

References

1. J. Seda, S. Horrall, Mosquito bites - StatPearls - NCBI bookshelf (National Center for Biotechnology Information, 2023)
2. A. D. Vander, A. Labib, G. Yosipovitch, Frontiers, Immunology **13** (2022)
3. H. Lee, S. Halverson, N. Ezinwa, Primary Care: Clinics in Office Practice **45**, 3 (2018)
4. F. J. Colón-González, et al., The Lancet Planetary Health **5**, 7 (2021)
5. C. Baril, et al., BMC Bioinform **16** (2023)
6. T. K. Yamana, E. A. Eltahir, Parasites & Vectors, **6**, 1 (2013).
7. A. Ford, M. Gearey, T. Acott, Int. J. Mosq. Res. **4** (2021).
8. Y. A. Halasa, et al., PLoS ONE, **9**, 3 (2014)
9. B. K. Prahani, I. R. Dawana, B. Jatmiko, Int. J. Emerg. Technol. Learn **18**, 10 (2023)
10. D. Solomon, *A different view: An inquiry into visualization of bibliometric data*, in ASEE Proceeding of Annual Conference & Exposition, Seattle, Washington (2015)
11. H. N. Hidaayatullaah, N. Suprapto, E. Hariyono, B. K. Prahani, D. Wulandari, J. Phys. Conf. Ser. **2110**, 012026 (2021)
12. B. K. Prahani, J. Alfin, A. Z. Fuad, B. Jatmiko, E. Hariyono, N. Suprapto, Int. J. Emerg. Technol. Learn. **17**, 17 (2022)
13. Nurhasan, B. K. Prahani, N. Suprapto, M. A. A. Ardha, Int. J. Instruction, **15**, 3 (2022)
14. B. Jatmiko, B. K. Prahani, N. Suprapto, S. Admoko, U. A. Deta, N. A. Lestari, M. N. R. Jauhariyah, M. Yantidewi, D. Muliyati, J. Phys. Conf. Ser. **2110**, 012018 (2021)
15. B. K. Prahani, M. Z. B. Amiruddin, N. Suprapto, U. A. Deta, T. -H. Cheng, *Int. J. Educ. Method*, **8**, 3 (2022)
16. B. K. Prahani, I. A. Rizki, B. Jatmiko, N. Suprapto, T. Amelia, Int. J. Emerg. Technol. Learn. **17**, 8 (2022)
17. N. Suprapto, A. Kholiq, B. K. Prahani, U. A. Deta, J. Phys. Conf. Ser. **012017** (2021)

18. B. Jatmiko, T. Sunarti, B. K. Prahani, E. Hariyono, Dwikoranto, F. C. Wibowo, S. Mahtari, Misbah, M. Asy'ari, J. Phys : Conf. Ser. **2110**, 012020 (2021)
19. B. K. Prahani, M. Z. B. Amiruddin, B. Jatmiko, N. Suprapto, T. Amelia, Int. J. Interactive Mobile Technologies, **16**, 8 (2022)
20. U. A. Deta, I. Laeliyah, N. A. Lestari, M. Yantidewi, M. N. R. Jauhariyah, B. K. Prahani, AIP Conf. Proc. **2595**, 020013 (2023)

Integration of independent learning and physics innovation in STEAM-based renewable energy education to improve critical thinking skills in the era of Society 5.0 for Sustainable Development Goals (SDGs) 2030

Sri Dwi Indahwati[1*], *Fida* Rachmadiarti[1], *Eko* Hariyono[1], *Binar Kurnia* Prahani[1], *Firmanul Catur* Wibowo[2], *Muhammad Abd Hadi* Bunyamin[3], and *Muhammad* Satriawan[1]

[1]State University of Surabaya, Postgraduate Science Education Program, Jl. Ketintang Cell. No. Kel, Ketintang, Kec. Gayungan, Surabaya, East Java 60231, Indonesia
[2]State University of Jakarta, Physics Education Program, Rawamangun, Kota Jakarta Timur, Daerah Khusus Ibukota Jakarta 13220, Indonesia
[3]Universiti Teknologi Malaysia, School of Education, 81310 UTM Skudai, Johor, Malaysia

Abstract. This study aims to integrate Free Learning and Physics Innovation in STEAM-based Renewable Energy Education to improve critical thinking skills in Era Society 5.0 in order to achieve the 2030 Sustainable Development Goals (SDG). This research uses a literature review method that collects and analyzes related articles. The results showed that integrating Free Learning and Physics Innovation in STEAM-based Renewable Energy Education is an effective approach to improving students' critical thinking skills. Renewable energy education is a relevant and important topic in SDG 2030, and the STEAM approach helps students understand and overcome this challenge. Era Society 5.0 emphasizes technology integration, so the educational approach must constantly be updated and adapted to the times. The results of this literature review can assist educators, policymakers, and researchers in developing educational strategies that align with the times' needs.

1 Introduction

This research aims to integrate the concept of "Freedom of Learning" and physics innovation [1-5] in STEAM-based renewable energy education [6] to improve students' critical thinking skills to support the achievement of Sustainable Development Goals (SDG) 2030 in the Era of Society 5.0 [7, 8]. The main challenge is to combine the flexibility of "Freedom of Learning" with innovative physics so that students not only understand the physics behind renewable energy but can also apply critical thinking to

* Corresponding author: sri.22006@mhs.unesa.ac.id

solve global energy problems [9-12]. It is hoped that the results of this research can help educators and stakeholders overcome educational challenges in this ever-changing era.

Several studies related to integrating the concept of "Freedom of Learning" and physics innovation in STEAM-based renewable energy education and its impact on student's critical thinking skills and the achievement of SDG 2030 in the Era of Society 5.0 have been carried out [13-16]. Febriansari [17] has developed a framework combining the STEAM learning model and design thinking that can be applied to learning renewable energy materials. Meanwhile, Sulisworo [14] examines the use of information and communication technology in STEM-based education as an alternative for adapting students' competencies to work and society, emphasizing the importance of critical thinking skills. Arifuddin [18] has integrated STEM in learning to encourage students to think critically, comprehensively, and innovatively in finding solutions to problems, using Educational Robots as an effective learning tool. In addition, Rahayu [16] developed an E-Student Worksheet integrated with STEAM on acid-base material in digital flip form, which can increase interactivity in learning. These studies provide deeper insight into how the integration of the concept of "Freedom of Learning" and physics innovation can be applied in the context of STEAM-based renewable energy education and its impact on student's critical thinking skills and the achievement of SDG 2030 in the Era of Society 5.0 [19]. This research focuses on integrating Independent Learning and Physics Innovation in STEAM-based Renewable Energy Education to Improve Critical Thinking Skills in the Society 5.0 Era for Sustainable Development Goals (SDG) 2030. To develop an innovative approach to integrating STEAM in renewable energy learning, students can work in teams to design, develop and implement renewable energy solutions. Researchers consider Project-Based Learning (PjBL) with learning designs that focus on real projects relevant to renewable energy [20-24].

This will encourage active engagement, independent exploration, and problem-solving related to developing new concepts or theories that did not exist before. For example, if your research on renewable energy produces a new concept or theory that has never been discussed, this can be considered new in research. The type of research used is literature study or library research, namely collecting data obtained from books, journals or scientific papers related to the research object.

This research aims to design, implement and evaluate an innovative educational model that helps students develop critical thinking skills in facing renewable energy challenges in the Society 5.0 era and contributes to achieving SDG 2030. This article highlights the importance of integrating independent learning, physics innovation, and the STEAM approach to renewable energy education. Through this approach, students can better understand the challenges and opportunities of renewable energy and are able to create creative solutions for a sustainable future [25, 26]. Thus, this article provides insight into how integrating these three elements can improve students' critical thinking skills in understanding and overcoming renewable energy problems and contribute to achieving sustainable development goals in the Society 5.0 era, especially related to renewable energy and SDG 2030.

2 Methods

This research examines the integration of independent learning and physics innovation in renewable energy education using the STEAM approach to improve critical thinking skills in the era of Society 5.0 for Sustainable Development Goals (SDG) 2030 [27]. This research is a type of qualitative research using a library study method with an art education concept approach. In qualitative research, the main focus is on an in-depth understanding of the phenomenon or topic being studied through the interpretation of the data and the meaning contained therein [28-30]. The concept approach to art education adds a dimension of creativity and the application of art approaches in teaching and learning. Data sources in this research include various journals, articles, books, and other references relevant to the research topic, especially those related to renewable energy education, independent learning, physics innovation, and the STEAM approach.

This literature review aims to present a comprehensive summary of previous research related to a particular topic [31, 32]. The aim is to inform what is already known about the topic and identify what knowledge is still lacking. Literature reviews can also provide a rational basis for research that has been conducted and inspire further research ideas [33]. In this research, we will use a literature review as the main approach to understand this topic more deeply.

Literature study involves collecting data from various sources such as journals, books, documents, the internet, and libraries. The method includes collecting library materials, reading, recording relevant information, and managing these materials for writing purposes [34]. This method allows researchers to access various sources to support understanding and analysis of research topics.

This research uses a descriptive qualitative research method, allowing researchers to describe and analyze existing phenomena in detail. Data is obtained from various sources to ensure the diversity and accuracy of the information required. The data collection technique used is literature study or library research, with Google Scholar as the main search platform for finding scientific sources. This research aims to understand the integration of Independent Learning and Physics Innovation in STEAM-based Renewable Energy Education and improve students' critical thinking skills in facing renewable energy challenges in the Society 5.0 era in order to achieve SDG 2030.

3 Results

Based on initial search results on Google Scholar with the keywords, namely "Integration, Independent Learning, Physics Innovation, Renewable Energy Education, STEAM, Critical Thinking Skills, Society 5.0 Era, SDG 2030," researchers found 784 relevant article titles. However, not all of these articles have themes that match the research objectives, and some articles are duplicates. Then, researchers carried out further searches on Google Scholar by limiting the search to only the title (in title) using the same keywords. As a result, 563 articles were found that met the research objectives. Meanwhile, 221 articles that did not match the research objectives and duplicates were excluded from the study.

Of the 563 articles found, researchers screened them to ascertain whether the articles had complete manuscripts. 530 articles were excluded because they were not available in full text or paid versions. Finally, the remaining 33 articles are available in full-text versions. Furthermore, of the 33 articles, screening was carried out again based on the research criteria, and 26 articles were excluded because they did not meet the criteria. Finally, the researcher found seven journals with similar research themes/contents. Research articles that best fit the research objectives are selected from the seven journals based on the most appropriate criteria. This is shown in the research flowchart in Figure 1.

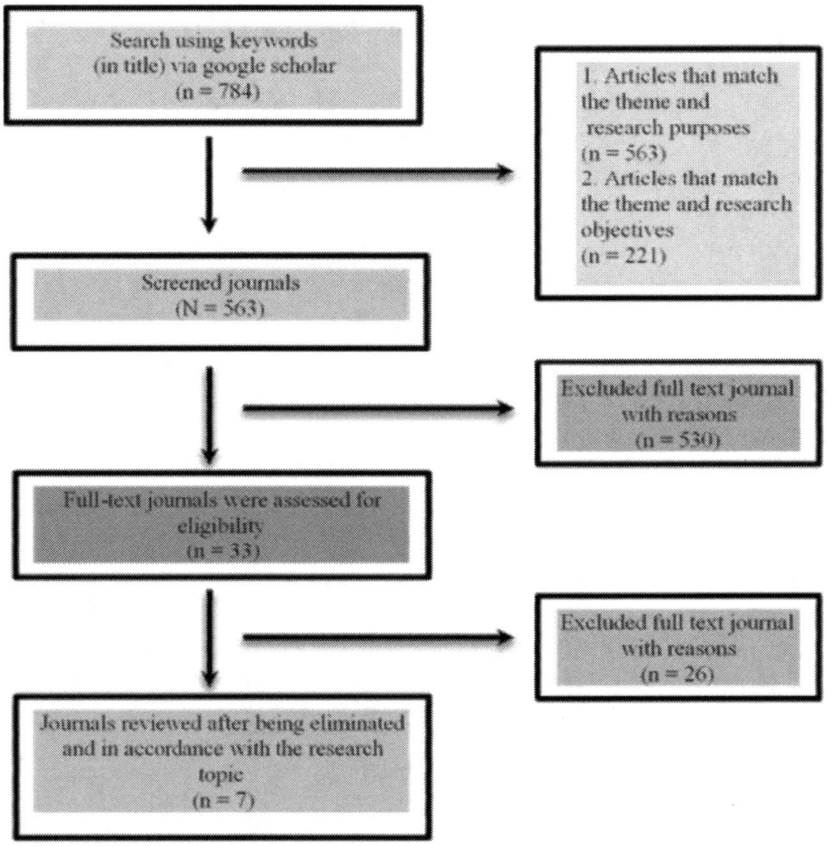

Fig. 1. Research flow chart.

The results of this research will provide valuable insight into the benefits and potential of integrating independent learning, physics innovation, and the STEAM approach in renewable energy education. Apart from that, it is also hoped that this research can provide practical recommendations for developing curriculum and learning methods that are more effective in improving students' critical thinking skills in the Society 5.0 era in order to achieve the 2030 Sustainable Development Goals (SDG) related to renewable energy. With a deeper understanding of how integrating these approaches can support sustainable development, educators and policymakers can play an active role in creating a highly competitive generation that is innovative in facing future renewable energy challenges [35]. In addition, this research can also contribute to a global understanding of the importance of renewable energy education in achieving SDG 2030 and creating a sustainable world for future generations.

No	Authors	Finding
1	Cahyorini Wulandani, Meida Afina Putri, Rahmalia Indah Pratiwi, Komareeyah Sulong (2022) [36]	Every industrial revolution has an impact that will give birth to change. Education is one of the fields that has an impact in every revolutionary era. To be able to produce students who are appropriate to the times, the curriculum and learning activities must always innovate and change to suit current developments. STEAM-Project Based Learning or project-based STEAM can be used to develop early childhood skills in the era of Industrial Revolution 5.0 by combining elements of science, technology, engineering, art and mathematics as well as applying strategies consisting of reflection, discovery, application and communication in implementing Learning Activities
2	Atiaturrahmaniah Atiaturrahmaniah, Ida Bagus Putu Aryana, I Wayan Suastra (2022) [37]	The STEAM (Science, Technology, Engineering, Mathematics) model can be applied to students in schools basic because it can improve critical thinking skills, collaboration, communication skills and able to analyze all forms of information that has been obtained, thus forming students with character and students are also able to increase their interest in scientific literacy in looking for various comparisons existing facts and realities so that they can find solutions to the problems they find.
3	Yose Indarta, Nizwardi Jalinus, Waskito, et al., (2022) [38]	The era of society 5.0 requires society to be able to solve various problems or dynamics social by utilizing technology such as the Internet of Things (IoT), Artificial Intelligence (AI), technology robots, even big data.
4	Andrea Lesková, Zuzana Uličná, Hedviga Tkáčová, Klodiana Leka, Daniel Alvarez Mateo (2023) [39]	It is necessary to guide the child from early childhood to build values, critical thinking, but also own self-esteem and empathy. Since the foundations for a child's education are laid in the family, it is precisely family education that can (un)prepare the child for entering the world of digital media and its optimal functioning in it.
5	Agus Sugiarto, Rahmat Mulyono (2022) [40]	The role of teachers in implementing STEM learning is very important. Teachers who have optimal knowledge and experience in preparing STEM learning will provide optimal learning experiences for students. However, many teachers face challenges in implementing STEM learning. Therefore, support from education stakeholders is very necessary to develop STEM learning and prepare the young generation to face challenges in the era of Society 5.0.
6	Lia Amelia Megawati, Budi Arief (2021) [41]	Education has an important role in creating the young generation as Agents of Change and Agents of Producers who can create real change. Good quality education is needed to support an Indonesian society that is productive, competitive, and able to compete at the international level in the 21st century. The young generation, as agents of innovation, must be able to make important contributions to implementing the concept of sustainable development or SDGs. Independent Learning Campus (MBKM) is relevant to

No	Authors	Finding
		21st-century skills and SDGs. However, its implementation at the study program level is still not optimal. Lack of socialization of MBKM policies, guidelines and operational procedures means that student participation in the MBKM program scheme is still low. It is necessary to increase online and offline socialization at the study program level so that all students can feel the benefits of this program.
7	Iim Halimatul Mu'minah (2021) [42]	The STEAM approach in 21st-century learning is key in preparing for the Society 5.0 era. With a focus on science, technology, engineering, arts, and mathematics, students are developed in critical, creative, and logical thinking as well as solving everyday problems. They are also capable of producing innovative products and designs. In the era of Society 5.0 which is full of change, the STEAM approach helps students be ready as future innovators.

4 Discussion

This research highlights the important potential of integrating independent learning, Physics Innovation, and the STEAM Approach in renewable energy education. The implications of this research can be used as a basis for developing curriculum and learning methods that are more effective in improving students' critical thinking skills in the Society 5.0 era in order to achieve the 2030 Sustainable Development Goals (SDG) related to renewable energy [36, 38, 41]. In addition, this research can also contribute to global understanding of the importance of renewable energy education in achieving SDG 2030 and creating a sustainable world for future generations. Apart from that, the results of this research can also provide benefits for educators and educational practitioners in designing teaching strategies that are more innovative and relevant to future renewable energy challenges. By understanding the importance of integrating independent learning, Physics Innovation, and the STEAM Approach, educators can adopt more interactive and fun learning methods to encourage active student involvement [35, 37]. The results of this research can also have an impact on developing human resources who are more competent and competitive in the field of renewable energy. Students involved in learning based on integrating independent learning, Physics Innovation, and the STEAM Approach will have strong critical thinking skills, the ability to innovate, and a deep understanding of renewable energy. This will prepare them to contribute to advancing renewable energy technologies, confronting climate change, and achieving the 2030 SDG sustainable development goals [41]. However, this research also has some limitations. The main limitation is using the literature study method as the only data source. Although literature studies provide valuable insights into the integration of Free Learning, Physics Innovation, and the STEAM Approach in renewable energy education, the data obtained from the literature may be limited and do not include experience and views from the field [42]. Therefore, to produce more comprehensive and representative findings, further research may involve other research methods, such as field research or interviews with experts and practitioners in renewable energy education.

5 Conclusion

Integrating Independent Learning, Physics Innovation, and the STEAM Approach in Renewable Energy Education has great potential to improve students' critical thinking skills in the Society 5.0 era to achieve the 2030 Sustainable Development Goals (SDG) related to renewable energy. Results from seven relevant articles consistently show that this approach is effective. By combining these approaches in renewable energy learning, students experience more meaningful and active learning, conducting experiments, integrating scientific disciplines, and applying their knowledge in authentic contexts. This increases their understanding of renewable energy and develops critical thinking skills essential in the Society 5.0 era. This integration also supports the achievement of SDG 2030 regarding renewable energy, equipping students to become agents of change in sustainable development. This approach is not just about academic learning but also preparing young people to play an active role in creating a sustainable future. To improve STEAM-based renewable energy education and students' critical thinking skills in the Society 5.0 era towards SDG 2030, independent learning, Physics Innovation, and the STEAM Approach must be integrated into the curriculum. Teachers need to receive adequate training, educational facilities must be improved, and cooperation with the renewable energy industry needs to be increased. Student assessments should focus on critical thinking skills, technology can be used as a supporting tool, and students should be encouraged to develop innovative projects. Environmental awareness also needs to be promoted. By following these suggestions, renewable energy education can more effectively prepare young people to achieve sustainable development goals.

References

1. Y. E. Ika, A. F. Laka, P. N. Supardi, D. Rendo, D. N. Birhi, 2022 Optika J. Pendidik. Fis. **6** 158.
2. F. Alatas, Implementasi dan Problematika Merdeka Belajar: Tantangan dan Peluang Penerapan Merdeka Belajar Kampus Merdeka (Akademia Pustaka, Tulungagung, 2021)
3. A, Purwanto, D. H. Putri, D. Hamdani, J. Kumparan Fis. **4**, 25 (2021)
4. I. Susanto, R. Daya, J. Darma Agung **30**, 59 (2022)
5. Z. F. Rozi, R. N. Sasongko, Bioedusains J. Pendidik. Biol. Sains **5**, 410 (2022)
6. S. Suwardi, Paedagogy J. Ilmu Pendidik. Psikol. **1**, 40 (2021)
7. S. H. Olsen, E. Zusman, M. Hengesbaugh, N. Amanuma, S. Onoda, Governing the Sustainable Development Goals in the Covid-19 Era: Bringing Back Hierarchic Styles of Governance? (Asian Development Bank Institute, Tokyo, 2021).
8. M. D. Tavares, G. M. Azevedo, Adv. Tour. Technol. Syst. **208**, 49 (2021)
9. R. Rosmiati, L. Liliasari, B. Tjasyono, et al., Reflective Practice 21, 4 (2020)
10. A. Amran, I. Jasin, M. Perkasa, et al., Journal of Physics: Conference Series **1521**, 4 (2020)
11. S. Sutarno, D. H. Putri, E. Risdianto, et al., Journal of Physics: Conference Series **1731**, 1 (2021)
12. A. A. Sihombing, S. Anugrahsari, N. Parlina, Y. S. Kusumastuti, Asian Journal of

University Education **17**, 4 (2021)

13. N. Khoiri, S. Ristanto, A. F. Kurniawan, Jurnal Pendidikan IPA Indonesia **12**, 2 (2023)
14. H. Retnowati, S. Subanti, International Online Journal of Education and Teaching **7**, 1 (2020)
15. H. Galindo-Dominguez, Educational Technology & Society **24**, 3 (2021)
16. M. I. Simeon, M. A. Samsudin, N. Yakob, International Journal of Technology and Design Education **32** (2020)
17. D. Febriansari, S. Sarwanto, S. J. Yamtinah, Inov. Pembelajaran **8**, 1 (2022)
18. R. Arifudin, A. Setiawan, Z. Abidin, D. A. Efrilianda, J. Jumanto, Abdimasku J. Pengabdi. Masy. **5**, 2 (2022)
19. M. L. Bernacki, L. Vosicka, J. C. Utz, C. B. Warren, Journal of Educational Psychology **113**, 6 (2021)
20. E. C. Miller, S. Severance, J. Krajcik, Journal of Science Teacher Education, **32**, 7 (2021)
21. N. Jalinus, S. Syahril, R. A. Nabawi, Journal of Technical Education and Training **11**, 1 (2019)
22. M. Baran, M. Baran, F. Karakoyun, A. Maskan, Journal of Turkish Science Education **18**, 4 (2021)
23. D. Triana, Y. U Anggraito, S. Ridlo, Journal of Innovative Science Education, **9**, 2 (2020)
24. M. Muhibbuddin, N. Yustina, S. Safrida, Ijaedu-International E-Journal of Advances In Education **6**, 16 (2020)
25. M. Maulidia, P. Dargusch, P. Ashworth, F. Ardiansyah, Renewable and Sustainable Energy Reviews **101** (2019)
26. A. Rahman, P. Dargusch, D. Wadley, Renewable and Sustainable Energy Reviews **144** (2021)
27. D. Demina, T. Rexy, A. Danyl, Journal International Inspire Education Technology **1**, 2 (2022)
28. F. Y. Sebele-Mpofu, Cogent Social Sciences **6**, 1 (2020).
29. N. A. Stahl, J. R. King, Journal of Developmental Education **44**, 1 (2020)
30. N. Harahap, Penelitian Kualitatif (Wal Ashri Publishing, Medan, 2020)
31. C. Vindrola-Padros, G. A. Johnson, Qualitative health research **30**, 10 (2020)
32. A. Mehrad, M. H. T. Zangeneh, International Journal For Research In Educational Studies **5**, 7 (2019)
33. A. S. Denney, R. Tewksbury, J. Crim. Justice Educ. **24** (2013)
34. M. Abdar, et al., Information fusion **76** (2021)
35. U. A. Izzati, B. S. Bachri, M. Sahid, D. E. Indriani, Journal for the Education of Gifted Young Scientists **7**, 3 (2019)
36. C. Wulandani, M. A. Putri, R. I. Pratiwi, K. Sulong, Indonesian Journal of Early Childhood Educational Research (IJECER) **1**, 1 (2022)
37. A. Atiaturrahmaniah, I. Bagus, P. Aryana, I. W. Suastra, JPGI (Jurnal Penelitian Guru Indonesia) **7**, 2 (2022)
38. Y. Indarta, N. Jalinus, W. Waskito, A. D. Samala, A. R. Riyanda, N. H. Adi, Edukatif: Jurnal Ilmu Pendidikan **4**, 2 (2022)

39. A. Lesková, Z. Uličná, H. Tkáčová, K. Leka, D. A. Mateo, Journal of Education Culture and Society **14**, 2 (2023)
40. A. Sugiarto, R. Mulyono, Didaktik: Jurnal Ilmiah PGSD STKIP Subang **8**, 2 (2022)
41. L. A. Megawati, B. Arief, *Relevansi antara Merdeka Belajar Kampus Merdeka dengan Keterampilan Abad 21 dan SDGs*. In Prosiding Seminar Nasional Merdeka Belajar Kampus Merdeka (MBKM), Bogor: Portal Prosiding MBKM Universitas Pakuan (2021)
42. I. H. Mu'minah, *Studi Literatur: Pembelajaran Abad-21 Melalui Pendekatan Steam (Science, Technology, Engineering, Art, and Mathematics) Dalam Menyongsong Era Society 5.0*. In Prosiding Seminar Nasional Pendidikan (2021, October)

Decision support system as an element of web-based integrated pest control on cabbage plants

Ni Wayan Surya Wardhani[1*], *Prayudi* Lestantyo[2], and *Nur Silviyah* Rahmi[1]

[1]Department of Statistics, Brawijaya University, Malang, Indonesia
[2]Management of Islamic Education, Maulana Maliq Ibrahim State University Malang, Malang, Indonesia

Abstract. Cabbage leaves are consumed as a healthy vegetable but the plant is very susceptible to attack by leaf caterpillars (P. xylostella). The residue from the periodic use of insecticides is very dangerous to human health. Hence Decision Support System (DSS) was developed which can be used as a reference when spraying insecticides is carried out. The system is intended to estimate the cabbage leaf damage based on web-based photo imagery that is used for detecting the percentage of pest attacks so that pesticides are only applied if the damage to the cabbage leaves is detected to be rather severe. Users are required to upload photos of cabbage plants to be detected. This research is carried out through literature studies, theoretical studies, data collection and location surveys of cabbage plant centers, statistical modeling, making a cabbage leaf hole display detection system with photo images, and making packages or libraries for models that have been developed, adding features that make it easier for users to run it. This developed system after passing the development process carries out 100% black box testing works according to the desired requirements. The implementation of pest experts and testing of this system gets an accuracy of about 87%, so the implementation is carried out on users. This developed system after going through the development process is carried out 100% blackbox testing to function according to the desired needs. The implementation of pest experts and testing of this system gets an accuracy of about 87%, so the implementation is carried out by the user. The results of this study have an analysis that can be stored in a database so that the utilization of recording analysis results can be directly stored in the database, can be printed, and can issue results from the Decision Support System for spraying cabbage pests.

1 Introduction

Agriculture plays a crucial role in achieving the Sustainable Development Goals (SDGs) set by the United Nations (UN). One of the relevant SDGs is Goal 2, namely "End Hunger, Achieve Food Security, Improve Nutrition, and Promote Sustainable Agriculture." Sustainable and productive agriculture is essential to ensure adequate food availability for the entire world's population. Indonesia is one of the world's agrarian countries, but ironically

* Corresponding author: wswardhani@ub.ac.id

in recent years Indonesia has imported foodstuffs from abroad [1]. However, the challenges faced in agriculture are pests and plant diseases that can threaten productivity and food security. One of the plants that are susceptible to this attack is the cabbage plant. Efforts to reduce the impact of pests and plant diseases on cabbage cultivation are a key aspect of achieving Goal 2 and several other goals, such as Goal 15 ("Living on the Land"). According to data from Statista [2], cabbage production in Indonesia experienced significant fluctuations. This happens because of the attack of cabbage pests and their management.

One approach that can be used to overcome this problem is to develop a Decision Support System (DSS) as a web-based Integrated Pest Control element. DSS is a tool that combines information from various sources to help farmers and agricultural managers make better decisions related to pest and plant disease control. A web-based DSS will give farmers easy and quick access to information on identifying pests and diseases, estimating the risk of infestation, and recommending appropriate controls. The use of this technology not only helps increase efficiency in decision-making, but also reduces the risk of crop loss due to pests and plant diseases. Research by Roche et al. [3], has researched decision support system based on Bayesian modeling for pests management: Application for wireworm risk assessment in corn fields, but still not publicly accessible. In the context of SDGs Goals 2 and 15, this approach supports productive sustainable agriculture, reduces potential losses, and contributes to global food security. Research by Subedi et al., [4] has found the Impact of climate change on the biology and ecology of insect pests: Implications for pest management strategies, crop production, and Food security.

This is the main part of controlling plant pests to achieve food security in accordance with the SDGs.The results of the study by Miluch et al. [5] showed that when monitoring with pheromones, the moths caught were not directly related to the density of the larvae. Therefore visual observations are better used to predict larval density. Most of the visual observations made in previous studies were observations of the number of larvae. However, observing the larvae is difficult because the size of the larvae is very small and tends to hide behind the leaves [6]. Visual observation of larvae to predict larval density can be replaced by observing the number of holes on the leaves [7]. Research on guidelines for the design of cabbage intercropping systems as a first step in a meta-analysis relay for vegetables becomes a benchmark in the development of an integrated system.

In this era, several combined methods to increase crop production are carried out on a large scale. The method used is not in the form of opening new agricultural land, but by applying multidisciplinary knowledge (intensification). One of them is the application to detect pests, plant diseases, or other plant growth disturbances. Research by Mahlein [8], explains that the trend in the detection of plant disturbances and diseases is using increasingly innovative information technology.

The trend of knowing the nature of plant diseases now refers to the study of visual patterns in a particular plant. Research by Patil et al. [9], provides advances in various methods used to study plant diseases/traits using image processing. The method studied is to increase yield & reduce human subjectivity in detecting plant diseases. Furthermore, research by Dong et al. [10] created an Automatic System for Plant Pests and Diseases Dynamic Monitoring and Early Forecasting. This can be used as a basis for dynamic monitoring of pests but many must be adjusted to the plant growth and development requirements.

Research by Arnal Barbedo [11], uses digital image processing techniques to detect, measure, and classify plant diseases from digital images in the visible spectrum. Although disease symptoms can manifest in any part of the plant, the method explores the symptoms seen on leaves and stems only. In this study, three parts were involved, namely detection, quantification severity, and classification. Each class, in turn, is divided according to the main technical solution used in the algorithm

Research development by Raza et al. [12], combines visible thermal light image data with detailed information and develops a machine learning system for plants infected with powdered tomato dew fungus Oidium neolycopersici. The method is to extract a new feature set from the image data using local and global statistics showing that by combining the depth information, greatly improves the detection accuracy of diseased plants.

Research on the implementation of the image processing method was also carried out by Pujari et al. [13], to detect diseases caused by fungi that attack various plants. Another study by Holsteen et al. [14], has detected stem disease of the flax plant which is one of the most important plants in several Asian countries. An automated system based on an Android application has been implemented to capture images of the disease-infecting hemp stems and send them to a dedicated server for feature extraction. On the server side, the affected portion of the image will be segmented using a threshold formula that is adjusted based on hue-based segmentation. Consequential feature values will be extracted from the segmented portions for texture analysis using the color methodology. However, this research is still in the development process and has not yet been socialized to the public.

Based on the latest research by Saradhambal et al. [15], introduced a defined color-based segmentation model to segment infected areas and assign them to relevant classifications. Plant diseases can be detected by image processing techniques. Disease detection involves steps such as image acquisition, image pre-processing, image segmentation, feature extraction, and classification. It shows the affected leaf part in percentage. Our previous research has carried out an analysis of cabbage plants using a statistical approach with an uneven distribution of data to determine cabbage pest attacks. This research will be developed into a website so that it can be analyzed more quickly [16].

The horticulture subsector is very prospective for development, especially cabbage, where cabbage plays an important role because it contains many vitamins that are useful for human health, as well as a source of income for farmers. In order to identify the level of attack by plant pests through the number of holes on the leaves, not much has been explored before the application of pesticides. The main challenge is how to reduce the use of pesticides to increase the quality and quantity of production. Therefore, research is focused on pests that attack cabbage leaves to be identified so as to create a system where farmers can make the right decisions to take action. The results of this study are also used as a supporting element for the cabbage IPM database.

By integrating aspects of information and communication technology (ICT) with agriculture, the development of DSS as an element of Integrated Pest Management in Cabbage offers innovative solutions that can support the achievement of SDG goals related to sustainable agriculture, food availability, and protection of terrestrial ecosystems. Through easy and widespread access to information related to agriculture and pest control, farmers can increase productivity, reduce environmental impact, and support overall sustainability.

2 Methods

Data collection was carried out on cabbage planting areas in the Batu and Tumpang areas. This location was chosen because it has a suitable height for the cultivation of cabbage plants. The collected data is then processed in a statistical computing laboratory for modeling. In addition, from the data obtained, the development of Plutella xylostella pest detection software was also carried out in the information technology laboratory. The object of research is photos of cabbage plants that have various conditions, both attacked by pests and not attacked by pests. Photo data will be analyzed and the level of pest attack will be grouped.

Some concepts need to be done in building an analysis system. Hartono (2004) states that there are seven important parts that must be considered:

a. Planning, the first step in building an information system is to make a plan (planning), namely making all plans related to the information system project;
b. Analysis, the intended analysis is an analysis of the information system workflow that is currently running and identifies whether the workflow is efficient and according to certain standards. The analysis is carried out by a Business Process Analyst (BPA) who has experience and/or understands the management system workflow in the area being analyzed;
c. Design is a very important step in the SDLC cycle because it defines the foundation of the information system. Errors in design can cause obstacles and even project failure. There are 2 types of designs made in this step, namely business process designs and programming designs. The programming design is carried out by the System Analyst (SA), namely making the necessary designs for programming based on the business process designs that have been made by the BPA. This design will be a guideline for programmers to write source code. Programming design includes database design, screen layout design, process diagram design, report layout design, and development;
d. Development, the development carried out is programming, namely the work of writing computer programs in a programming language based on certain algorithms and logic;
e. Testing, this process is to test the quality of information systems, to identify discrepancies in the results of an information system with the expected results;
f. Implementation, this process is to implement an information system that has been created so that users use it to replace the old information system. Implementation process includes notifying the user, training user, installing system, data entry/conversion, prepare user ID;
g. Operation and Maintenance, there are several routine procedures performed on the information system, namely system maintenance, backup and recovery, and data archive.

The stages of system development can be seen in the figure 1.

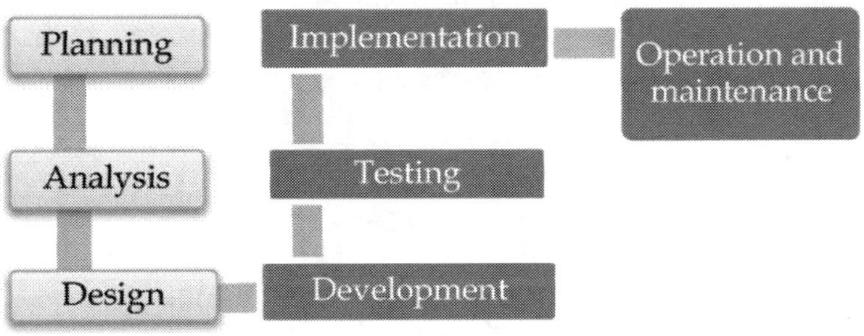

Fig. 1. System Development Stages Flowchart

The development offered is a website-based information system and open access for the user community. This can be done with website-based development and a user-friendly design. Need development for a fast process. Algorithms and databases must be the basis for the development of this system. The development activities carried out are in the form of website design development (Front End), Database Development (Back End), Domain and Hosting Creation, Integration, and Local Migration to Open Access Websites. The research stages are divided into 3 stages, namely testing, implementation, and operation and maintenance.

Basically, system testing aims to ensure that the system created can run well as expected. There are two methods for testing that are often carried out by developers, namely the Black box or Black box Testing method, and the White box or White box Testing method. At this stage, it consists of system testing activities, system whitebox testing, system blackbox testing, testing by IT experts, and repair and finishing. The main implementation of this system is for cabbage farmers to find out the level of pest attack quickly with photo images of cabbage. For this stage, besides cabbage farmers, it is also implemented in the biostatistics laboratory for input on the implementation of an open-access website-based information system. This stage consists of implementing the system as a whole, implementing the system for farmers, and implementing the system for experts. The operation of the system is an important part of the website-based information system that is launched. Apart from that, it analyzes bugs/errors from the user so that it can operate properly. As for maintenance, there needs to be periodic maintenance of the information system.

The research utilizes the Gray Level Co-occurrence Matrix (GLCM) method as an extraction method. Research by Zhang et al. [17] has provided evidence of results using GLCM provides good results and can be applied to plant types. Gray Level Co-occurrence Matrix (GLCM) is used for texture analysis/ feature extraction. Moreover, GLCM will describe the frequency of the appearance of pairs of two pixels with an intensity is in a certain distance and direction from the image [18]. The coordinates of a pair of pixels have a distance d and an angle orientation θ. Distance is represented in pixels, while angle in degrees. Angle orientation is formed based on four corner directions, namely, 0°, 45°, 90° and 135°, and the distance between pixels is 1 pixel [19]. The steps in the GLCM calculation are presented as follows: (i) Formation of the initial GLCM matrix from pairs of two pixels aligned in the direction 0°, 45°, 90° or 135°; (ii) Form a symmetrical matrix by adding the initial GLCM matrix with its transpose value; (iii) Normalize the GLCM matrix by dividing each matrix element by the number of pixel pairs.

3 Result and discussion

Website-based information system and open access for the user community. This can be done with website-based development and a user-friendly design. Need development for a fast process. Algorithms and databases must be the basis for the development of this system. Making the application that has been done is making an analysis of pest attacks on cabbage plants. The application work process is shown in Figure 2.

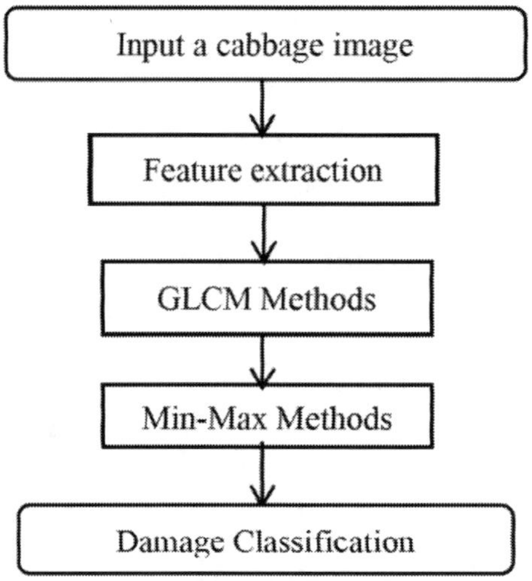

Fig. 2. Application work process

The initial appearance of the system is like this, Analysis is on the left while the right side is used to issue analysis results. Input the image for the extraction analysis of cabbage damage. The initial appearance of the system can be seen in Figure 3.

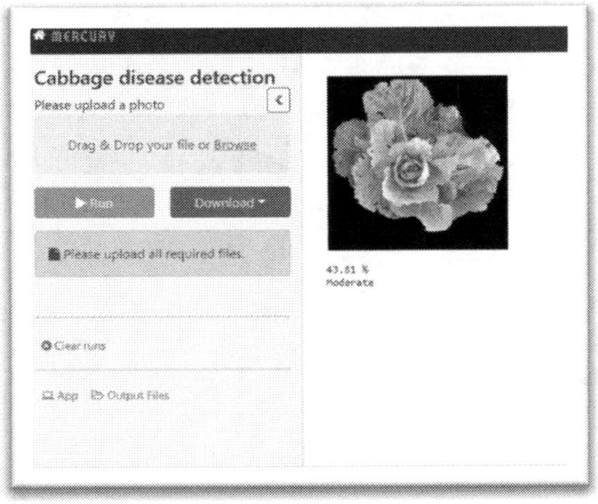

Fig. 3. Initial display of the system

After inputting the system, a display will appear as shown in Figure 4. This indicates that the input image is received.

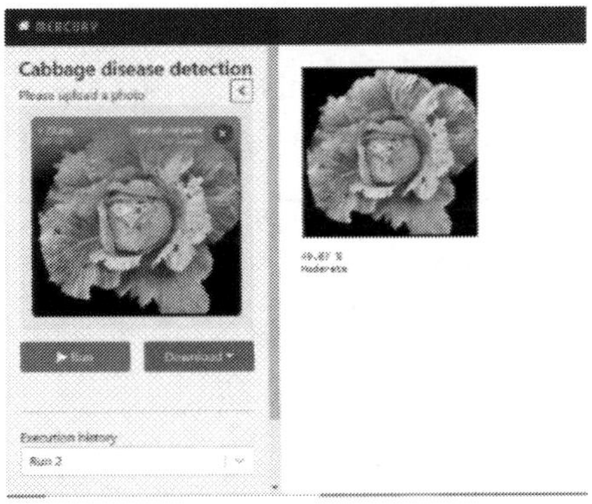

Fig. 4. Input image is received

To be able to perform an analysis, click the run button as shown in Figure 4. Aplikasi akan mulai menganalisis. Proses analisis ditampilkan pada gambar 5.

Fig. 5. analysis process

The results of the analysis process are shown in Figure 6.

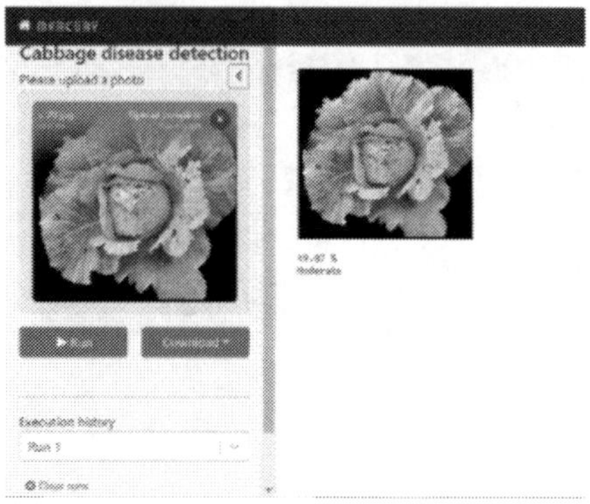

Fig. 6. Results of the analysis carried out

The next stage is testing, system testing aims to ensure that the system created can run as well as expected. There are two methods for testing that are often carried out by developers, namely the Black box or Black box Testing method, and the White box or White box Testing method.

Table 1. Test results for IT expert systems

Test Scenario	Test Case	Test Step	Result
User uploads files	Users upload files in jpeg, png, jpg or other image formats.	The user is on a web page. The user enters an image via the browse/drag button and places the file in the available field.	File is uploaded and ready to be executed.
	files in formats other than jpeg, png, jpg or other image formats.	The user is on a web page. The user enters an image via the browse/drag button and places the file in the available field.	The file remains uploaded and ready to be executed.
Classification process	Process of classifying files in jpeg, png, jpg or other image formats that have been uploaded.	1. The user is on a web page. 2. The user presses the run button for the classification process. 3. Users are waiting for the results of clarification	The web page will display the results of the classification as well as the files that have been uploaded.
	the process of classifying files in jpeg, png, jpg or other formats other than uploaded images	1. The user is on a web page. 2. The user presses the run button for the classification process. 3. Users are waiting for the results of clarification	The web page does not display any results.

Furthermore, the main implementation stage of this system is for cabbage farmers to find out the level of pest attack quickly with photo images of cabbage. For this stage, apart from cabbage farmers, it is also implemented in the biostatistics laboratory for input on the implementation of an open-access website-based information system.

The next stage is operation and maintenance. The operation of the system is an important part after the website-based information system that is launched. Apart from that, it analyzes bugs/errors from the user so that it can operate properly. As for maintenance, there needs to be periodic maintenance of the information system. After analyzing as many as 225 images were taken twice the analysis. The first uses this system and the other uses experts to determine the accuracy of this system. Of the 225 experiments carried out, 182 trials were the same between the expert and this system. So the accuracy of this system is above 80%.

Previous research conducted by Wardhani, et. al. [16], shows the lack of comprehensive evaluation metrics to measure classification performance on imbalanced data. The authors mention that imbalanced data is a common problem in classification, especially in image classification tasks such as medical or agricultural fields. However, they highlight that commonly used metrics such as accuracy, F1 score, and g-mean may not accurately reflect the performance of classifiers on imbalanced data. The because accurate evaluation metrics are critical for assessing classifier performance on imbalanced data. Without robust metrics, determining the effectiveness of various classification techniques and making informed decisions in real-world applications will be challenging. Therefore, further research is needed to explore and compare the performance of different metrics in the context of imbalanced data classification.

In research conducted by Dong, et. al. [10], a system is needed that can provide timely and spatially sustainable information regarding the occurrence and development of pests and diseases in a wide area. Current ground monitoring and meteorological-based forecasting methods are insufficient to meet these requirements. Additionally, this research highlights the need for improved database construction and the incorporation of communication technologies, deep learning, and artificial intelligence to efficiently extract and mine information from massive multi-source data. Additionally, there is a need to optimize system automation and intelligence to support green crop protection in China.

Research conducted by Carillo-Reche, et. al. [7] has limited understanding and utilization of competition control strategies, such as root pruning, in intercropping systems. The authors question whether this strategy is fully utilized to increase productivity and product quality in intercropping systems. This indicates the need for further investigation and exploration of competition control strategies to optimize the benefits of intercropping in agricultural systems.

Based on research conducted by Roche et. al. [3], concrete means are needed to protect maize field plants from wireworms. Although there are several data-driven models used to evaluate wireworm infestation risk, such as the VFF-QC web application developed and used in Quebec, there is no widely available DSS for wireworm infestation risk assessment yet.

4 Conclusion

From these results, it can be concluded that by using the feature extraction method from the analysis carried out in photo images taken to detect pest attacks by observing the number of holes in cabbage leaves can be applied in mathematical and statistical system modeling. Mathematical and statistical modeling can be used as a basis for classifying the distribution of P. xylotella pests in a cabbage field using the Bagging and Boosting methods. The development of a photo-image-based system application can be used as data to support decision-making for the use of pesticides in tackling P. xylotella pests.

Acknowledgement

We would like to thank the Brawijaya University. This research activity was funded by a professor research grant from the Faculty of Mathematics and Natural Sciences, Universitas Brawijaya.

References

1. A. Fatchiya, I. Muflikhati, T. Soedewo, urnal Penyuluhan **14**, 2 (2018)
2. Statista, Production of Chinese cabbage in Indonesia from 2012 to 2022 (Statista research departement, 2023)
3. J. Roche, M. Plantegenest, P. Larroudé, J. B. Thibord, L. C. Ronan, S. Poggi, Technol. **4** (2022)
4. B. Subedi, A. Poudel, S. Aryal, J. Agric. Food Res. **14** (2023)
5. C. E. Miluch, L. M. Dosdall, M. L. Evenden, Crop Prot. **45** (2013)
6. P. B. Baker, A. M. Shelton, J. T. Andaloro, J. Econ. Entomol. **75**, 6 (1982)
7. J. Carrillo-Reche et al., Agric. Ecosyst. Environ. **354** (2023)
8. A. K. Mahlein, Plant Dis. **100**, 2 (2016)
9. S. B. Patil, T. W. Hurd, A. K. Ghosh, C. A. Murga-Zamalloa, H. Khanna, PLoS One **6**, 6 (2011)
10. Y. Dong et al., IEEE J. Sel. Top. Appl. Earth Obs. Remote Sens. **13** (2020)
11. J. G. Arnal-Barbedo, Springerplus **2**, 1 (2013)
12. S. Raza, S. I. Bozhevolnyi, M. Wubs, N. Asger Mortensen, J. Phys. Condens. Matter **27**, 18 (2015)
13. J. D. Pujari, R. Yakkundimath, A. S. Byadgi, Procedia Comput. Sci. **46** (2014)
14. A. L. Holsteen, S. Raza, P. Fan, P. G. Kik, M. L. Brongersma, Science **358**, 6369 (2017)
15. G. Saradhambal, R. Dhivya, S. Latha, R. Rajesh, Int. J. Pure Appl. Math. **119**, 14 (2018)
16. N. W. S. Wardhani, M. Y. Rochayani, A. Iriany, A. D. Sulistyono, P. Lestantyo, *Cross-validation Metrics for Evaluating Classification Performance on Imbalanced Data*, in 2019 Int. Conf. Comput. Control. Informatics its Appl. Emerg. Trends Big Data Artif. Intell. IC3INA (2019)
17. G. Zhang et al., Agronomy **10**, 1 (2020)
18. I. Setiawan, W. Dewanta, H. A. Nugroho, H. Supriyono, Jurnal Media Infotama **15**, 2 (2019)
19. R. A. Surya, A. Fadlil, A. Yudhana, J. Inform. J. Pengemb. IT. **2**, 2 (2017)

Design and analysis of heat transfer on a customized eco-print batik steamer with automatic temperature and time control using Computer Fluid Dynamic (CFD)

Handini Novita Sari[1*], *Rizal* Mahmud[2], *Nur Aini* Susanti[3], *Imami Arum Tri* Rahayu[4], *Susi Tri* Umaroh[3], and *Agung Prijo* Budijono[1]

[1]Universitas Negeri Surabaya, Mechanical Engineering Department, Surabaya, Indonesia.
[2]Meiji University, Mechanical Engineering Informatics Department, Tokyo, Japan.
[3]Universitas Negeri Surabaya, Mechanical Engineering Education Department, Surabaya, Indonesia.
[4]Universitas Negeri Surabaya, Fashion Education Department, Surabaya, Indonesia.

Abstract. The process of making eco-print batik carried out by eco-print batik industry players still uses conventional methods with steaming pans. The conventional method used has a quite crucial impact on the equipment used, the bottom of the pan is susceptible to damage because it is used continuously for quite a long time. To overcome this problem, a steamer design will be created specifically for eco-print batik using Solid Work software. Following the completion of the design, this research investigates how the heat transfer process occurs, the circumstances of the heat flux created, and the temperature during the heating process using computer fluid dynamics (CFD) with simulation methods as an approach to real conditions. Observations were made in 300s, 600s, 900s, 1200s. The steamer material employed in this simulation is stainless steel 304, and the simulation results show that the steamer's heating temperature is. Based on the simulation results of high pressure in closed system conditions and the influence of materials with good thermal conductivity on the steamer, which have an impact on increasing temperature and heat flux during the process, through these considerations, the steamer is designed to be equipped with temperature control and a safety valve which functions to release pressure. steam when excessive. In this way, the steamer will be more effective and efficient in terms of time, heating, and the quality of the batik results that will be produced later when compared to the pan used previously.

1 Introduction

This eco-print batik is one type of batik whose manufacturing method utilizes natural dyes from tannins or dyes of leaves, roots or stems placed on a piece of cloth, then the cloth is boiled. Boiling this cloth takes approximately two hours to get perfect results. So

* Corresponding author: handinisari@unesa.ac.id

far, the eco-print batik boiling process carried out by eco-print batik industry players still uses conventional methods, namely using ordinary boiling pots. During the boiling process, temperature, pressure, and time conditions are not controlled when using conventional methods. The conventional method used has a crucial impact on the equipment used, for example, in the pan, the bottom of the pot is vulnerable to damage because it is used continuously for a long duration of time. In addition, the fabric bending treatment so that all the cloth put in the pot is sufficiently accommodated will damage the pattern that appears on the batik, the impact of this results in crease marks from the bend. In this conventional method, one pot can only hold one to two batik cloths, so in this case, it is inefficient in terms of time and quantity which will also have an impact on the consumption of fuel used during the production process.

The steamer design made specifically for batik eco-print will be equipped with temperature and time control so that the process of boiling batik eco-print fabric can run without interference from uncontrolled temperature and time conditions so that users can easily adjust the conditions according to their individual needs. The steamer will be designed with material that is sturdier and more reliable using materials that have high heat resistance. The capacity or capacity of the steamer will also be calculated according to needs so that it will be more effective and efficient in terms of time with a larger input quantity than before.

The first step taken in this research is to make an eco-print batik steamer design with solid work software. After the design is completed, this study will also examine how the heat transfer process takes place, how the heat flux conditions are produced, the maximum temperature obtained, and the convection coefficient during the heating process. In this case, using computer fluid dynamics (CFD) with simulation methods as an approach to real conditions. One of the most effective methods for assessing and managing superheater models is computational fluid dynamics (CFD), which enables study of extremely complex geometries while taking into consideration the intricacies of three-dimensional geometric models [1] [2]. It is also feasible to measure the heat transfer between exhaust gases and steam during superheater operation, as well as the heat transfer coefficient on both the steam and the hot side of the superheater [3].

Research related to the phenomenon of heat transfer, condensation, evaporation and material treatment of tools that use stainless steel material has been carried out both experimentally and simulated, such as research conducted by [4]. The purpose of this research is to determine how the initial temperature of the substrate affects the heat transmission, microstructure, hardness, and residual stress of AISI 316 L stainless steel deposits on AISI 304 steel substrates. Perhaps the methods presented here might be applied to further additive manufacturing research in alloys and other metal equipment, bringing reliability and durability to additive manufacturing (AM). Wei Li described condensation heat transfer in annuli outside the enhanced horizontal stainless steel tube, which he claims is commonly used to improve heat transmission in various scenarios [5]. One of the primary processes that occurs in energy conversion systems is boiling heat transfer. Arcasi research et. al. [6] discuss the experimental examination of hysteresis phenomena during boiling stream heat transfer in horizontal stainless steel. The relative influence of neighboring cavities is still a topic of theoretical debate in the literature today [7], from numerical [8], and experimental [9] points of view.

Tests performed under the same operating conditions but with different thermal histories (heat flux charged greater in the transient phase) are thought to be influential in giving the wall a lower superheat value and hence higher heat transfer performance. The boiling theory eventually enforced the experimental trend, taking into account the effect of hysteresis as the activation of surface cavities subjected to higher temperature boundary conditions. In the study of H.E. Zhang et al., a CFD model was built to mimic the entire process of directional energy deposition of stainless steel. To investigate physical phenomena associated with this process, such as temperature behavior, melt flow characteristics, and layer growth, systematic simulations of melt ponds and deposited layers are used [10]. Process parameters, sample characteristics, and DED process modeling are all examples of DED process modeling. Particle ejection studies reveal that a number of elements, such as coaxial nozzle shape, inert gas flow rate, and particle characteristics, do have a significant impact on powder flow [11-15] With a steady stream of powder, the ejected particles heat the focused laser beam falling into the melt pond and are considered an additional mass [16-18]. In addition, clad formation is also directly affected by the size and temperature of the melt pool [19, 20]. As a result, it is vital to better understand how the melt pool evolves and how it influences clad formation. According to Asghar Hozoorbakhsh et al.'s study on An investigation of heat transfer and fluid flow on laser micro-welding upon the thin stainless steel sheet (SUS304) using computational fluid dynamics (CFD) [21].

2 Methods

The method used in this study is computer fluid dynamic (CFD) which is a simulation method to analyze heat transfer studies that occur when the steamer is used for the batik fabric steamer process. The complete design of the steamer is shown in Figure 1.

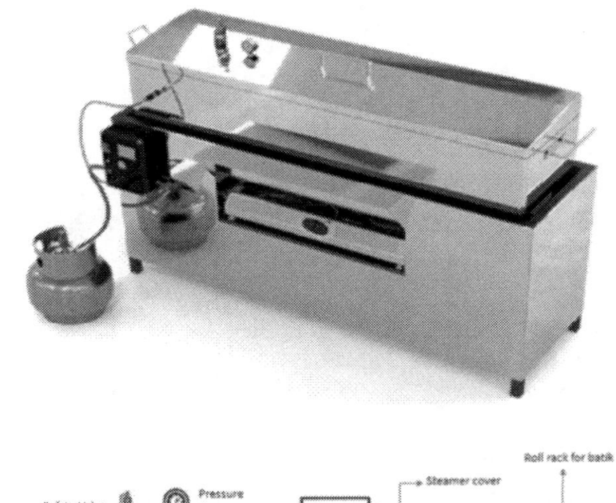

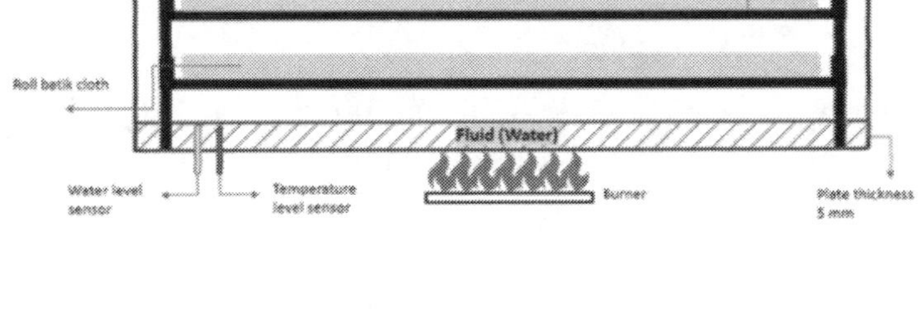

Fig. 1. Eco-print batik steamer desi

2.1 **Modelling heat and mass transfer**

2.1.1 Heat transfer

Heat transfer in fabrics is assumed based on the equation of Eq. (1) [22, 23].

$$\rho_m \; c_{pm} \frac{\partial T}{\partial T} + \nabla \left(-k_m \nabla T\right) + \rho_w \; c_{pw} \; u_w \; \nabla T = 0 \qquad (1)$$

Where ρ_m and ρ_w are the densities of cloth and water (kg/mp_w3), respectively. k_m is the thermal conductivity of the fabric (W/(m °C), and is the heat capacity of the fabric and water (J/(kg °C), u_w is the fluid velocity (m/s), T is the temperature (°C) and t is time (s).

2.1.2 Mass transfer

Based on the mass conservation formula and based on the equation of water transport with product given by equation (2) [22]

$$(\partial C)/\partial t \; + \nabla \; (Cu\_w) \; = \; \nabla (D \nabla C) \qquad (2)$$

where C represents the moisture concentration (kg of water/kg of sample), t is time(s), u_w is the velocity of the fluid (m/s) and D is the moisture diffusion coefficient (m^2/s)

2.2 **Boundry and initial condition**

The convective heat flux from the hot air to the surface of the product is given by eq. (3) [22].

$$q = h \; (T_{steamer} - Ts) \qquad (3)$$

Where q is the heat flux convective flux (W/m2), h is the heat transfer coefficient average (W/(m2°C)), T steamer is the temperature of the steamer (°C) and T s is the surface temperature of the product (fabric) (°C).

2.2.1 Heat transfer boundary condition

The heat flux boundary condition is given by eq. 4

$$(1\text{-}f) \; h \; (T_{steamer} - T_s) = -n \; (k_m \nabla T + \rho_w \; c_{pw} \; u_w T) \qquad (4)$$

f is the fraction of energy lost by evaporation of water on the surface. The value of f is between 0 and 1. The parameter f is viewed as an empirical switching function that determines how much of the provided energy is utilized for evaporation. Ts ≥100 °C predominates during evaporation, and f quickly climbs to values closer to 1 than to 0. The precise value of f will be determined by the conditions of heat transmission and mass loss, and it should be calculated at this stage through trial and error. Mass transfer boundary condition. The mass flux boundary condition is given by eq. 5.

$$\text{n}.\ (-D\nabla C + u_w C) = f\,h\,\frac{(T_{steamer} - T_s)}{H_{evap}\,\rho}\,(C - C_{eq}) \quad (5)$$

2.2.2 *Initial condition*

T= To and C = Co

2.3 **Experiment setup**

The material used as the basic material of the steamer is stainless steel 304, the use of this material is based on the consideration that stainless steel type 304 is a versatile type of stainless steel. Its chemical composition, mechanical strength, weldability, and corrosion resistance are excellent at a relatively affordable price. Stainless steel type 304 is widely used in the industrial world and small scale. The complete input parameters of the material are shown in Table 1, as well as the water properties shown in Table 2.

Table 1. The input parameters material steamer SS304

Properties of Outline Row 5: Steel Stainless 304

	A	B	C
1	Property	Value	Unit
2	Material Field Variables	Table	
3	Density	8055	kg m^-3
4	Isotropic Thermal Conductivity	16,2	W m^-1 C^-1
5	Specific Heat	500	J kg^-1 C^-1

Table 2. Water property

Properties of Outline Row 7: Water Fresh

	A	B	C
1	Property	Value	Unit
2	Material Field Variables	Table	
3	Density	997,4	kg m^-3
4	Isotropic Thermal Conductivity	0,604	W m^-1 C^-1
5	Specific Heat	4179	J kg^-1 C^-1

Table 3. Material of Burner: Brass

Properties of Outline Row 3: Brass

	A	B	C
1	Property	Value	Unit
2	Density	8600	kg m^-3
3	Isotropic Thermal Conductivity	111	W m^-1 C^-1
4	Specific Heat	162	J kg^-1 C^-1

The image below is an illustration of the boundary conditions of an eco-print batik steamer which is simulated running in closed system conditions.

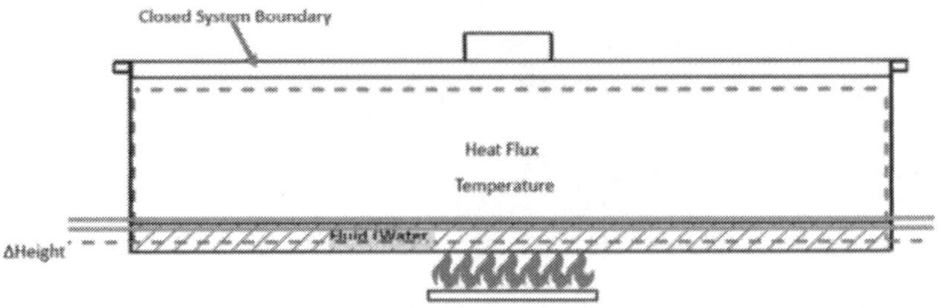

Fig. 2. Boundary layer condition

Input for Simulation Parameters:
1. Temperature of Heat Source
2. Film Coefficient of Water
3. Bulk Temperature
4. Time Variation [300s, 600s, 900s, 1200s, 1500s]
(to check water evaporation/5 minutes)
Objective:
1. Temperature Distribution (T)
2. Heat Flux Distribution (Q)

2.4 **Meshing**

The meshing process is one of the important processes in CFD, the quality of the mesh greatly determines the output results of the simulation. Meshing is the process of dividing a domain into several cells. In general, mesh is divided into 2, namely mesh with a structured grid and mesh with an unstructured grid. The general scalar equation used in the discretization process is. Place the figure as close as possible after the point where it is first referenced in the text. If there is a large number of figures and tables, it might be necessary to place some before their text citation.

$$\underbrace{\frac{\partial(\rho\phi)}{\partial t}}_{\substack{transient\\term}} + \underbrace{\nabla\cdot(\rho\mathbf{u}\phi)}_{\substack{convection\\term}} = \underbrace{\nabla\cdot(\Gamma\nabla\phi)}_{\substack{diffusion\\term}} + \underbrace{Q}_{\substack{source\\term}}$$

$$\underbrace{\frac{\partial(\rho\phi)}{\partial t}}_{\substack{transient\\term}} + \underbrace{\frac{\partial(\rho u\phi)}{\partial x} + \frac{\partial(\rho v\phi)}{\partial y}}_{\substack{convection\\term}} = \underbrace{\frac{\partial}{\partial x}\left(k\frac{\partial\phi}{\partial x}\right) + \frac{\partial}{\partial y}\left(k\frac{\partial\phi}{\partial y}\right)}_{\substack{diffusion\\term}} + \underbrace{Q}_{\substack{source\\term}} \quad (6)$$

By dividing the work area into cells, the discretization process can be carried out, so that the CFD calculation process can be carried out, but in CFD the approach can be carried out with finite differences or finite volumes, after the calculation/iteration process is carried out until convergent results are obtained, the respective values The points will be known, so that the values of these points can be displayed, either through an attractive GUI visualization, or simply numerical data in a table.

Figure 3 is the result of meshing from an eco-print batik steamer. The average skewness value of the meshing steamer is 0.47865 while the minimum skewness is 1.3057e-010, meaning that the meshing quality can be said to be very good.

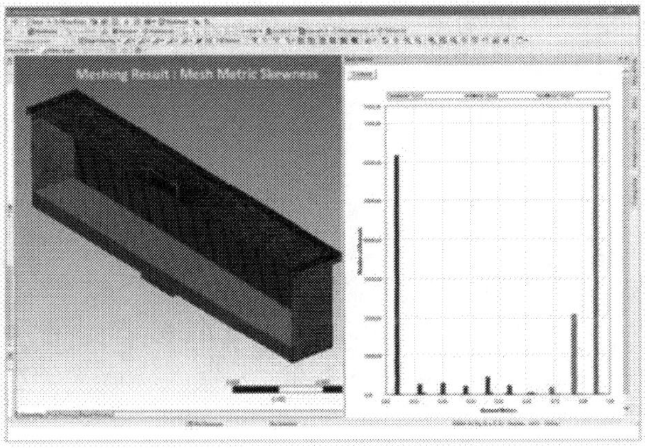

Fig. 3. Meshing of steamer

3 Result and discussion

3.1 Temperature distribution

Figure 4 depicts the modeling results of the temperature distribution of the eco-print batik steamer made of 304 stainless steel. Where significant temperature differences are depicted, at the heat source the temperature was set at 120 °C, and simulation observations were carried out at t=300s to t=1200s, As time goes by, overall heat propagation occurs both vertically and horizontally. At t=300s to t=1200s, the bottom of the steamer temperature shows a figure of around 118.79-120.5 °C, where the steamer is filled with 28.32 liters of water with a fluid height of around 59 mm. The temperature distribution in the steamer section containing fluid is at t=300s to 900s in the temperature range of 102.93-117.1 °C, with an average temperature of 109.878 °C. At t=1200s the average fluid temperature shows an increase of 110.174 °C. This means that at t=300s the fluid has reached the boiling point of water. In theory, water will boil at a temperature of 100 degrees Celsius at a pressure of 1 atmosphere. Because the steamer is made of thick and strong stainless steel and has a tight lid, the water vapor produced during the boiling process cannot possibly escape and only collects in the steamer. This collected water causes the water pressure in the steamer to rise, which causes the boiling temperature to also rise to 100 °C.

At t 300s, 600s, and 900s the average fluid temperature is the same, this is because the observation time distance is too close, and the cross-sectional area and water volume are quite large which also influences the convection heat transfer process that occurs in the fluid. The surface that has a higher heat load will transmit heat, in this case, the density of the lower water decreases, so the water particles are pushed to the surface of the steamer, and their place is automatically replaced by water which falls to the bottom of the lower surface until the water boils.

https://doi.org/10.1051/e3sconf/202345002002

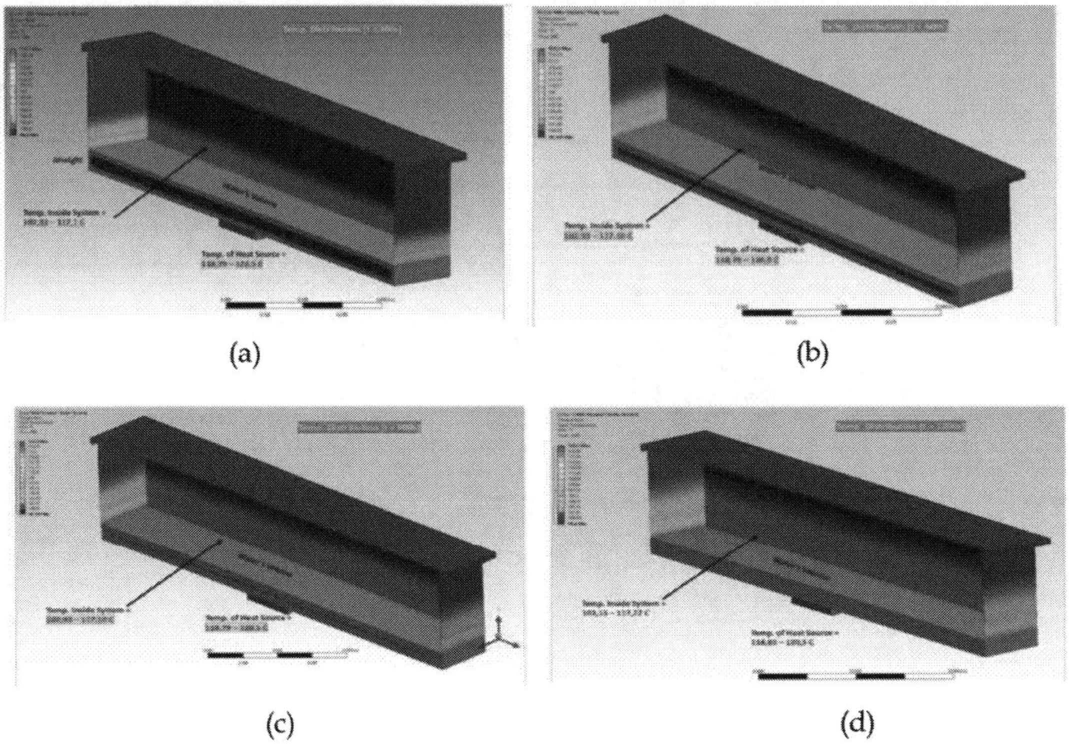

Fig. 4. Temperature distribution at time (a) 300s, (b) 600s, (c) 900s, and (d) 1200s

3.2 *Heat Flux*

The heat flux condition at t=300s in the fluid is in the range 2.133-357.11 W/m^2 with the water volume reduced to 28.32 liters from the initial volume of 28.77 liters. At t=600 to 900s, the heat flux value in the fluid is in the range 8,467-718,920 W/m2, and at t=1200s the heat flux value in the fluid is 12,867- 876.96 W/ m^2. From Figure 5 we can see that the heat flux increases with increasing temperature and time. Meanwhile, the volume of water also decreased frequently with increasing temperature and time, respectively amounting to 27.63 liters, 26.88 liters, and 26.16 liters.

Because this steamer works in a closed system, as a safety measure, the steamer is provided with a safety valve that functions to release steam pressure when it is excessive. This is because the steamer uses a tight lid so that water vapor cannot escape. At a certain time, this tool reaches the threshold or standard for the start of calculating the cooking time, which is marked by a hissing sound due to water vapor passing through the safety valve. So it can be concluded that the process of steaming eco-print batik cloth using this steamer can be faster than using a regular pan because the heat absorbed by the trapped steamer cannot come out again (closed system) so the pressure inside the steamer rises along with the temperature rise and no change in volume occurs. Init (isochoric process) or there is only a slight reduction in water due to the presence of the safety valve. The aim of the steaming process for eco-print batik cloth is to bring out the basic color of the leaves so that it can produce a beautiful pattern of fine cloth.

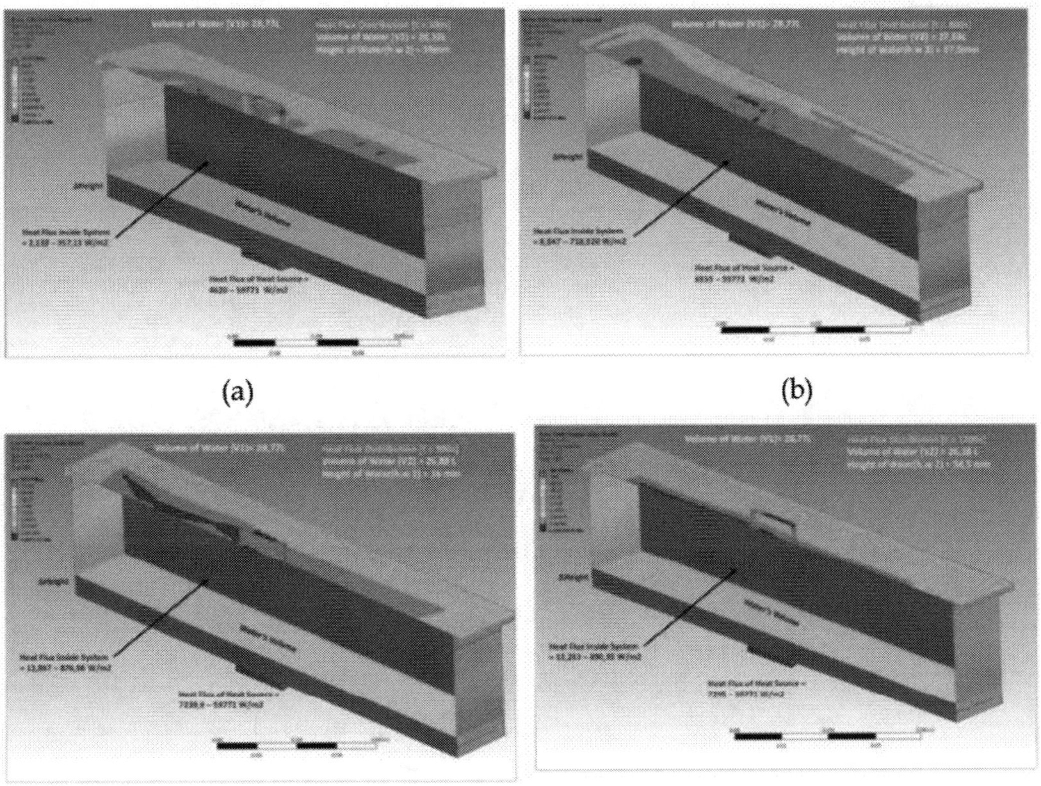

(a) (b)

(c) (d)

Fig. 5. Heat flux of steamer at t (a) 300s, (b) 600s, (c) 900s, and (d) 1200s

The high pressure in closed system conditions and the influence of materials with good thermal conductivity on the steamer, have an impact on increasing temperature and heat flux during the process, so through these considerations, the steamer is designed to be equipped with temperature control. This temperature control functions to make it easier for users to adjust temperature control. Based on research conducted by Entien Darmawati and Sutopo who examined the effect of steaming temperature for eco-print batik cloth on color sharpness and fastness, it was concluded that the most optimal temperature was in the range of 70-80 °C [24]. With this temperature control, the temperature conditions to get the best quality from eco-print batik can be maintained.

4 Conclusion

Based on results and discussion, it can be concluded that : (1) The temperature distribution in the steamer section containing fluid is at t=300s to 900s in the temperature range of 102.93-117.1 °C, with an average temperature of 109.878 °C. At t=1200s the average fluid temperature shows an increase of 110.174 C. This means that at t=300s the fluid has reached the boiling point of water ; (2) Heat flux at t=300s in the fluid is in the range 2.133-357.11 W/m^2 with the water volume reduced to 28.32 liters from the initial volume of 28.77 liters. (3) At t=600 to 900s, the heat flux value in the fluid is in the range of 8,467-718,920 W/m^2, and at t=1200s the heat flux value in the fluid is 12,867-876.96 W/m^2.The high pressure in closed system conditions and the influence of materials with good thermal conductivity on the steamer, have an impact on increasing temperature and heat flux during the process, so through these considerations, the steamer is designed to be equipped with temperature control and a safety valve which functions to release steam pressure when excessive.

References

1. T. D. Granda, M. Trojan, Energy **199**, 117423 (2020)
2. P. Ludowski, D. Taler, J. Taler, Appl Therm Eng. **58** (2013)
3. P. Madejski, D. Taler, A. Korzeń, Arch. Energ. **3** (2012)
4. D. V. Isquierdo, R. H. M. Siqueira, S. M. Carvalho, M. S. F. Lima, J. Mater. Res. Technol. **18** (2022)
5. W. Li, H. Chen, J. Wang, Y. Gao, X. Wang, Int. J. Therm. Sci. **177** (2022)
6. A. Arcasi, R. Mastrullo, A. W. Mauro, L. Viscito, Int. J. Heat Mass Transf. **164** (2021)
7. S. Ming-Heng, J. Ma, W. Bu-Xuan, Int. J. Heat Mass Trans. **36** (1993)
8. N. B. R. Mallozzi , R. L. Judd, Int. J. Heat Mass Transf. **43** (2000)
9. R. L. Judd, J. Heat Transf. **110** (1988)
10. Y. M. Zhang, C. W. J. Lim, C. Tang, B. Li, Int. J. Therm. Sci. **165** (2021)
11. S.Y. Wen, Y.C. Shin, J.Y. Murthy, P. E. Sojka, Int. J. Heat Mass Transf. **56**, 25–26 (2009)
12. A. J. P. J. Ibarra-Medina, Phys. Procedia **5** (2010)
13. H. Liu, X. L. He, G. Yu, Z. B. Wang, S. X. Li, C. Y. Zheng, W. J. Ning, Sci. China Physics, Mech. Astron. **58** (2015)
14. G. F. J. Yan, I. Battiato, Int. J. Adv. Manuf. **91**, 1–4 (2016)
15. O. B. Kovalev, I. O. Kovaleva, I. Y. Smurov, J. Mater. Process. Technol. **249** (2017)
16. J. Ibarra-Medina, Development and Application of a CFD Model of Laser Metal Deposition (University of Manchester, Manchester, 2012).
17. I. Tabernero, A. Lamikiz, S. Martínez, E. Ukar, Phys. Procedia **39** (2012)
18. I. Arrizubieta, A. Lamikiz, F. Klocke, S. Martínez, K. Arntz, Int. J. Heat Mass Transf. **115** (2017)
19. X. Wang, et al., *Research on remelting phenomena and metal joining of fused-coating additive manufacturing,* in Journal of Physics: Conference Series, 1063, IOP Publishing (2018)
20. J. Du, X. Wang, H. Bai, G. Zhao, Y. Zhang, International Journal of Thermal Sciences **114** (2017)
21. A. Hozoorbakhsh, M. I. S. Ismail, A. A. D. M. Sarhan, A. Bahadoran, N. B. A. Aziz, Int. Commun. Heat Mass Transf. **75** (2016)
22. E. Bird, W. Stewart, E. Lightfoot, Transport phenomena (John Wiley and Sons, New York, 2001).
23. R. G. M. Van der Sman, Meat Sci. **76**, 4 (2007)
24. E. Darmawati, Berkala Penelitian Teknologi Kulit, Sepatu, dan Produk Kulit **20**, 1 (2021)

E3S Web of Conferences **450**, 02003 (2023) https://doi.org/10.1051/e3sconf/202345002003
ICoSBi 2023

The effect of using a fried rice machine on product effectiveness and productivity

Any Sutiadiningsih[1*], *Ila Huda Puspita* Dewi[1], *Hafizhuddin Zul* Fahmi[1], *Wahyu Dwi* Kurniawan[1], and *Muhammad Izzat Nor* Ma'arof[2]

[1]Universitas Negeri Surabaya, Surabaya, Indonesia
[2]INTI International University, Negri Sembilan, Malaysia

Abstract. In general, fried rice is a dish that is popular with the public, conventionally prepared using a set of conventional equipment with problems: (1) its operation requires a lot of physical effort, is tiring and sweaty, not intensive, (2) the equipment is relatively unhygienic, the quality of the results is less stable. The SEGORO fried rice-making machine is the solution to this problem. The SEGORO fried rice unit consists of: a set of heaters with gas energy, equipped with an electric fire starter, a 12 Volt DC motor that drives the stirrer and rotates the Wok, equipped with a relevant wok, equipped with a gas handle valve. Analysis results: (1) one processing batch for 15-20 portions; (2) processing time from 12 minutes to 7 minutes, productivity increased up to 2x; (3) conventionally workers work tiringly and produce a lot of sweat. 'SEGORO' fried rice machine, equipped with an electric stirrer motor and Wok turning motor does not require high physical effort, and is not tiring. In conclusion, the Segoro fried rice machine uses little electricity, low voltage, saves gas and physical energy, its operation is driven by a stirrer motor, and is equipped with a wok turning motor whose speed can be adjusted. The Segoro fried rice machine is made from stainless steel 304, it is more hygienic and meets food-grade standards, and the quality of the processed product is more stable.

1 Introduction

From the many foods spread throughout Indonesia, the majority of the country's population. This makes rice a staple food. One of the processed foods made from Rice is the most popular and sold by MSMEs, street vendors, and restaurants namely fried rice. Making fried rice still requires skill basic hands, a long time, and a lot of energy.

* Corresponding author : anysutiadiningsih@unesa.ac.id

1.1 Background

In the cooking process, especially in the manufacture of food that involves manual steps such as fried rice, several problems arise in traditional methods that can impact the efficiency and quality of the final product. The manual method of making fried rice generally involves intensive work and requires considerable physical effort. The following are some of the problems encountered in the manual method: (1) Worker fatigue: The manual method of making fried rice tends to require a lot of hand and physical movement. This can cause workers to experience high levels of fatigue. Excessive fatigue can reduce worker productivity and even impact their welfare; (2) Long Time: Because the manual method still relies on conventional equipment and steps are carried out sequentially, the time needed to complete one fried rice order can be longer. This can affect kitchen efficiency and customer service time; (3) Poor Quality of Food: Fatigue experienced by workers can have an impact on the quality of the food produced. Especially when workers sweat, sweat dripping onto pans or equipment can affect the temperature and quality of the food produced. This can reduce the taste and final quality of the dish served; (4) Lack of Cleanliness and Hygiene: The manual method of making fried rice tends to make the ingredients come into contact with workers' hands more often. This can cause hygiene and hygiene problems, as there is a risk of transferring microbes from workers' hands to the food being processed. Lack of hygiene in the cooking process can threaten the health of consumers; and (5) Varied Order Levels: In a restaurant or food stall environment, order levels can fluctuate rapidly. Workers often have to handle large volumes of orders in a short amount of time. This can overwhelm workers and lead to reduced efficiency and increased risk of misrepresentation.

Therefore, there is a need to consider using more efficient, hygienic, and ergonomic methods in the fried rice process to overcome the problems described above. Modern methods or the right tools can help increase productivity, quality, and worker welfare as well as maintain higher standards of hygiene in the cooking process. The SEGORO, an innovative semi-automatic fried rice machine can produce high-quality products and answer the problems above. In addition, the purpose of creating this machine innovation is to make it easier to make rice that requires a frying process, register product certification and quality management so that it is ready for commercialization, share knowledge and information on fried rice-making machines, and be able to improve services to the community so that income increases.

1.2 Previous research

1.2.1 Stirring automatic

In a study on [1] the effect of different stirring speeds on acid refolding of rice glutelin, researchers found that 1% (w/v) rice glutelin hydrocolloid throughout the acidification process. A significant decrease in the size of hydrated rice glutelin particles was observed when the stirring speed was increased to 750 rpm, while further increasing the speed had no significant effect on the particle size. Samples stirred at 750 rpm showed the highest solubility, and samples stirred at 0 rpm showed the lowest solubility. Furthermore, both surface hydrophobicity and molecular weight initially increased and then decreased, reaching a minimum value for the sample stirred at 750 rpm. Principal component analysis (PCA) was used to identify the correlation between changes in different properties such as solubility, particle size, β-sheet content, surface hydrophobicity, and zeta potential and different agitation treatments. In summary, the selection of stirring speed significantly influences various aspects of the refolding process of rice glutelin during acidification. Therefore, selecting the appropriate stirring speed is important to maintain quality control in the production of protein hydrocolloids.

Huber et al. [2], showed the possibility of achieving a more uniform distribution of U-235 enrichment in downblend castings by using zirconia crucibles instead of graphite crucibles. Using graphite as a container, the electromagnetic field generated by the induction heating device interacts directly with the graphite. Zirconium oxide, on the other hand, is not conductive in this situation and therefore does not interact with the field. The induced field interacts directly with the metal, creating electromagnetic stirring (EMS) within the molten metal bath. The results showed that the enrichment uniformity of castings using zirconia crucibles was approximately 8 times better than that of castings using graphite crucibles, as measured by plate enrichment area and coefficient of variation of enrichment.

Based on the study of using a catalytic micromotor as a self-stirring microreactor for efficient dual-mode colorimetric detection [3], we successfully developed a colorimetric detection system using a catalytic micromotor (MIL-88B @Fe3O4). With responsive rotational motion and excellent catalytic performance, this new catalytic MOF-based micromotor not only opens a new dimension of nanotechnology in colorimetric analysis, but also has applications in precision manufacturing, biomedical analysis, environmental management, etc. It shows great potential in various fields. These micromotor-based microreactors can be easily used for various chemical microreactions in various applications.

Based on a study focused on parametric optimization and experimental validation of multiple fertilizer mixing using air bubbles and knife stirring [4], and a study based on discrete element simulation, multiple fertilizer mixing was performed to improve overall utilization. The goal is to improve the uniformity of the fertilizer mix. The simulations were performed using his EDEM and Fluent software, and the results were confirmed by experiments. Simulation results show that the ideal mixing speed range for individual blade groups is 600 to 900 revolutions per minute (r·min-1), with the best mixing when the blades are positioned at an angle of 20 degrees (α). It shows that uniformity is achieved. Bench test results show that the mixing uniformity of air injection and double knife stirring is better than the double knife group, and the double knife group is better than the single knife group, which is consistent with the simulation results. Moreover, the deviation between the experimental and simulated values is in the range of 2%, confirming the reliability of the simulation model. The results of this study serve as a valuable reference in the design of variable-speed fertilizer mixing equipment.

Based on research [5] pertaining to the proposal of a novel method for creating continuous rheological materials utilizing rotating barrels equipped with stirred screws and the subsequent evaluation of their microstructure, a fresh process for fabricating rheological materials has been introduced. This innovative method employs specialized rotational barrel-type equipment designed for use in rheo die-casting and rheoforming processes.

The specially designed rotational barrel-type equipment has the unique capability of producing rheological materials continuously while also offering control over shear rate and temperature. The findings suggest that as the rotation speed increases, the rotation time lengthens, and the cooling rate decreases, the equivalent diameter and roundness of the primary particles become smaller. However, it has been verified that grain agglomeration and an increase in grain size occur when the rotation time exceeds 1800 seconds.

This study presents a new thermo-solid model based on the work on a thermo-solid coupled model [6] that simulates the heating and agitation of particles in a rotating drum using disk discontinuous deformation analysis. This model combines Discontinuous Deformation Analysis (DDDA) and Discrete Thermal Element Method (DTEM) and is called the DDDA-DTEM model. Its purpose is to reproduce the process of heating and stirring particles in a rotating drum. The results show that a rotation speed of 15 revolutions per minute (RPM), 2 to 6 medium stirring blades, and a load rate of 20% are helpful in achieving high-quality processing results. The model performance shows that the DDDA-DTEM model established here serves as a viable tool for simulating the heating and stirring of particles in a rotating drum. This is of great importance in refining granular material processing strategies and industrial plant designs.

In light of research [7] focused on utilizing ultrasound to create gluten-free dough for automated dumpling production, it's important to consider the growing prevalence of celiac disease. Celiac disease is an autoimmune genetic condition that leads to intestinal damage as a reaction to specific proteins in the diet. Individuals with celiac disease must avoid foods containing gluten, such as dough made from grains containing this protein. Unfortunately, gluten-free dough typically exhibits subpar rheological properties and is unsuitable for the automated shaping of dumplings. This article introduces an innovative approach, combining ultrasonic technology with gluten-free dough preparation to enhance its rheological qualities, making it suitable for the automated dumpling molding process.

The application of ultrasonic treatment, specifically at a 35 kHz frequency, plays a pivotal role in homogenizing the dough's structure and modifying its rheological characteristics. Sonication results in a twofold increase in dough volume and subsequently enhances dumpling mass yield by 2 to 10% per kilogram of dough. Beyond the economic benefits, this technology also provides an added sterilization effect during dough preparation.

Based on research [8] conducted on the creation of a real-time automated citrus fruit sorting device, this machine is meticulously designed and utilizes a computer vision-based adaptive deep learning model for optimal performance. Traditional automation methods for postharvest tasks often suffer from time and data inefficiencies, resulting in less-than-optimal outcomes. These automated systems typically necessitate the involvement of highly skilled software experts for calibration and reconfiguration, which can inflate costs significantly.

In this study, modern sensors and intelligent devices capable of conducting deep learning image analysis have been harnessed to design automated machines for postharvest operations, including washing, vision-based sorting, and weight-based grading of citrus fruits. This approach minimizes the need for human intervention and significantly reduces attrition, while achieving outstanding performance in executing specific tasks. The accuracy of the machine's performance is ensured through the use of optimal mechanical component design, which is achieved through kinematic synthesis and dimensional analysis.

Based on the research [9] conducted to develop an automated pest monitoring system for rice fields using machine vision, it is essential to effectively implement integrated pest management. Light traps are widely used around the world for pest monitoring in the agricultural field. This paper introduces the development of an automatic monitoring system for pests captured in rice light traps using image processing technology. To prevent the accumulation of captured insects, a vibrating plate and rotating conveyor belt are used to disperse the insects. This study shows a strong correlation (r=0.92) between automatic and manual identification methods based on daily pest counts from single light trap images over a one-year period. Field experiments highlight the effectiveness and accuracy of automatic light trap monitoring systems.

Built upon research [10], this study concentrates on creating and assessing an automated control system for rice polishing machines that utilizes computer vision and fuzzy logic. The main goal is to establish an intelligent Automatic Control System (ACS) that utilizes machine vision and fuzzy logic methods to regulate rice polishing machines. The evaluation results show that the developed ACS can accurately determine the desired operating conditions of the polishing machine with an accuracy rate of 89.2%. On average, each monitoring cycle takes approximately 14.73 seconds, with 6.4 seconds for kernel sampling and transportation into the imaging chamber, 7.33 seconds for capturing three kernel images, image processing, and running the fuzzy inference process, and the remaining 1.5 seconds dedicated to pressure level adjustments within the mechanism.

Based on research [11] focused on the development of a control system for automatic mini plug seed planting machines, the context in China's vegetable production landscape primarily revolves around small and medium-sized farms, rather than traditional open field cultivation. These facilities often contend with low ceilings and limited dimensions, making it challenging to employ planting machinery. Consequently, many vegetables farmers resort to manual planting due to these spatial constraints. To address the automation needs of small and medium-sized agricultural setups, a compact, remote-controlled, wireless automatic planting machine was conceived. This device can swiftly transplant seedlings within greenhouses while also prioritizing environmental sustainability. It leverages wireless communication to govern all of its operations. Within the agricultural production process, the control system of agricultural machinery plays a crucial role in determining the level of automation. Therefore, this study focuses on designing an innovative control system for the automatic planting machine. The control system hardware is divided into three essential components: sensors for signal collection, programmable controllers for signal processing, and actuators for specific actions.

Based on research [12] focused on the supervisory control of automatic pouring machines, this paper introduces a supervisory control approach for an entire molten metal pouring system aimed at enhancing plant productivity, ensuring worker safety, and improving product quality. The process involves the use of a pouring process model to calculate the forward tilt control input using an adaptive feed-forward control system. This input is designed to maintain a consistent liquid level in the sprue cup while accounting for variations influenced by the build-up of slag in the ladle.

The back tilt input is determined through a hybrid shape approach applied to mitigate the tilting action. The supervisory control system smoothly transitions from forward tilting to backward tilting by employing model predictive control to achieve precise control over the quantity of the pour. Experimental tests validate the effectiveness and practicality of the proposed comprehensive control system.

Based on a study [13] focused on reducing pressure drops in the design of spiral impeller static mixers through additive manufacturing, a novel static mixer design referred to as the SSM Impeller has been introduced. Experimental measurements of pressure drop and mixing characteristics were conducted and subsequently verified through computational fluid dynamics (CFD) analysis. The study examined five different SSM Impeller designs, each featuring variations in taper, discharge, and attack angles. In comparison to a standard SSM of equivalent size, the SSM Impeller demonstrated a reduction in pressure (and power) of up to 18.2%. These reductions have the potential to yield substantial energy savings and allow for downsizing in industrial applications employing static mixers. Moreover, the experimental results underscore the capability of additive manufacturing to produce custom SSM Impellers without the need for costly machining and joining methods.

In a study [14] concerning counter-current mixer-settler chromatography utilizing a spiral disk assembly modified with glass beads as barriers, a novel spiral disc design was developed. This design involved the strategic placement of barricades at 6 mm intervals along the central spiral channel to partition it into discrete sections. Within alternate sections, glass beads were introduced, promoting a planetary motion that led to repeated mixing and deposition of the polymer phase system. The efficacy of this spiral disc assembly within the mixer-settler setup was evaluated in the context of separating lysozyme and myoglobin using a polymer phase system. The most favorable outcomes were achieved when employing spiral discs equipped with barricades featuring openings ranging from 1.2 to 0.4 mm on each side, especially at high rotational speeds of up to 1200 rpm.

1.2.2 Stainless steel 304 food grade

According to research [15] that investigated the effect of allium extract inhibitors on the localized corrosion behavior of 304 stainless steel through electrochemical noise analysis showed that adding the inhibitor up to the optimum concentration (8 cm^3 L^{-1}) was effective. Improvements increase immunity to spectral noise. Shot noise results show that in the presence of Allium Sativum (garlic) inhibitor, the frequency distribution of events shifts to a higher frequency range, indicating reduced localized corrosion of stainless steel. Analyzes of hole formation and hole growth showed that although the antioxidant properties of allium extract increased the hole formation rate, the formation of stable holes with a radius greater than 30 μm was significantly reduced due to metastable hole filling. It became clear that it had been thwarted. Phosphate compounds. Microscopic examination of the corrosion morphology supported the hypothesis that phosphate compounds formed within the pit core and the steel surface was subsequently repassivated.

In a study entitled "Exploring surface topographical properties and bacterial adhesion: a case study investigating the adhesion of Listeria monocytogenes scotia on type 304 stainless steel" [16], using a type 304 stainless steel surface, A detailed investigation was carried out and the aim of the study was to determine the differences in bacterial adhesion to this particular material by examining two common surface types: extrusion and soil. This study involved investigating 15 surface topography parameters at three different processing temperatures and identifying possible relationships between these parameters and microbial attachment on these surfaces. To achieve this, the surfaces were analyzed qualitatively and quantitatively using a variety of techniques, including scanning electron microscopy, energy-dispersive X-ray spectroscopy, and confocal microscopy. Using statistical techniques such as analysis of variance and multivariate regression analysis, L. Monocytogenes on stainless steel surfaces Scott A. The results showed that there is a strong correlation between bacterial adhesion and surface properties, especially surface isotropy, average surface roughness, surface spacing, and processing temperature in the case of 304 stainless steel material.

Drawing from research on [17] stainless steel corrosion within the food and pharmaceutical sectors, stainless steel finds extensive application in these industries, primarily owing to its exceptional resistance to corrosion and its superior mechanical attributes. These attributes hold immense significance since the substances utilized in food and pharmaceutical manufacturing must adhere to stringent purity and quality standards. The prudent selection of an appropriate stainless steel grade plays a pivotal role in averting corrosive occurrences that can have adverse repercussions on the overall production processes. In this discussion, we will delve into an examination of the stages within the food and drug production procedures, with a specific focus on potential corrosion mechanisms affecting stainless steel under specific operational conditions.

2 Methods

This method uses reverse engineering, then is registered for certification and commercialized. This design emerged based on the background of existing problems and was supported by several supporting literature so that a suitable design concept was obtained to solve the problem. Then make working drawings of several components and determine the tools according to the design. After that, prepare to make tools. The steps are to create a framework, and then the process of making tools and tool systems. After going through this, a trial is carried out to determine whether the tool works according to the research objectives or not. If it meets the research objectives, research results and data can be obtained so that the performance of the SEGORO machine can be analyzed whether it is running effectively or not. Then conclusions can be drawn based on the data that has been found, then submitted to small and medium businesses, then the tools are implemented for the performance of small and medium businesses and the research is completed. The research flow can be depicted in Fig. 1 below.

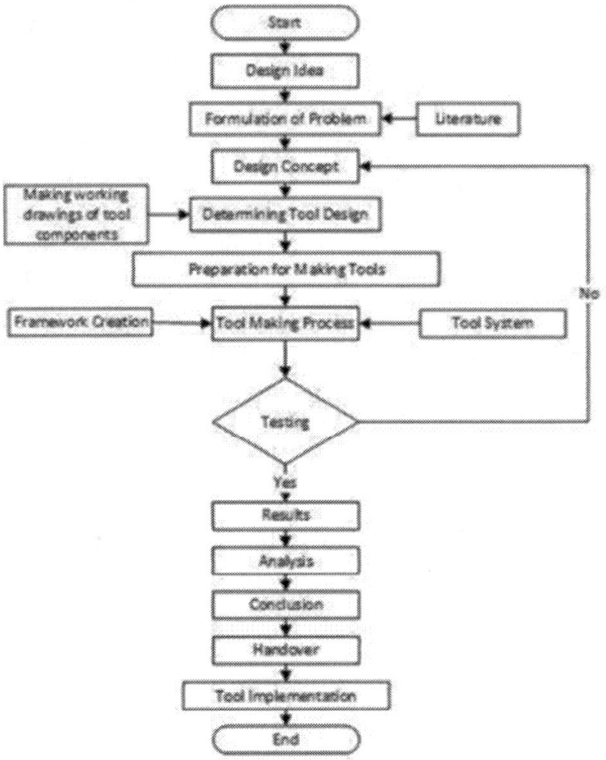

Fig. 1. Research Flow of Segoro Machine

This research compares cooking fried rice with a fried rice machine and conventional methods in 15 portions/batch. This research aims to find out the comparison of cooking fried rice using a fried rice machine and conventional methods.

3 Result and discussion

3.1 Engine specification

The SEGORO machine has been created based on analysis and working drawings. Based on the working drawings, a prototype is then created. The specifications for the fried rice machine are in Table 1 below.

Table 1. The SEGORO fried rice machine specifications

Aspect	Information
Ingredients	Stainless Steel 304 standard Foodgrade
Dimension	80 x 70 x 70 cm
Corner turning drive	Motor wiper
Shape of stirrer,	The shape of the stirring tool, spiral with medium - large - medium sizes in harmony with the curved surface of the wok
The stirring tool speed	Can be adjusted according to needs
Stove	Type 203 Burner Size 9cm
Wok dimension	Diameter 50 cm
Capacity	15-20 portion/ batch
Stirring rotation speed	30 rpm (can be adjusted according to needs)
Distance between stove and bottom wok	5 cm
be equipped	fire protection device from wind
Power	100 watt

The SEGORO fried rice machine has several components: (1) a stove set equipped with a gas valve handle to regulate the size of the fireplace; (2) one unit of spiral-shaped stirrer whose curve is adjusted to the curve of the Wok surface. On the mixing spiral, there are 8 pairs of tools for breaking down rice chunks 'resembling two fingers' (specially designed). For operation, it is equipped with a stirring speed control device; (3) the driving motor is a 12 V DC motor that is used to drive the stirrer and rotate the Wok electrically (semi-automatically); (4) The wok is equipped with 4 hooks to maintain its stable position during the process. With the operation of all these components, it is capable of maximizing the process of breaking down rice grains and mixing spices more quickly. This way it does not require a lot of physical effort, and you can even do other work, for example, tidying and cleaning the work area. The SEGORO fried rice machine is made from stainless steel (SS) 304. SS 304 is a type of food-grade stainless steel, easy to shape and resistant to very high corrosion, the surface of SS 304 is straight and not wavy, easy to clean. The working drawing of the fried rice machine can be seen in the following image (Fig. 2).

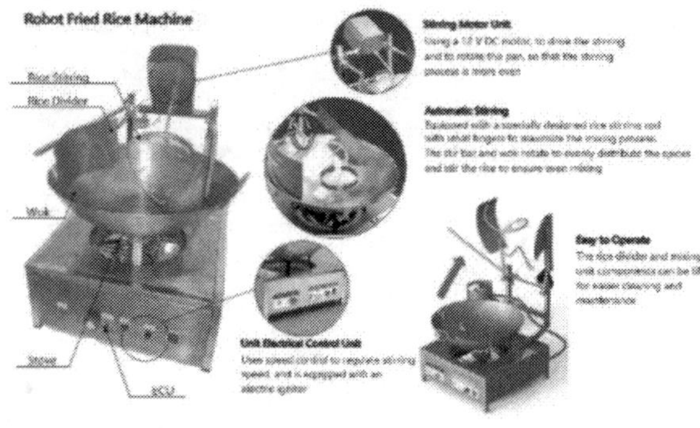

Fig. 2. Design model for the fried rice-making machine

The spiral stirrer has claws for scraping the rice at the bottom of the pan so that it is mixed perfectly and completely, with a length of 5 cm and an angle of 90°. The claw image can be seen in detail in the image below (Fig. 3).

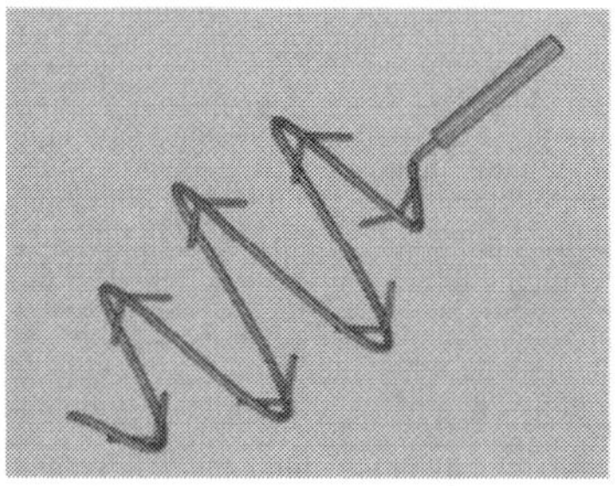

Fig. 3. The claw of stirring SEGORO Machine

3.2 Result

The comparison results of testing the fried rice machine with the conventional method can be seen in Table 2.

Table 2. The comparison results of testing the fried rice of conventional method with the fried rice by Machine (the SEGORO)

Based on Table 2. 1500 grams of rice with a gas requirement of 80 grams, the quality of the rice grains is crushing chunks of rice is more difficult, requires a lot of energy, takes longer, leaving a few pieces of rice, and produces 15 servings in one cooking process in 12 minutes. Meanwhile, trial II with a quantity of 1500 grams of rice and a gas requirement of 70 grams showed that the quality of the rice grains was crushing chunks of rice is easier, doesn't involve much human effort, the rice grains are split evenly, and the time required to cook the fried rice was 7 minutes to produces 15 portion/ batch. In terms of effectiveness, cooking large portions of fried rice is more effective using a fried rice machine compared to conventional methods.

Information	Test I (Conventional)	Test II (with the SEGORO)
Amount of rice (gr)	1500	1500
Gas requirements (gr)	80	70
Rice quality	Crushing chunks of rice is more difficult, requires a lot of energy, takes longer, leaving a few pieces of rice.	Crushing chunks of rice is easier, does not involve much human effort, the rice grains are split evenly.
Taste and color	Uneven	Equally
Time (minutes)	12	07
Physical load (interview and observation results)	The perpetrator (student) feels tired, his face shows tension (serious)	The perpetrator (student) feels relaxed, his face shows relaxation and he can tidy up his work desk

Fig. 4. Cooking with conventional methods

Figure 4 shows the conventional cooking process for fried rice. Throughout the processing process, physical effort is required to crush the rice chunks and mix the rice with the spices. From the beginning to the end of processing, workers continuously do physical work. Apart from that, the spatula used is made from 304 stainless steel which is quite heavy, thus adding to the load in the continuous stirring process. The time required is relatively long, namely 12 minutes. The processed rice grains have not been completely decomposed, there are still some lumps of rice (small lumps).

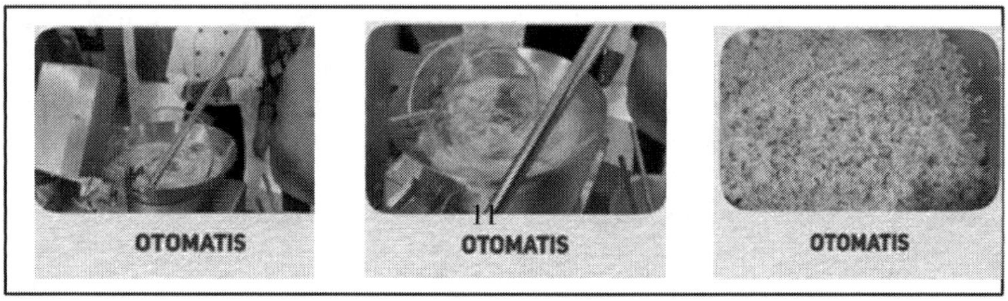

Fig. 5. Cooking with the SEGORO machine

In Figure 5 the cooking process with the SEGORO fried rice machine. At the beginning of the process, stirring the eggs is done manually, then after the rice is added, the stirring is done with a semi-automatic stirrer which has been specially designed and with the Wok rotating at medium rotation. Crushing chunks of rice is done more easily, very little physical effort is required, and the grains of rice are split evenly. The overall quality of the processed product shows that the color of the rice shows that the spices and rice are mixed very evenly throughout. Overall, 15 portions are prepared in 7 minutes. The equipment is easy to clean.

3.3 Discussion

3.3.1 Stirring Machine Automatic

This SEGORO machine is a semi-automatic mixing machine with 100 watts of power for cooking fried rice in large portions, where one cooking process produces 15 portions. This machine is equipped with a spiral stirrer with a claw length of 5 cm and an angle of 90° with a pan diameter of 50 cm so that it can accommodate large portions.This way it doesn't require a lot of physical effort, and you can even do other work, for example, tidying and cleaning the work area. The time needed to start heating the pan is no more than one minute so the cooking process is very short for large portions, namely 7 minutes. This machine can cook to the bottom of the pan so it can cook optimally with even taste and color and the rice grains remain intact. For a more detailed machine design, see the following image (Fig. 6).

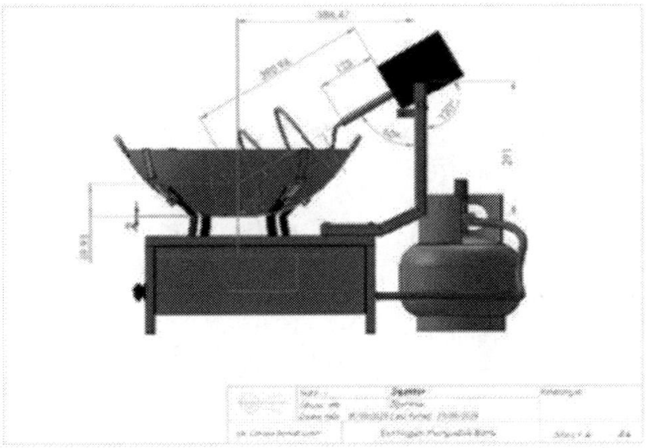

Fig. 6. Detailed SEGORO design

Based on the research that has been carried out, it can be concluded that the SEGORO machine is very effective for use in cooking large portions and is also very good for reproduction and commercialization. This is in line with research [1]. Research for automatic rotation mixers was obtained from research literature [2], [3], [4], [5], [6], [7], [8], [9], [10], [11], [12]. The spiral stirrer research is in line with research [13] and [14].

3.3.2 Stainless steel 304 food grade

This SEGORO machine is easy to operate from preparing the equipment to starting cooking, which takes 7 minutes. This machine is very easy to clean and rust-proof because it uses food-grade 304 stainless steel. The 304 stainless steel material can be proven in research [15], [16], and [17].

4 Conclusion

Using a fried rice machine can speed up the frying process from 12 minutes to 07 minutes/batch (15 portions) so that productivity can be increased up to 2x. The Segoro machine only uses a small amount of electricity with low voltage and saves gas usage. Apart from that, it can save energy so it doesn't get tired easily because it is done by machine, the product is more hygienic because the machine material is made from food grade standards, namely stainless steel 304, the quality of the processed product is more stable.

References

1. C. He, L. Song, Z. Liu, H. Xiong, Q. Zhao, Int. J. Biol. Macromol. **228** (2023)
2. Z. Huber, M. Athon, S. Shen, E. Conte, K. McCoy, C. Lavender, J. Nucl. Mater. **555** (2021)
3. H. Zhao, H. Zeng, T. Chen, X. Huang, Y. Cai, R. Dong, J. Colloid Interface Sci. **643** (2023)
4. X. Xi et al., Comput. Electron. Agric. **210** (2023)

5. P. K. Seo, S. M. Lee, C. G. Kang, J. Mater. Process. Technol. **209**, 1 (2009)
6. G. Huang, C. Tong, S. Zhang, X. Chen, Powder Technol. **402** (2022)
7. S. A. Ulasevich et al., Ultrason. Sonochem. **68** (2020)
8. S. K. Chakraborty et al., Eng. Appl. Artif. Intell. **120** (2023)
9. Q. YAO et al., J. Integr. Agric. **19**, 10 (2020)
10. H. Zareiforoush, S. Minaei, M. R. Alizadeh, A. Banakar, B. H. Samani, Comput. Electron. Agric. **124** (2016)
11. Q. Yang et al., Comput. Electron. Agric. **169** (2020)
12. K. Yano and K. Terashima, Control Eng. Pract. **18**, 3(2010)
13. M. Hildner, J. Lorenz, B. Zhu, A. Shih, Chem. Eng. Process **191** (2023)
14. Y. Ito et al., J. Chromatogr. **1151**, 1–2 (2007)
15. M. Peirow Asfia, M. Rezaei, Mater. Chem. Phys. **274** (2021)
16. J. Das, J. A. Chase, M. L. Partyka, E. R. Atwill, B. Linke, J. Food Prot. **83**, 3 (2020)
17. A. Zaffora, F. Di Franco, M. Santamaria, Curr. Opin. Electrochem. **29** (2021)

Automatic shoe embossing machine integrated with real-time data cloud system to improve SMEs footwear productivity

Iskandar[1*], *I Made* Arsana[1], *Rizdana Galih* Pambudi[1], *Yuli Sutoto* Nugroho[2], and *Muhammad Febrilian Dwi* Syahputra[3]

[1]Departement Mechanical Engineering, State University of Surabaya, Surabaya, Indonesia.
[2]School of Electronic Engineering and Computer Science, Queen Mary University of London, London.
[3]Master Student of Information System, Sepuluh Nopember Institute of Technology (ITS), Surabaya.

Abstract. The primary function of the automatic shoe embossing machine integrated with a real-time data cloud system is to enhance the shoe manufacturing process by imprinting designs onto leather shoes, while simultaneously collecting and transmitting production data for monitoring and analysis. This research project aims to design and develop a cutting-edge shoe embossing machine. To achieve this, we employ a Raspberry Pi-based cloud system to gather and store production data in a database format, allowing for real-time monitoring and operator feedback. The research methodology encompasses several stages, including design, manufacturing, assembly, operational testing, performance evaluation, and data analysis. In conclusion, this research project demonstrates the successful development of an automatic shoe embossing machine integrated with a real-time data cloud system. This innovation streamlines the shoe manufacturing process, ensuring high-quality embossing while providing valuable real-time production data for enhanced control and decision-making.

1 Introduction

After several centuries of industrial progress, the era of Industry 4.0 has arrived in modern times. In 2011, the idea of Industry 4.0 was first put forth to help the German economy grow [1]. The first industrial revolution was typified by mechanical production plants using steam and water power at the end of the 18th century. The second industrial revolution was typified by mass labor production using electrical energy at the beginning of the 20th century. The third industrial revolution was typified by automatic production using electronics and internet technology in the 1970s. Currently, the fourth industrial revolution, known as Industry 4.0, is underway, with the features of cyber-physical systems (CPS) production based on heterogeneous data and knowledge integration [2]. CPS's primary responsibilities are to increase the efficacy and efficiency of the whole sector while meeting the flexible and dynamic demands of production [3]. Cloud-based manufacturing, the Internet of Things

* Corresponding author: iskandar@unesa.ac.id

(IoT), RFID, ERP, cloud-based radio frequency identification, and social product development are just a few of the technologies and related paradigms that are included in Industry 4.0 [4].

The manufacturing industry is becoming more competitive than it has ever been, and manufacturing organizations are facing a variety of challenges as a result of sophisticated markets, shifting consumer preferences, and international rivalry [5]. Multinational corporations and imports compete with Indonesian industry in the country's markets. Manufacturers must possess a variety of skills in the globalized market, including quality, delivery, flexibility, and pricing [6]. Companies must focus on this crucial component if they want to maintain a competitive edge in the manufacturing sector, which is the efficient and effective management of resources and operations. Manufacturing companies are beginning to recognize and use productivity improvement (PI) as a competitive strategy. Productivity is a concern for organizations involved in manufacturing [7]. Consequently, manufacturing organizations are placing a greater emphasis on productivity and PI as a means of competing [8].

Small and medium-sized businesses, or SMEs, are the backbone of the manufacturing sector in developed nations like our own [9]. As of right now, SMEs are thought of as the foundation of the Indonesian economy. 99% of all businesses in Indonesia are SMEs, and they make up a significant portion of the country's revenue. This accounts for up to 97% of workers and more than 50% of Indonesia's GDP overall [10]. According to data from SME (2011) and the Ministry of Cooperation, there are 55.530.000 SMEs in Indonesia. Thus, maintaining the SME sector's competitiveness is crucial for the nation's economy as a whole as well as the manufacturing sector's overall growth. For SMEs, globalization offers both opportunities and challenges [11]. SMEs must become more competitive to thrive in an environment where tariff protection is fast eroding. SMEs that are internationally competitive enough to gain admission into the global value chain can take advantage of growth opportunities in a global marketplace [12]. If the others are to survive, they will have to change how they operate and become more competitive to face the difficulties [13].

An increasing number of SMEs are businesses that sustain the Indonesian economy on an annual basis. The Indonesian Ministry of Cooperatives and SMEs report that from 2015 to 2019, SMEs' growth rate increased [14]. The data does not accurately represent the total number of SMEs because the number of SMEs listed is an approximation. This is because there are a lot of SMEs and it is challenging to record them because the majority of them have not yet registered their businesses.

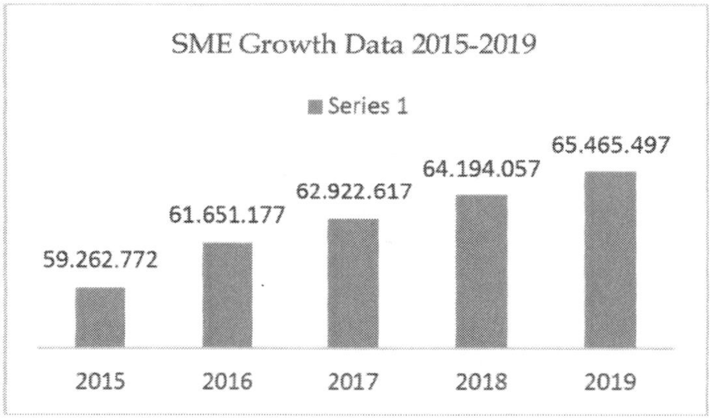

Fig. 1. SME Growth Data 2015-2019 Ministry of Cooperatives and SMEs of the Republic of Indonesia

In 2021, the government launched the Online Single Submission - Risk Based Approach (OSS RBA) platform which can be accessed via www.oss.go.id. This website is used as a medium for registering business permits in Indonesia for business actors. By 2022, the number of SMEs that have registered their businesses on the OSS platform has reached 8.71 million units. For the East Java province region, there are already 1,153,576 registered SME units. Indonesian SMEs still face numerous challenges today [15], such as funding scarcity, inadequate technology, and a lack of expertise in production and company management. Despite the existence of SMEs, this is a crucial component of economic growth. It appeared that the Indonesian government was having trouble keeping SMEs operating continuously. Although the government has launched a program to assist small and medium-sized enterprises (SMEs) with soft loans, the initiative has not yielded significant results thus far, primarily because of incomplete and inaccurate information. The research of Utomo and Dodgson (2001), which explains why the adoption of information technology in Indonesia is so slow, supports the theory that this explanation may be due to information technology hurdles that occurred in Indonesia [16]. The reason for this is that not all locations have received equal coverage from the infrastructure, which is still operating slowly.

To succeed in an environment that is changing quickly, small businesses with tight budgets need specific "tools" to be developed [17]. These tools need to be easy to use and, most importantly, they need to change the companies in a big way when they are put into use. Industry-wide technical advancements open up new avenues for businesses to undergo digital transformations and provide cutting-edge goods and services.

One of the SMEs in East Java is Footwear. This SME still lacks adequate production technology support and poor business management, especially financial management, causes the quality and quantity of products produced to be low, so that their products are less popular with consumers.

Fig. 2. Brand label manual process on the insole

One of the production aspect problems that need to be resolved immediately is the process of labeling leather shoe brands. The process of labeling shoe brands is currently done by cutting brand labels screen printed on fabric and stitching them onto the inner lining of the shoe (insole), this takes a long time, starting from cutting, sewing, and inserting the insole into the shoe takes ± 2.5 minutes, the results are not good, do not provide added value in terms of appearance, fade quickly and even fall off. This method is ineffective, inefficient, and requires higher production costs. To increase the quality and quantity of production effectively and efficiently, innovation in production technology for leather shoe brand label printers is needed, namely a semi-automatic electrical energy leather embossing press machine. To print a brand with this semi-automatic electric energy leather embossing press machine only takes ±5 seconds, the quality is very good, beautiful, and attractive, provides added value in terms of appearance, elegance, luxury, and classy, and does not fade for a long time. This can increase product competitiveness and consumer satisfaction. By using a semi-automatic leather embossing press machine with electrical energy, the production capacity for shoe brand label printing can be increased 30 times faster, with better quality.

2 Methods

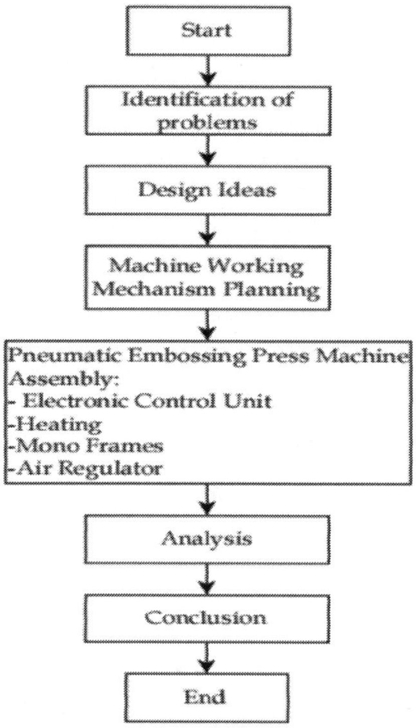

Fig. 3. Research Flowchart

The planning for making this embossing press machine was carried out in stages. This machine utilizes pneumatic power as a mattress pressure calculator whose pressure is adjusted. There is a control box to turn the heater on and off. The following is the flow of how an embossing press machine works with pneumatic power.

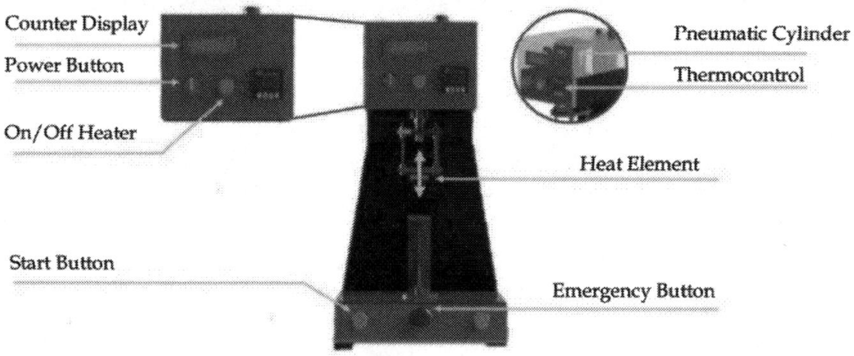

Fig. 4. Machine Mechanism

1. The Heater button turns on the heating element according to the set temperature (+/-10 minutes before use)
2. The start button is used to activate the actuator (cylinder pusher) to press down and hold it for the pre-set time, then the pressing profile will return to its initial position.
3. The heating element can move up and down driven by a pneumatic cylinder. The number of 1x embossing processes will be calculated and displayed on the counter display screen

Our focus is on the problems related to collecting, delivering, and analyzing critical machine-embroidered data. We suggest that the time-consuming manual note-based solutions are in use today. Furthermore, it creates a barrier to real-time data access, making monitoring and troubleshooting more difficult. Additionally, we have provided a method that uses the Raspberry Pi cloud to automate this process, from data gathering to information display. The Raspberry Pi serves as the entry point for processing the sensor data [18, 19]. All of the current equipment is linked to sensors. Afterwards, exchange services are connected to these devices. When the data is accessible on the cloud, expert systems can process it or send it to the LED Display for review.

3 Results and discussion

3.1 Design and specification of emboss press machine

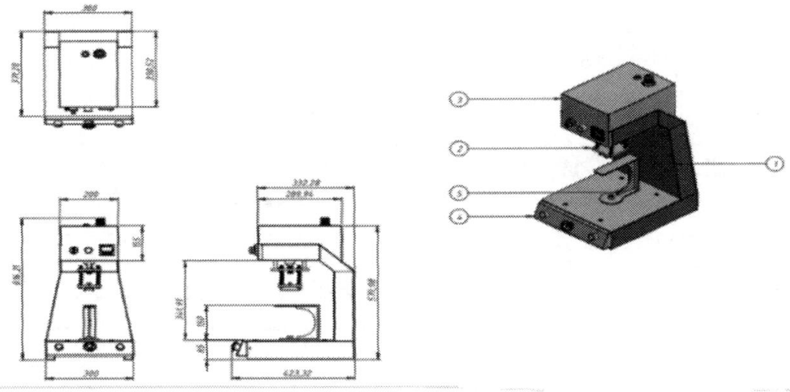

Fig. 5. Drawing of Emboss Press Machine

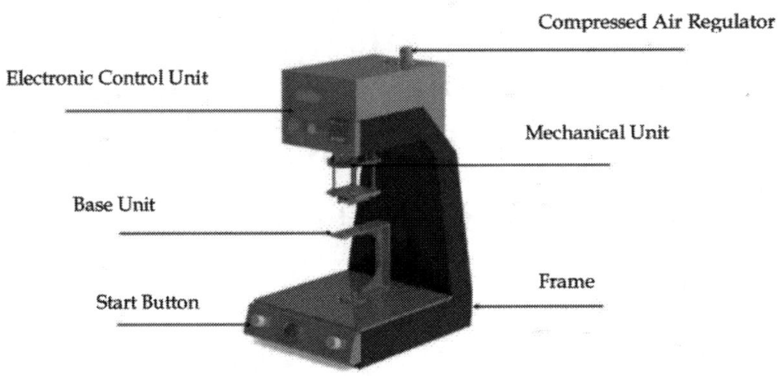

Fig. 6. Specification Machine

1. Air Regulator: Regulates pressure and separates water content from compressed air input
2. Electronic Control Unit: Regulates the temperature and time of the embossing process
3. Mechanical Unit: Heating profile pressing actuator that can be adjusted for pressing distance
4. Base Unit: Place to place the material that will be embossed
5. Frame Unit: Mono frame as a mount for each machine component
6. Start Button: Active button for the embossing process

The following are specifications for engine dimensions, weight, capacity, working temperature, engine power and air pressure:

Table 1. Specifications

Nu	Description	Grade	unit
1	Dimension	425x300x620	mm
2	Weight	19.5	kg
3	Print Capacity	+/- 12	Print/min
4	Temperature	Up to 210	°C
5	Engine Power	250	Watt
6	Air Pressure	<1	Bar

The Internet of Things, or IoT, is a network of electronically connected physical items. The term "thing" in the Internet of Things refers to objects that have been given an IP address and are able to gather and send data across a network without the need for human aid or intervention [20]. Examples of such devices include people wearing heart monitors or cars with built-in sensors. Decisions are influenced by the items' ability to interact with their internal states and external surroundings thanks to embedded technology. Here, the notion of the Internet of Things is combined with two distinct ecosystems: Arduino and LED Display.

Ultimately, these are to be used in the development of a digital color controller. The goal of this project is to create a smart mirror that can display information such as displays the number of times the machine has embossed the leather. Use of the Raspberry Pi is required for all these clever features. The use of a smart mirror allows for time savings. 1.2GHz quad-core 64-bit Broadcom BCM2837 CPU with 1GB RAM This BCM43438 features an on-board 40-pin extended GPIO (General Purpose I/O) with wireless LAN and Bluetooth Low Energy (BLE); four USB 2.0 ports, four-pole stereo output, and a composite video port for analog video transmission; full-size HDMI (High Definition Multimedia Interface); a CSI (Camera Serial Interface) display port for connecting a Raspberry Pi touchscreen display; a

Micro SD port for storing data and loading your operating system; upgraded to a 2.5A switched Micro USB power source.

3.2 SMEs productivity

To anticipate complaints of musculoskeletal disorders (MSDs) among MSME footwear workers, prepare tools that meet comfort or ergonomic standards to carry out the work process. Ergonomic approaches and evaluations have been applied in many ways. An example is in designing products, laying out layouts, work facilities, and workplaces with the main aim of being more effective, comfortable, safe, healthy, and efficient (ENASE). With this embossing press machine, workers do not need a long time in the label-making process. The increase in the number of assets obtained from the addition of 1 unit of embossing press machine means that the assets of partner shoe SMEs will increase/increase. Turnover increases. By using a manual system (the label needs to be sewn on first and then placed on the shoe) you can only print a label on one shoe in ± 3 minutes or 10 pairs of shoes per hour. The production capacity of this embossing press machine is 12 labels per minute in one embossing process or can emboss 360 shoe labels/hour. So, this automatic embossing press machine is able to increase production capacity ± 36 times greater than the manual labeling method with a sewing machine.

4 Conclusion

A type of heat press machine for printing brands on shoe leather, the electric energy semi-automatic leather embossing press machine (also known as a shoe stamping machine) has a pneumatic cylinder that is outfitted with a temperature controller (Thermocontrol) and an embossing heating element (heat embossing). With a 220 Volt electrical source, the temperature of the analog thermometer used in this machine can be raised to a maximum of 210°C. For example, this Semi-Automatic Leather Embossing Press Machine uses electric energy to print brand labels in ±5 seconds, which is 36 times faster than screen printing, cutting, and sewing using a sewing machine to create shoe brands. This works well and efficiently.

References

1. G. Reischauer, Technol. Forecast. Soc. Change **132** (2018)
2. Z. You, L. Feng, IEEE Access **8** (2020)
3. R. Davies, T. Coole, A. Smith, Procedia Manuf. **11** (2017)
4. Y. Lu, J. Ind. Inf. Integr. **6** (2017)
5. T. Wattanapruttipaisan, Asia-Pacific Dev. J. **9** (2002)
6. M. A. Vonderembse, M. Uppal, S. H. Huang, J. P. Dismukes, Int. J. Prod. Econ. **100** (2006)
7. N. Slack, Int. J. Oper. Prod. Manag. 7 (1987)
8. A. Arshad Ali, A. Mahmood, A. Ikram, A. Ahmad, J. Open Innov. Technol. Mark. Complex **6**, 154 (2020)
9. N. Yoshino, F. T. Hesary, Major Challenges Facing Small and Medium-Sized Enterprises in Asia and Solutions for Mitigating Them (The Asian Development Bank Institute, Tokyo, 2016)
10. Kristiyanti, Maj. Ilm. Inform. **3** (2012)

11. H. S. Ng, D. M. H. Kee, Entrepreneurial SMEs Surviving in the Era of Globalisation: Critical Success Factors, in Global Opportunities for Entrepreneurial Growth: Coopetition and Knowledge Dynamics within and across Firms 75–90 (Emerald Publishing Limited, 2017)
12. E. Genc, M. Dayan, O. F. Genc, Ind. Mark. Manag. **82** (2019)
13. H. Lim, F. Kimura, The Internationalization of Small and Medium Enterprises in Regional and Global Value Chain (The Asian Development Bank Institute, Tokyo, 2010)
14. A. Anastasya, MSME data, number and growth of micro, small and medium enterprises in Indonesia (Ukmindonesia, Indonesia, 2023).
15. M. I. Darfaz, D. Kurniansyah, K. Febriantin, J. Ilmu Sos. dan Pendidik, **7** (2023)
16. B. Furuholt, F. Wahid, *E-Government Challenges and the Role of Political Leadership in Indonesia: The Case of Sragen*, in Proceedings of the 41st Annual Hawaii International Conference on System Sciences (HICSS), IEEE, 2008)
17. Q. R. Hamidah, A. T. Sejati, A. Z. Mujahidah, Soc. Humanit. Educ. Stud. Conf. Ser. **2**, 345 (2019)
18. T. Tavade, P. Nasikkar, *Raspberry Pi: Data logging IOT device*, in 2017 International Conference on Power and Embedded Drive Control (ICPEDC), IEEE, 2017)
19. A. Alarcón-Paredes, et al., Appl. Sci. **9**, 3046 (2019)
20. L. Atzori, A. Iera, G. Morabito, M. Nitti, Comput. Networks **56** (2012)

Development of pneumatic shoe press machine with integrated real-time cloud database system for SMEs footwear productivity

Nur Kholis[1], Yunus[2*], *Rizdana Galih* Pambudi[2], *Ade Priyo* Widhikdho[3], and *Theodorus Wiyanto* Wibowo[2]

[1]Department Electrical Engineering, State University of Surabaya, Surabaya, Indonesia.
[2]Departement Mechanical Engineering, State University of Surabaya, Surabaya, Indonesia.
[3]School of Mechanical and Manufacturing Engineering, The University of New South Wales, Sydney NSW, Australia

Abstract. Abstract. This innovative effort focused on advancements in the shoe-making process by introducing a state-of-the-art pneumatic shoe press integrated with a real-time cloud database system tailored to the needs of small and medium-sized enterprises (SMEs). This research aims to design and engineer a state-of-the-art pneumatic shoe press capable of printing complex designs on leather footwear. This machine can collect, send, and store production data in a cloud-based database, so as to monitor and analyze their manufacturing operations in real-time. The research methodology utilizes a structured approach from the design, manufacturing, assembly, operational testing, performance evaluation, and data analysis stages. The application of a pneumatic shoe press integrated with a real-time cloud database system provides various benefits for SMEs in the footwear industry. The system's advantage lies in the use of a real-time cloud database that allows SMEs to gain immediate access to production data. The machine is capable of printing complex designs with high precision in one minute per pair of shoes. The product quality is good, so it can meet the standards of the shoe industry. This research successfully developed a pneumatic shoe press integrated with a real-time cloud database system, which can improve the productivity and quality of SMEs in the shoe industry. This technological innovation simplifies the manufacturing process, ensures the best product quality, and provides real-time production data.

1 Introduction

Fashion is a very interesting and profitable business [1]. In the era of the revolution industrial 4.0, all humans need fashion that can reflect themselves, because in this era, first impression is very important [2]. A person who gives a good first impression will make himself accepted, appreciated, and recognized by others [3]. One of the important fashion products to support your appearance is shoes [4]. Sidoarjo is one of the districts

* Corresponding author: yunus@unesa.ac.id

that has many micro, small and medium enterprises (MSMEs), almost 306,481 [5]. UD. Bajarsari is one of the MSMEs in Sidoarjo that produces shoes. The shoe products produced include school shoes, office shoes, loafers, and safety shoes. UD. Bajarsari in producing shoes still uses simple equipment and is done manually, so the quality and quantity of production are still low [6]. However, UD. Banjari still has to carry out production because it already has many customers who order from the place and the number of shoe users is increasing [7]. The process of making shoes at UD. Bajarsari can be seen in Fig. 1.

Fig. 1. Shoe production process with manual tools

The production equipment used in general still uses conventional tools, making it difficult to produce goods effectively and efficiently. The condition of the existing equipment is quite poor and it is necessary to update the tools and machines to improve the quality and quantity of production [8]. The process of gluing the sole and upper is done by gluing the upper to the sole, then followed by beating and pressing using a hammer so that the upper and sole can adhere firmly as shown in Fig. 2.

Fig. 2. Shoe gluing process

The manual gluing process is not very effective and efficient and has many weaknesses. The pressing process given cannot be evenly distributed throughout the base and sides of the shoe circumference, so the bond strength is uneven [9]. The manual gluing process results in less pressure applied because the sole of the shoe will be deformed due to hammer blows and pressure. The adhesion of the results of manual gluing is less strong and the gluing process can only be done one by one, so it requires a large enough emphasis that makes labor tired quickly so that work productivity decreases. The time required for one shoe gluing process takes one minute for one shoe which includes the base and the perimeter of the shoe. The production capacity is only one pair of shoes every two minutes for one worker [10].

Shoes produced by UD. Banjari when viewed from the aspect of physical appearance quality, shoe products produced by this small shoe industry already look good, namely shiny, trendy, and impressive like classy and expensive shoes so that they make confidence for those who wear them. However, from the aspect of bonding strength between the sole and upper, the quality is still not good, the bonding strength is still uneven for the entire circumference of the glued shoe. The low quality and quantity of production is a challenge for MSMEs in developing their business [11]. This is because many consumer orders cannot be fulfilled in accordance with the amount and time requested by consumers. As a result, many consumers are disappointed and cannot wait according to the time period promised by MSMEs. Some consumers even choose to cancel their orders or look for products from other MSMEs that can meet their needs. Shoe products can still be improved in terms of strength, evenness of bonding strength, and uniformity through the application of appropriate production technology that is effective and efficient [12]. Based on the events in the field that occur, it is necessary to apply appropriate technology to overcome or help improve the quality of shoes.

The appropriate technology is to design and engineer a state-of-the-art pneumatic shoe press capable of printing complex designs on leather footwear. The machine is designed to collect, send, and store production data in a cloud-based database. Advanced pneumatic shoe presses can monitor and analyze their manufacturing operations in real-time. The shoe press uses a pneumatic system, applying intense pressure to all parts of the glued shoe, resulting in a very strong and even bonding quality of the leather shoes.

2 Methods

The method that will be carried out in the implementation of the first research is a survey or field observation of several MSMEs to see firsthand the shoe production process and interviews. The results of surveys and interviews were analyzed to find out the problems faced by shoe MSMEs. The results of the analysis found that there are problems that are important to be resolved immediately, namely the process of pressing the upper glue with the sole which is done manually using hammer pressure and screw press, requires a lot of energy, a long time, and the resulting adhesive strength is not optimal, because the pressure is uneven. If these problems are not resolved immediately, they can make shoe MSMEs lose in trade competition. This research activity is carried out through several stages, which include the design, manufacturing, machine testing, and evaluation stages.

2.1 Design

The design stage starts with designing the design and description of the machine to be developed. When making a design assisted with 3D Modeling software. After the design has been completed, then proceed in making a blueprint.

2.2 Manufacturing

Selection of materials that match the standards used to design the machine. After all the materials are collected, continue assembling the machine according to the predetermined design. The main components are first made, after completion at this stage then proceed to testing the machine.

2.3 Testing

When manufacturing is complete, it is then continued by testing the function of using this automatic pneumatic system shoe press machine to find out the performance in the partner's small shoe industry.

2.4 Evaluation

After the pneumatic shoe press machine with an integrated real-time cloud database system is completed, an evaluation is carried out during the process of making the machine until it is used by partners.

3 Result and discussion

Based on observations at MSMEs and information provided related to the operational conditions, the process of pressing the upper gluing with the sole carried out by MSMEs uses a conventional system. The process of pressing the upper gluing with the sole is done manually using hammer pressure and screw press which requires a lot of energy and a long time. However, the adhesion strength produced is not optimal because the pressure applied is uneven.

To overcome these problems, this research initiated the manufacture of a leather shoe sole and upper adhesive press machine with an automatic pressure sensor integrated with a real-time cloud database system. Before the manufacturing process, the first step is to design a drawing of a shoe-pressing machine using an automatic air pressure (pneumatic) system. The machine design is made with the help of 3D modeling applications. The results of the press machine design can be seen in Fig. 3.

Fig. 3. Design the design of the pressing machine

Based on Fig. 3, the press consists of a pneumatic press control, retaining lever, serrated arm, compressor, and electric motor. This press machine is also equipped with a real-time cloud system. The press machine can apply strong pressure to all parts of the glued shoes, resulting in a very strong and evenly distributed quality of leather shoes. The provision of air pressure can be controlled by a pneumatic press control equipped with a timer or timer that can be adjusted as needed. This press machine is also integrated with a cloud database storage system that aims to facilitate data collection of production that has been produced based on daily production results [13].

This Pneumatic Shoe Press Machine with an Integrated Real-time Cloud Database System has an engine motor used for the compressor uses a 1 PK engine motor. This machine is also equipped with a shoe rack that can be used to store shoes that have been pressed. The resulting automatic pneumatic system shoe press machine and machine trials in shoe SMEs can be seen in Fig. 4.

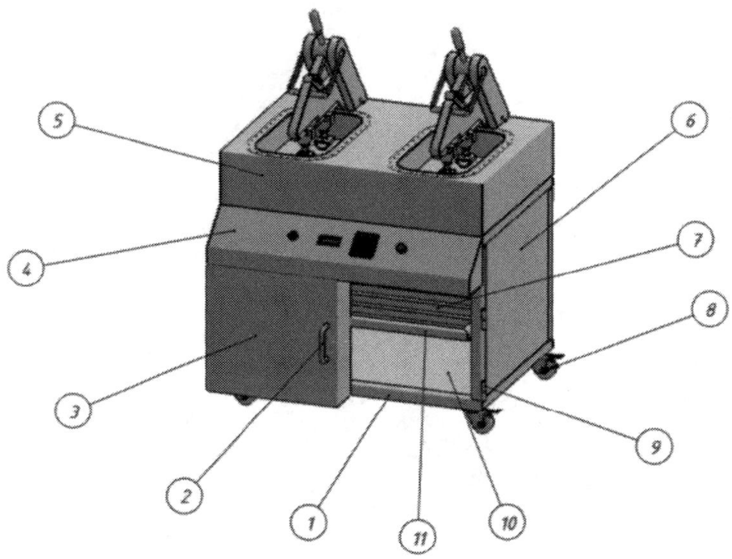

Fig. 4. Detail machine

Table 1. Machine description

No	Description	Specification
1	Frame Unit	Stainless stell 201
2	Door Handle	
3	Door	
4	Control Panel	Acrilic
5	Top Cover	Stainless stell 201
6	Side Cover	
7	Rear Cover	
8	Wheels	Plastic
9	Hinge	Stainless stell 201
10	Bottom Cover	
11	Shelf	

The working principle of this automatic pneumatic system shoe press machine is (1) connect the compressor electric motor cable to the electric power source; (2) press the ON / OFF button on the machine in the ON position; (3) set controls, including the desired pressure and the length of time of pressing; (4) the compressor turns on to accommodate air; (5) place the shoes on the leather membrane; (6) lock the shoe holder arm; (7) press the start button to start the air-filling cycle automatically; (8) holding the pressure according to the input time, the faucet opens automatically so that the air can escape; (9) lift the shoe holder arm and take the shoes that have been pressed from the membrane; (10) do the next pressing starting from the fifth step to the ninth step; and (11) turn off the machine if the shoe pressing process is complete. The machine specifications can be seen in Table 2.

Table 2. Specification machine

No	Description	Grade
1	Dimension	940cm x 840cm x 1300cm
2	Engine Power	0.5 PK
3	Capacity	2 shoes
4	Pressure	3-4 Bar
5	Counting	Raspberry PI
6	Control box	1. Emergency 2. ON/OFF 3. Buttom counter
7	Display	LCD

The function test experiment of using this automatic pneumatic shoe press was conducted to determine the performance in a small shoe industry [14]. The machine function test process was carried out by: 1) gluing the sole and upper that has been installed in the sulas; 2) preparing a pair of shoes to be pressed; 3) placing a pair of shoes on top of two leather membranes on the pressing machine; 4) lock the shoe holder arm; 5) press the start button to start the air filling cycle automatically; 6) hold the pressure according to the inputted time, the faucet opens automatically so that the air can escape; 7) lift the shoe holder arm and take the finished pressed shoe from the membrane; 8) lift the shoe holder and take the finished pressed shoe from the membrane, 9) the shoe pressing process is complete.

Based on the machine performance testing carried out, the data obtained is that it takes ± 20 seconds to press a pair of shoes with details: 1) the time for placing shoes on the pressing membrane is approximately five seconds, the pressing time is held for 10 seconds to produce a strong bond, and for taking shoes from the membrane is 5 seconds, so that the production capacity is three pairs of shoes/minute, twice as large as the manual gluing process of a pair of shoes/minute. The results of the function test comparing manual shoe pressing with using an automatic pneumatic shoe press are shown in Table 3.

Table 3. Results of manual shoe pressing trials using an automatic pneumatic shoe pressing machine in pairs/minute.

Test to -		1	2	3	4	5	Average
Manual	Number of shoes	25	25	25	25	25	25
	Pressing time	25'39"	28'03"	26'11"	30'12"	25'43"	27'9"
Machine	Number of shoes	25	25	25	25	25	25
	Pressing time	12'35"	12'37"	13'23"	12'55"	13'02"	12'54"

Based on Table 3, it can be seen that the results of the total shoes pressed using an automatic pneumatic shoe pressing machine are collected in cloud data storage which makes it easier for MSMEs to collect data on production that has been produced based on daily production. The results of shoes from the automatic pneumatic system shoe press have good quality. This machine can be said to have passed the criteria of the research. The machine can also increase production to meet demand from consumers who also do not forget about quality [15]. Evaluation of the development of this press machine still cannot be fully controlled by the machine and still needs human intervention.

4 Conclusion

This research successfully developed a pneumatic shoe press integrated with a real-time cloud database system, which can improve the productivity and quality of SMEs in the shoe industry. This technological innovation simplifies the manufacturing process, ensures the best product quality, and provides real-time production data. This machine can almost double shoe production with an average time required of only 12'54". The press results have a strong and even bond.

References

1. M. Pradana, Reventiary, Jurnal Manajemen **6** (2016)
2. N. Maryanti, R. Rohana, M. Kristiawan, M. International Journal of Educational Review **2**, 1 (2020)
3. M. R. Leary, Self-presentation: Impression management and interpersonal behavior (Routledge, 2019)
4. S. Supardi, et al., Jurnal Ekonomi Manajemen Sistem Informasi **4** (2023)
5. A. Sudaryanto, I. A. Maruta, SOCIETY: Jurnal Pengabdian dan Pemberdayaan Masyarakat **1** (2020)
6. M. Suyanto, M. Abdulrahim, Journal Community Service Consortium **1** (2020)
7. I. Novizar, Z. Mahjali, A. Agustiar, Jurnal Bisnis Tani **1** (2015)
8. L. Holbeche, Aligning human resources and business strategy (Routledge, 2022)
9. S. Anwar, E. Wahjudi, Yunus, Jurnal Pengabdian Pada Masyarakat **1** (2022)
10. K. Alvarez, D. Aldas, J. Reyes, *Towards lean manufacturing from theory of constraints: a case study in footwear industry*, In 2017 International conference on industrial engineering, management science and application (ICIMSA), IEEE, (2017, June)
11. S. Bhat, E. V. Gijo, A. M. Rego, V. S. Bhat, TQM Journal **33**, 2 (2021)
12. J. Patnaik, B. Bhowmick, Technol Anal Strateg Manag. **34** (2022)
13. C. Xu, et al., Int J Digit Earth **15** (2022)
14. M. S. Gad, K. A. A. Yehia, A. A. Abdelhakeem, Fullerenes, Nanotubes and Carbon Nanostructures **26,** 11 (2018)
15. M. Font-i-Furnols, L. Guerrero, Meat Sci. **193**, 108941 (2022)

https://doi.org/10.1051/e3sconf/202345002006

The efficiency improvement of the shoe polishing machine for SMEs footwear industry

I Made Arsana[1], *Nur* Kholis[2*], *Yuli Sutoto* Nugroho[3], *Theodorus Wiyanto* Wibowo[1], and *Catur Surya* Saputra[1,4]

[1]Departement Mechanical Engineering, State University of Surabaya, Surabaya, Indonesia.
[2]Department Electrical Engineering, State University of Surabaya, Surabaya, Indonesia.
[3]School of Electronic Engineering and Computer Science, Queen Mary University of London, London.
[4]Department Science Education, State University of Surabaya, Surabaya, Indonesia

Abstract. The Shoe Polishing Machine has become an essential device in household and commercial environments to restore shine and extend the life of shoes while saving time and effort. The machine is equipped with a soft brush and polish dispenser, providing an efficient and convenient way to revitalize shoes without the need for manual polishing. Various types of shoe materials such as leather, synthetics, and suede can be treated with this machine, which comes with adjustable shine and polish intensity settings. The advantages of the Shoe Polishing Machine include saving time and effort, suitable for use in hotels, airports, and shoe shops. It also improves hygiene by reducing the spread of polish residue and germs common with manual polishing. The safety of Shoe Polishing Machine users is also a priority with the built-in safety mechanism. Maintenance is easy, making the machine reliable in the long run. Overall, the Shoe Polishing Machine is essential for maintaining the appearance and quality of shoes. With its automatic operation, versatility, and convenience, it is a valuable asset in personal and commercial shoe care. This machine can be used as an efficient and effective solution to achieve flawlessly shiny shoes.

1 Introduction

The younger generation now has a tendency to follow the latest trends or up to date, they have a strong appeal, dare to express themselves, have confidence in their appearance, are afraid of being out of fashion, and have a relaxed yet fashionable style [1]. Fashion is a reflection of the style of dress, accessories, hairstyle, and other elements that can enhance one's appearance [2]. Fashion is supported by several things that can increase the slickness of the clothes worn, one of which is shoes. Shoes are quite important when it comes to dressing style [3]. Various shoe models include flat shoes, boots, oxfords, heels, sneakers, and others that have different wearing functions, depending on fashion needs and fashion styles [4].

* Corresponding author: nurkholis@unesa.ac.id

In line with the significant use of shoes based on data, Indonesia is among the top four footwear-producing countries in the world. Indonesia is behind China, India, and Vietnam in shoe production [5]. The number of footwear industries in Indonesia is recorded at 18,687 business units, including 18,091 small-scale business units, 441 medium-scale business units, and 155 large-scale business units. Of the tens of thousands of business units, labor absorption has reached 795,000 people [5]. Starting from the data, the need for the use of shoes is very high. The amount of use of shoes in daily activities allows a person to have more than one shoe [6]. So, the maintenance of two to three shoes also requires a lot of time and energy [7]. If the shoes look dirty when worn or not shiny, it will reduce the attractiveness and confidence in appearance. So, shoe polishing is needed to keep the shoes looking clean and shiny.

Shoe polishing of shoe soles is intended to sparkle and avoid shoe dust. Manual shoe polishing requires a long time and a lot of energy [8]. Self-polishing if not using the correct technique can damage the shoe and shorten the life of the shoe. Along with the rapid advancement of technology and the development of increasingly modern times, humans are faced with pressure to continue to innovate. In the era of globalization, time is considered precious and it is important to manage activities well [9].

Micro, small and medium enterprises (SMEs) generally have shoe polishing machines using open electric motors with the help of polishing burrs [10]. Machines in existing MSMEs on average come from their own assemblies and are not equipped with adequate security and also pay less attention in terms of health, where the coarse grinding wheels used will produce waste dust that has an impact on air pollution and breathing (health). In addition, the quality of conventional polishers has the potential to produce poor products due to unstable hand conditions in the polishing process. This machine can be crucial because it can save time and effort. Shoe polishing machines can also restore the shine of shoes and can extend the life of shoes. shoe polishing machines that can save users time and effort, and can improve shoe hygiene. A simple shoe polishing tool can be seen in Figure 1.

Fig. 1. Simple polishing tool

Based on the observation of the problem, it is necessary to apply appropriate technology to help facilitate shoe polishing. One of the appropriate technologies needed is an automatic shoe polishing machine equipped with a soft brush and polish dispenser, which can provide efficiency and comfort in revitalizing shoes [11]. The polishing machine will also be designed with adjustable shine and polish intensity settings. The safety of the user is also a priority so there is a built-in safety mechanism. [12]. Prakoso et al. [13] made a shoe polishing machine to help MSMEs, but the machine is not yet equipped with a Raspberry PI for calculations. This developed polishing machine is equipped with Raspberry PI for the calculation process of shoe productivity accurately, easily, and with many other advantages. The next advantage is easy maintenance so that it can be relied on in the long run. This shoe polishing machine can reduce polish residue and germs in manual polishing. In addition, this machine is equipped with a blower that functions to suck dust from the shoe sole polishing process. The filter on this machine will filter dust so that the air that comes out becomes clean, reduces pollution, and is environmentally friendly. The machine is also equipped with polishing eyes that are used to polish shoe soles that are not neat after gluing. With automatic machine operation and easy maintenance, this machine can be relied on personally or commercially.

2 Methods

This research activity has been carried out for five months in Sidoarjo Regency, East Java Province. The implementation of this research activity was carried out through several stages, namely:

2.1 Field observation

Observation activities are one of the vital stages in order to identify problems that actually occur in the field. At this stage, researchers found problems in the form of low quality and quantity of shoe production.

2.2 Design

Designing activities begin with making a design. The design-making process uses the help of SolidWorks software. The next stage is manufacturing by making the main components first and then continuing with other supporting components.

2.3 Polishing machine application

Procurement and application of a press machine equipped with a Raspberry Pi, suction blower, and Hepa filter equipped with purified water. The polishing machine will provide convenience to users to smooth and polish the perimeter of the shoe.

3 Result and discussion

The research was conducted from May to September 2023. Prioritized problems exist in production that must be resolved immediately. The problem is obtained by collecting data and various information on an existing problem. Based on the problems in the field directly, there are problems that need to be solved. The problem that needs to be solved is the process of selep shoes using an open electric motor with the help of coarse grinding eyes, the results of the selep process of used selep garbage flying and scattering have a bad impact on health. In addition, the quality of the process is not good due to unstable hand conditions and lack of ergonomics, which cause operators to tend to get tired quickly.

The solution to solve the existing problems will be to design a semi-automatic shoe polishing and buffing machine integrated with Raspberry PI. This machine is designed to have two roles, which can polish shoes and shine shoes, so it is expected to be used for polishing and buffing it is equipped with a suction blower to suck up flying dust as well as to improve the quality of shoes. The design of the machine begins with making a design first. The machine design is made with the help of the SolidWorks application. The results of the polishing machine design can be seen in Figure 2.

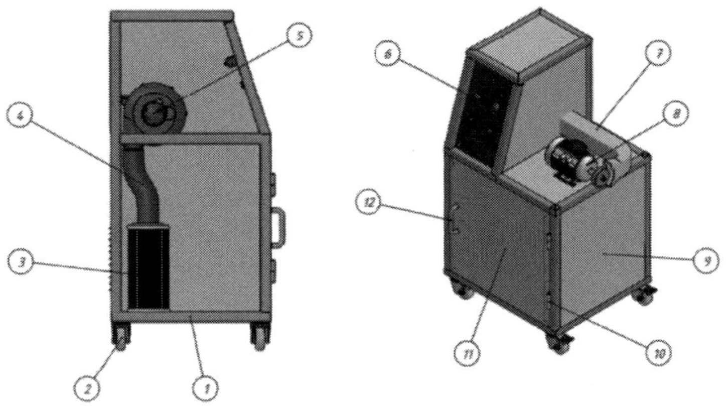

Fig. 2. Detail machine

Table 1. Image description

No	Description	Specification
1	Frame Unit	Stainless steel
2	Wheels	Plastic
3	Hepa Filter	30cm x 20cm
4	Hose	2"
5	Blower	350 watt
6	Control Panel	1. Emergency 2. ON 3. OFF 4. Counter 5. LCD
7	Suction Line	2"
8	Electric Motor	0.5 HP
9	Cover Unit	Stainless stell
10	Hinge	
11	Door Unit	
12	Door Handle	

Based on Figure 2, it can be seen that the semi-automatic shoe polishing and buffing machine integrated with Raspberry PI consists of many main components. The first component starts with the vacuum cleaner blower, hipa filter, and Raspberry Pi. Semi-automatic shoe polishing and buffing machine integrated Raspberry PI is designed to smooth the circumference of shoes driven by a high-speed electric motor equipped with a vacuum blower, so that it is able to polish the circumference and skin of very smooth shoes, very fast time and healthy working environment. [14]. A suction blower serves to suck dust so that it does not fly away [15]. The engine is also equipped with a hepa filter that functions to filter the air so that the air released from the engine is no longer dirty [16]. The rear view of the machine with the suction blower visible can be seen in Figure 3.

Fig. 3. Rear view of engine

The semi-automatic shoe polishing and buffing machine is equipped with a Raspberry PI. Raspberry Pi is one of the Internet of Things (IoT) components that can be applied to remote control with internet networks that can be applied to electronic equipment. Raspberry Pi functions to calculate the number of shoes that have been polished using a polishing machine [17]. An example of the front view of the machine can be seen in Figure 3 and the specifications of this tool can be seen in Table 2.

Fig. 4. Machine front view

Table 2. Machine specifications

No	Description	Specifications
1	Dimensions	p: 65cm x l: 52cm x t: 110cm
2	Electric motor	0.5 hp
3	Material	Stainless steel
4	Round	1500 rpm
5	Power	750 Watt

Procedures or stages to operate the semi-automatic shoe polishing and buffing machine integrated Raspberry PI, are as follows: first plug the machine power cable into the power source. After the cable is connected press the on button to turn on the machine. Polish shoes as needed. When each shoe is finished polishing, press the "Count" button to enter the number of shoes that have been obtained to the server. Then press the off button after the machine is not in use. This machine has been tested to determine the speed in polishing the circumference and very fine shoe leather. The test results of this tool can be seen in Figure 5.

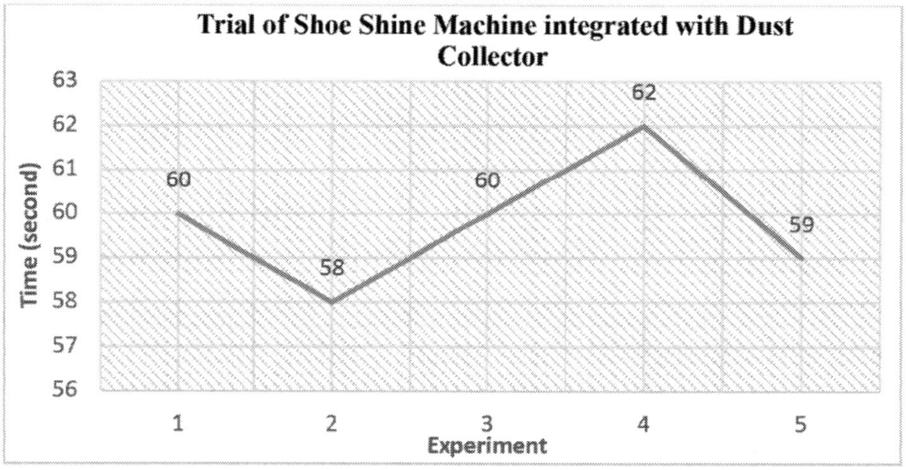

Fig. 5. Test results of shoe shine machine integrated with dust collector

The figure explains the results of machine trials conducted to polish the circumference and skin of very fine shoes with each trial tested on six pairs of shoes and the time required. The average time required to polish the circumference and skin of a pair of very fine shoes is 60 seconds or each pair of shoes only takes 10 seconds. The results of this trial are in line with research conducted by Jiao et al. [18] that the polishing machine has a simple structure and high processing efficiency, and can effectively polish such flat or round surfaces when combined with a dust collector. This machine also prioritizes safety in work, besides that it is also relatively easy to maintain. This polishing machine can be used as an efficient and effective solution to get shiny, flawless shoes.

4 Conclusion

The advantages of the Shoe Polish Machine include saving time and effort, suitable for use in hotels, airports, and shoe shops. It also improves hygiene by reducing the spread of polish residue and germs that are common with manual polishing. The safety of Shoe Polish Machine users is also a priority with the built-in safety mechanism. Maintenance is easy, making this machine reliable in the long run. Overall, a Shoe Polishing Machine is essential for maintaining the appearance and quality of shoes. This machine can improve quality and production in MSMEs. It can also be used as an efficient and effective solution to get shiny, flawless shoes.

References

1. J. A. Friedman, R. Zeckhauser, Intelligence and National Security **31** 2016)
2. L. Maghfiroh, S. S. Iriani, Jurnal Ilmu Manajemen **9** (2021)
3. A. M. Rahmawati, et al, Co. Management, and Industry (JEMI) **3** (2020)
4. L. K. Yew, J. D. Tan, South East Asia Res **30** (2022)
5. K. P. R. I. Kemenperin, The Fourth Largest Footwear Manufacturer in the World (Ministry of Industry, Indonesia, 2019)
6. R. Aznar-Gimeno, et al. Entropy **23**, 777 (2021)
7. S. Asano, et al., Sens Actuators A Phys. **309** (2020)
8. E. Chanty, JISO: Journal of Industrial and Systems Optimization **2**, 2 (2019)
9. P. Ramesh, M. Anish, J. B. Sundar, A. D Raj, International Research Journal of Multidisciplinary Technovation **1** (2019)
10. M. Niam, Machine control and electrical systems (PT. Nasya Expanding Management, Indonesia, 2021)
11. M. Alfian, H. L. Wijayanto, K. Kadriadi, M. Jafar, Jurnal Ilmiah Universitas Batanghari Jambi **23**, 1080 (2023)
12. J. Braband, H. Schäbe, Safety and Reliability **36** (2016)
13. A. F. Prakoso, D. Suwito, Yunus, Jurnal Pemberdayaan Masyarakat Madani (JPMM) **6** (2022)
14. D. R. Mandel, T. L. Hendriks, D. Irwin, Intelligence and National Security **37** (2022)
15. S. K. Sun, X. H. Jia, L. F. Xing, X. Y. Peng, Engineering Applications of Computational Fluid Mechanics **12** (2018)
16. D. A. Newcomer, P. LaPuma, R. Brandys, A. Northcross, J Occup Environ Hyg. **15** (2018)
17. K. Žáková, M. Rábek, IFAC-PapersOnLine **51** (2018)
18. C. Jiao, Y. Shu, Y. Chen, Z. Zhang, Optik (Stuttg) **224**, 165436 (2020)

Application of alacrity as a means to improve athlete's agility

Nur Ahmad Arief[1*], *Testa Adi* Nugraha[1], *I Gede Agung Sri* Sidhimantra[1], Nurhayati[1], *Saiful* Anwar[1], *Susi Tri* Umaroh[1], and *Yuli Panca* Asmara[2]

[1]State University of Surabaya, Surabaya, Indonesia.
[2]INTI International University Nilai, Negri Sembilan, Malaysia.

Abstract. Agility and speed are two important factors that determine the performance of athletes in various sports. Currently, agility and speed measurements still rely on the use of a stopwatch and human response speed. Although this method has been used in various sports settings for many years. There are several underlying problems, including low levels of accuracy and consistency, the potential for trainer or assessor fatigue, time-consuming data processing, and the need for proper agility measurement tools. By considering the problems from this background, this study aims to present the development of a touchless sensor-based agility and speed measurement tool that can help improve accuracy, consistency, and efficiency in measuring athlete agility and speed, namely "The Alacrity". With this tool, athletes are expected to be able to plan appropriate training programs to improve their performance and process data manually because the results are presented in a special Windows application and can be viewed in real-time.

1 Introduction

Agility and speed are two important factors that determine the performance of athletes in various sports. The faster the athlete performs, the better the results, in this case, timeliness is very important in determining the success of tests and competitions [1]. Accurate and consistent measurement of these two factors is essential for evaluating athlete progress, designing effective training programs, and improving competition results.

1.1 Background of the problem

Currently, agility and speed measurements still rely on the use of a stopwatch and human response speed. Although this method has been used in various sports settings for many years. Some of the underlying problems, among others, are as follows: (1) Low level of accuracy and consistency, the use of a stopwatch and human response speed as a tool for

* Corresponding author: nurarief@unesa.ac.id

measuring agility and speed has a low level of accuracy and inconsistent results. Factors such as delays in human reactions and human errors in timing can lead to inaccurate results. This can make it difficult for coaches or researchers to carry out proper monitoring and analysis of athlete progress. (2) Potential trainer or assessor fatigue, when testing many people or participants, trainers or assessors often have to perform tasks that require high focus and concentration. Manual use of a stopwatch and data analysis requires precision and patience, which can cause the trainer or assessor to experience mental fatigue and impact measurement accuracy. (3) Time-consuming data processing, the data generated from the stopwatch needs to be processed again in order to provide meaningful measurement results. Processing this data can be time-consuming and effort-consuming, especially when scoring is done on a large number of athletes. (4) The need for proper agility measurement tools, for both professional and amateur athletes, a proper understanding of their agility and speed lays the foundation for designing an effective training program. Therefore, there is an urgent need to develop measurement tools that can provide accurate and consistent data regarding the agility and speed of athletes.

By considering the problems from this background, this study aims to present the development of a touchless sensor-based agility and speed measurement tool that can help improve accuracy, consistency, and efficiency in measuring athlete agility and speed, namely "The Alacrity". With this tool, athletes are expected to be able to plan appropriate training programs to improve their performance and process data manually because the results are presented in a special Windows application and can be viewed in real-time.

1.2 Previous research

The relevant research is the development of an android-based physical fitness test measurement application [2] with a tool effectiveness level of 87%, sports measurement test experts 89%, and practitioners 92% so that it is declared good and becomes a new development. [3] also developed an agility measurement test tool using an infrared sensor with product effectiveness having results of 85% material experts, 96% media experts, and 85% by practitioners so that an agility measuring tool is effective to use.

Development of an agility test measuring instrument based on infrared sensor technology by Ramadhan et.al [4] as an update to an agility test measuring instrument with a stopwatch with validation results by experts an average of 98%. The next research is the application of a fitness test for cardiovascular endurance in the application in the form of a new product for measuring heart and lung endurance levels. This product contains general fitness information using the VO2max test method, the process of calculating values by classification, and research results with feasibility values. of 88.41% [5]. Aji [6] also developed a ladder tool for coordination, agility, and power exercises obtained from the validation results from material experts of 86% or feasible; media expert by 80% or appropriate; small group trials of 84.42% or appropriate; large group trials of 82.12% or appropriate. Based on the findings from Baskoro et.al [7] the initial feasibility test of a motion sensor-based side step agility assessment tool in Stage I, the sample achieved an assessment score of 90.93%. Expert trainers assessed it at 90.40%, IT media experts provided an assessment of 88.80%, and test and measurement experts rated it at 91.20%. Moving on to the Stage II feasibility test of the same tool, the sample's assessment score improved to 93.20%, expert trainers rated it at 92.00%, IT media experts at 91.20%, and test and measurement experts at 94.40%. The research data collected led the researchers to the conclusion that the development of this motion sensor-based side-step agility assessment tool is highly viable for measuring athletes' agility levels.

In the study conducted by Putra [8], the impact of ladder drills and zig-zag run exercises on improving agility in futsal athletes was investigated. The research findings indicated that both ladder drill exercises and zig-zag run exercises produced similar outcomes in terms of enhancing agility in futsal athletes. The creation of a desktop program-based tool for assessing physical and skill tests in futsal sports, as reported by Gumantan et.al [9], underwent extensive expert evaluation. The results showed that sports lecturers gave it a rating of 90%, computer experts in the media field evaluated it at 87%, and Asian licensed trainer practitioners scored it at 90%. With consistent performance of over 80% across all three stages, including the initial trial, small-scale testing, and large-scale evaluation, it can be inferred that this research and development effort has yielded a novel product for measuring sports tests within the field of futsal. As for the study and design of agility measurement system technology with wireless sensor network (WSN) technology as an athlete selection method by Hidayat [10] discusses the study of human agility in running movements so that femur and tibial speed and acceleration data are obtained at different speed variations, namely at walking speed 0.6 m/s (2 km/h) to 1.11 m/s (4 km/h).

Setya [11] designed a spinning iron test kit for enhancing the striking speed in UPGRI martial arts UKM. The final validation results from experts indicate its high feasibility: material experts provided an 87.5% rating (considered very feasible), media experts rated it at 95% (deemed very decent), and field test athletes scored it at 90.2% (also considered very decent). Consequently, it can be firmly concluded that the spinning iron tool for practicing pencak silat stroke speed is highly suitable for use as a training instrument in pencak silat. Firmansyah [12] created an agility training tool for futsal athletes that utilizes both light and sound, and the research outcomes indicate its effectiveness as a training aid for agility. The validation data from material and media experts placed it within the "feasible" category, achieving an 88.8% rating. Furthermore, evaluations from trainers and players in small-scale trials resulted in an 88.24% rating, also categorizing it as "feasible." Subsequently, in large-scale trials, the tool demonstrated a performance of 92.38%. Syahputri [13] developed a Pencak Silat kick reaction speed training tool with the results of a small-scale trial involving 7 athletes from the Garuda Sakti College showing that the kick reaction speed training tool met the criteria to continue in a large-scale trial because the percentage score of each aspect was between 87.5%-92.42%. The results of a large-scale trial of 15 athletes from the SPDKK college showed that the kick reaction speed training tool met the criteria to continue in the manufacture of mass products because the percentage of each aspect was between 91.67% -95%. It can be concluded that this training tool is urgently needed to make kick reaction speed training more effective and efficient, because it is very important in supporting athlete achievement, besides that the exercise is not boring and becomes input material for pencak silat trainers throughout Indonesia.

Meanwhile, Devi [14], in 2021, a training tool was created to enhance the reaction speed of punches and kicks in karate athletes. The initial trial involved three experts, and the results were subsequently validated, yielding an average validation score of 87.14%. Another trial was conducted with the second tool, which engaged six experts, and the results were validated, resulting in an average validation score of 90.4%. Drawing from the validation conducted by the six experts, it can be deduced that the development of these training aids to boost reaction speed is valid and suitable for training purposes aimed at improving reaction speed. Ashadi [15] also made a shuttle run tool with data stored in an Arduino-based database with the results of the research showing very good accuracy and accuracy of the tool. The accuracy of the tool is 0.808 (80.8%), with an error presentation of 4.74%. The accuracy of the tool is also very good, namely 0.995 (99.5%), with a standard deviation of 0.33. Meanwhile, the good practicalization test was obtained with a speed of 10.52 seconds, and the less good one was obtained with a speed of 16.39 seconds.

2 Methods

In implementing the tool, the step position that the athlete must achieve is according to the layout of the sensor panel placement shown in the following image in Fig.1.

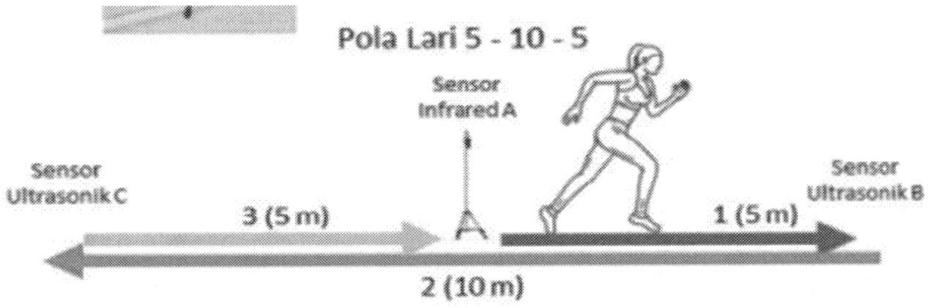

Fig. 1. Illustration of an athlete's footsteps during agility training

The implementation of the Alacrity measuring tool is carried out by setting the displacement of the athlete's foot movement based on the touchless sensor in the following order.The athlete's starting position is in the center (sensor A)
Step 1: Run 5 meters towards sensor B
Step 2: Run back 10 meters towards sensor C
Step 3: Run 5 meters to the starting position (sensor A)
So that the travel time of the athlete's position between sensors can be known and it will be recorded by the system in real-time.

3 Result and discussion

In the implementation of The Alacrity measuring instrument with the second variation, 10 people were used, consisting of 5 students majoring in sports coaching education, FIK Unesa and 5 students from sports clubs. The results of implementing The Alacrity measuring instrument are seen in the following Table 1.

Table 1. Results of implementation of The Alacrity measuring instrument on a limited scale

Atlet	L1 (5m)	L2 (10m)	L3 (5m)	Total
1	2.32	4.13	2.22	8.67
2	1.89	3.76	1.90	7.55
3	1.98	3.98	1.75	7.71
4	2.12	3.81	1.95	7.88
5	1.93	3.95	1.89	7.77
6	1.85	3.65	1.72	7.22
7	2.17	3.82	2.05	8.04
8	2.14	4.15	1.92	8.21
9	2.02	3.91	1.89	7.82
10	2.12	3.99	1.95	8.06
Mean	2.054	3.915	1.924	7.893
Item Varians	0.0212044	0.025205556	0.01996	0.06637
Total Varians				0.153157
Alpha Cronbach				0.849979

Based on the results of reliability calculations using Cronbach's Alpha method, Cronbach's Alpha coefficient value (α) = 0.85. Referring to the r_{table} value with df (0.05, n-2) of 0.632, then $\alpha > r_{table}$ so that it can be said that the results of the Alacrity measurement are reliable (trustworthy) and classified as very high (good) criteria because it is included in the range $0.800 \leq \alpha \leq 1{,}000$. Based on the description above, it can be said that the data generated by Alacrity is valid (accurate) because all item score correlation values > r_{table} (0.632). This Alacrity is also equipped with an athlete's movement time tracker while moving from one point to another so that the athlete's performance can be measured according to the desired standards. This makes the coach more focused on observing the athlete's agility movements carefully. Another advantage of this tool is that it is multi-functional because it can be used for athletes of all sports, especially for training agility.

4 Conclusion

Referring to the r_{table} value with df (0.05, n-2) of 0.632, then $\alpha > r_{table}$ so that it can be said that the results of the Alacrity measurement are reliable (trustworthy) and classified as very high (good) criteria because it is included in the range $0.800 \leq \alpha \leq 1{,}000$. Based on the description above, it can be said that the data generated by Alacrity is valid (accurate) because all item score correlation values > r_{table} (0.632). This Alacrity is also equipped with an athlete's movement time tracker while moving from one point to another so that the athlete's performance can be measured according to the desired standards.

References

1. M. Mariyono, S. Rahayu, E. R. Rustiana, Journal of Physical Education and Sports **6**, 1 (2017)
2. A. Gumantan, Jurnal Ilmu Keolahragaan **19**, 2 (2020)
3. A. Gumantan, I. Mahfud, Jendela Olahraga **5**, 2 (2020)
4. A. A. R. Ramadhan, I. Iyakrus, W. I. Bayu, S. Syafaruddin, H. Yusfi, Jurnal Program Studi Pendidikan Jasmani dan Kesehatan **10**, 4 2022
5. A. A. Putro, A. Suherman, K. Sultoni, J. Terap. Ilmu Keolahragaan **3**, 1 (2018)
6. G. P. Aji, Pengembangan alat ladder untuk latihan koordinasi, kelincahan dan power (Universitas Negeri Yogyakarta, Yogyakarta, 2016)
7. W. C. Baskoro, Pengembangan Alat Bantu Tes Kelincahan Side Step Berbasis Sensor Gerak (Universitas Negeri Medan, Medan, 2020)
8. P. D. A. Putra, J. Kesehat. Olahraga **10** (2022)
9. A. Gumantan, I. Mahfud, R. Yuliandra, JOSSAE J. Sport Sci. Educ. **6**, 2 (2021)
10. N. H. C. Hidayat, J. Siliwangi Sains Teknol. **3** (2017)
11. R. A. Setya, J. Phys. Act. Sport. **2**, 3(2021)
12. R. G. Firmansyah, Development of Cones Using Light and Sound as Agility Training Tools for Futsal Athletes (Universitas Jambi, Jambi, 2022)
13. Y. Y. Syahputri, Development of Pencak Silat Kick Reaction Speed Training Equipment (Universitas Negeri Medan, Medan, 2020)
14. K. Devi, " Development of Training Aids to Increase Reaction Speed of Punches and Kicks in Karate Athletes (Universitas Negeri Medan, Medan, 2021)
15. K. Ashadi, Making a Shuttle Run Tool with Data Stored in an Arduino Based Database (Universitas Negeri Padang, Padang, 2021)

Automatic patient bed comfort structure analysis based on the RULA method

Dzulkiflih[1*], *Wahyu Mulyo* Utomo[2], *Sista* Paramita[1], *Lilik* Anifah[1], *Agung Prijo* Budijono[1], and *Susi Tri* Umaroh[1]

[1]State University of Surabaya, Surabaya, Indonesia
[2]University Tun Hussein on Malaysia, Batu Pahat, Malaysia

Abstract. To improve patient care, reduce medical staff injuries, and increase the efficiency of the patient care process, automated patient beds are a very important innovation in the world of healthcare. However, there needs to be a study to find an ergonomic automatic bed design so that it is safe and comfortable when used by patients. This research concerns the study of the ergonomics of patient bed design using the Rapid Upper Limb Analysis (RULA) method. The patient bed that has been designed is then analyzed using RULA to obtain initial ergonomic data from a designed patient bed. There are 3 observations made, namely the patient's condition when sitting, the patient's condition when lying down, and the patient's condition when the bed can be adjusted low. When in the sitting position, the average RULA result is obtained at 3, when lying down, the average RULA result is obtained at 3, and when the bed is set low, the average RULA result is obtained at 3. This ergonomic value is an accumulation of the assessment of the structure of the upper arms and forearms, lower body, muscles, neck, torso, legs, wrists, and arms.

1 Introduction

The health and well-being of patients is a very important aspect of health services. One of the key factors in creating an optimal care environment is the ergonomic design of the patient bed. Ergonomics is an applied scientific discipline that focuses on the design and development of tools, including workplace settings, and optimal work strategies to maintain the health of the human body. Ergonomics aims to reduce fatigue, discomfort, risk of injury, and emotional stress [1]. Apart from that, ergonomics is a discipline that examines the limitations, strengths, and characteristics of humans, and utilizes this information in designing products, without ignoring the health, safety, and comfort aspects of the user [2]. A well-designed patient bed not only increases patient comfort but also makes it easier for medical staff to provide quality services. Patient beds are critical medical devices in hospitals and can be used by private individuals in caring for sick or recovering patients. The quality of the patient's bed is very important in ensuring

* Corresponding author: dzulkiflih@unesa.ac.id

patient comfort and ease of care. In hospitals, patient beds also play a role in providing the support necessary for optimal medical care. In addition, inappropriate work posture or position will result in work-related musculoskeletal disorders (MSDs)[3]. Therefore, designing and evaluating a good patient bed is very crucial. When someone is sick or in the process of recovery, comfort while lying in the patient's bed is an important factor in the healing and treatment process. A patient bed that is ergonomic and fits the patient's needs can help reduce the risk of pressure on the body, improve blood circulation, and reduce the discomfort that the patient may experience. Therefore, patient beds are needed that are designed with a structure that takes into account aspects of patient comfort and health as well as ease of care.

In this study, to analyze the comfort structure of the automatic patient bed, the RULA (Rapid Upper Limb Assessment) method was used. Even though there is a lot of previous research related to this field, the analysis of the comfort structure of automatic patient beds using the RULA method is still not optimal. This research contributes to this gap by developing an automated patient bed system in which a comfort structure analysis is obtained based on the RULA method. Results related to the automatic bed structure analysis in terms of comfort will be displayed. With this method, it will be shown how the system can be used to identify and measure factors that influence patient comfort and the ergonomics of the patient's bed. In this regard, this research can make an important contribution to improving the design of patient beds both for use in hospitals and for individual users.

2 Methods

This research method uses Rapid Upper Limb Assessment (RULA) which examines the ergonomics of the design of automatic patient beds. RULA is a method used to assess posture, strength, and movement of work activities related to upper limb activities[4, 5]. Apart from that, it is also a technique for evaluating and assessing body layout, strength, and movement transfer in work activities that touch the upper part of the body. This technique was also developed as a means of identifying possible risks of interference or injury experienced by workers as well as ergonomic factors related to upper extremity MSDs when carrying out certain activities[6]. The RULA ergonomic assessment instrument includes biomechanical loads, and task demands from the posture of the neck, trunk, and upper extremities which are measured and calculated using CATIA software. CATIA software is a tool that can be used for ergonomic analysis and design of workstations and environments. The steps in calculating RULA or the RULA assessment list are shown in Figure 1 below.

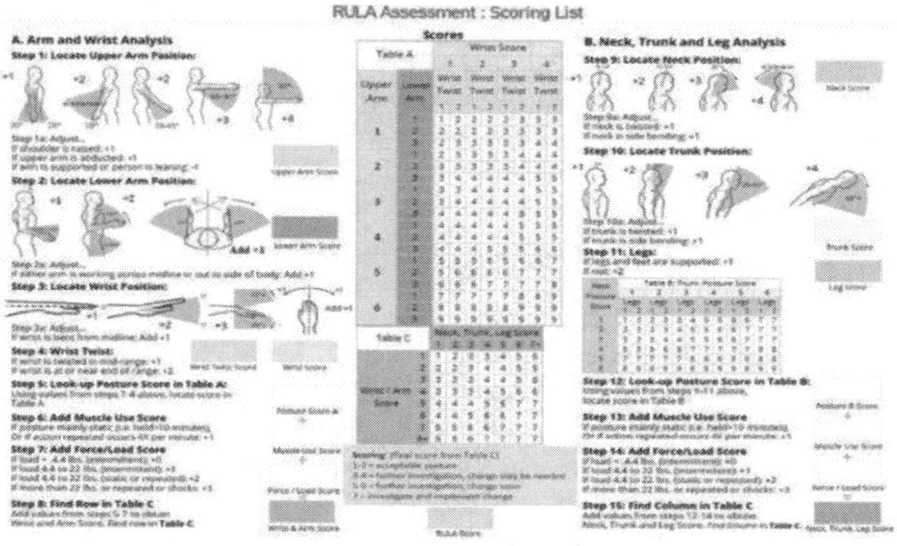

Fig. 1. RULA Assessment: Scoring List

RULA analysis was carried out on the upper limbs of the mannequin by taking into account factors such as weight, distance, and frequency. RULA is a survey method designed as a means of investigating human factors or ergonomics in the work environment regarding work disorders in the upper extremities. RULA is a screening tool that assesses the biomechanical load and posture of the entire body with a particular focus on the neck, trunk, and upper limbs. RULA is intended to be used as part of a more comprehensive ergonomics study [7].

In this case, RULA was developed and used specifically to evaluate the risks for patients who are treated using a patient bed that involves sitting, lying, right or left tilt, such as patients in hospitals who experience mobility problems. as well as risks for patients where patients lie down for long periods of time.Input data in RULA analysis includes body posture (head, torso, and upper limbs), force used, and types ofmovements or actions performed repeatedly. This RULA analysis is used for various aspects of a mannequin's posture automatically from a combination of visible variables and user data. In this study, some parts of the mannequin's body were adjusted according to the design of the patient's bed. assessment of the structural position of the upper arm, forearm, muscles, neck, trunk, legs, wrists, and arms. The patient bedthat has been designed is then analyzed using RULA to obtain initial ergonomic data for the designed patientbed. There are 3 reviews carried out, namely the patient's condition when sitting, the patient's condition whenlying down, and the patient's condition when the bed can be set low. The following is a simulation scenario of a patient bed for RULA assessment as seen in Figure 2.

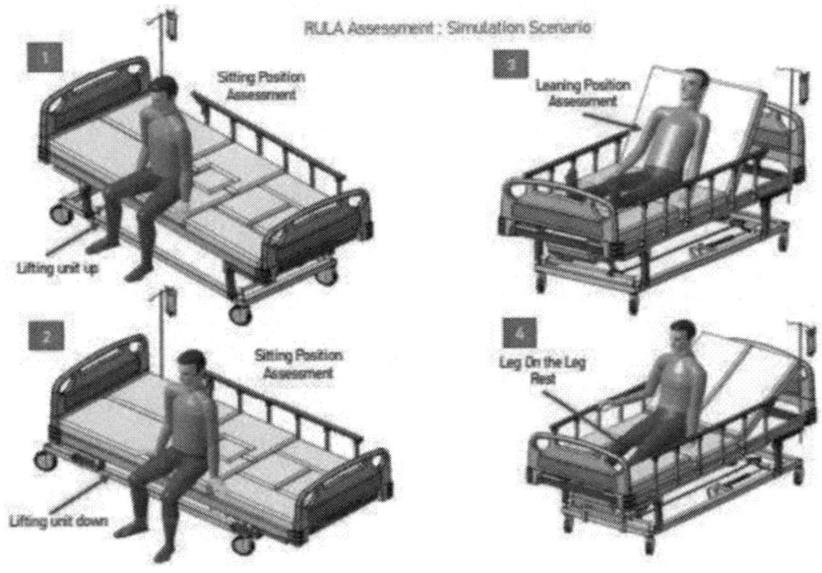

Fig. 2. RULA Assessment Simulation Scenario

Table 1. RULA Scoring and descriptions

Action Levels	RULA Score	MSD Risk Level	Descriptions
1	1-2	Negligible risk, no action required	The individual is operating in an optimal posture, with no potential risk of injury from their work position.
2	3-4	Low risk, some adjustments may be needed	The individual is working in a posture that could pose some risk of injury due to their work position, and this score is likely due to one specific aspect of the task.
3	5-6	Medium risk, further investigation, changes needed soon	The individual is working in a suboptimal posture, which carries a medium risk of injury from their work position. It's crucial to investigate and implement changes in the near future to prevent injuries.
4	7+	Very high risk, immediate changes required	The individual is operating in the poorest posture, with an imminent risk of injury from their work position. Immediate

			investigation and changes are necessary to prevent injury.

Table 1 displays the RULA action levels along with their respective descriptions. The score is represented on a numerical scale, ranging from 1 (the lowest score) to 7 (the highest score). Each action level is accompanied by its own set of descriptions, which elucidate the potential outcomes when the analysis aligns with the assigned score [5].

3 Results and discussion

Based on the results of the automatic patient bed design, the dimensions and geometry of the patient bed product are obtained as shown in Figure 3 below:

3.1 Product geometry & dimension

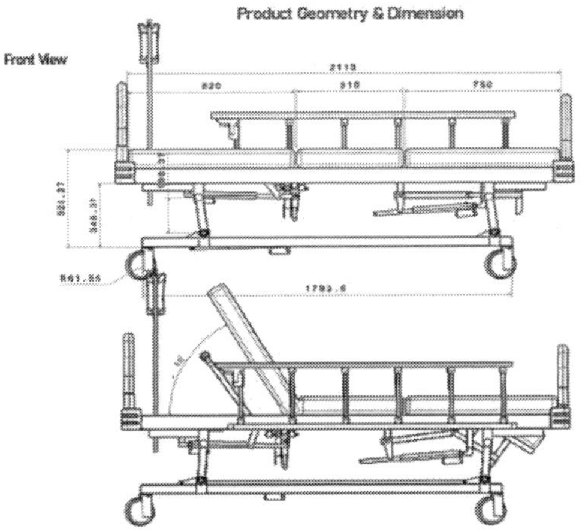

Fig. 3. Product Geometry and dimension of the patient bed

3.2 Structural study designed patient bed lifting unit

Based on the results of a structural study of the designed lifting unit for the patient's bed, in this simulation scenario the bed will be given a load of 1471 N (-150 kgf), where the results show that the stress distribution in the frame is less than 250 MPa or does not exceed the Yield Strength Then, the safety factor value is 15 so it can be said that this patient's bed frame is safe. In Figures 4 to Figure 7, the results of the structural analysis of the patient bed lifting frame unit are presented.

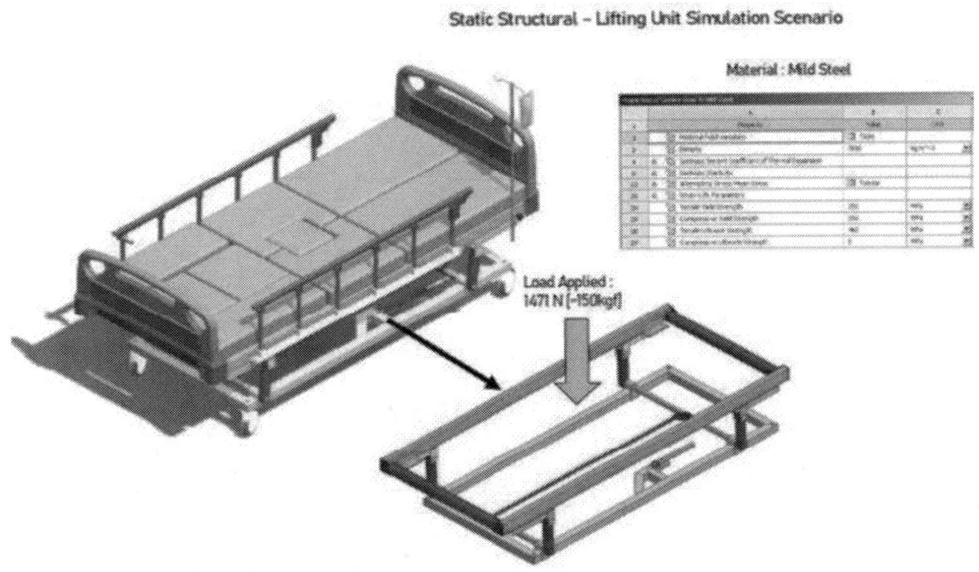

Fig. 4. Static Structural – Lifting unit simulation scenario

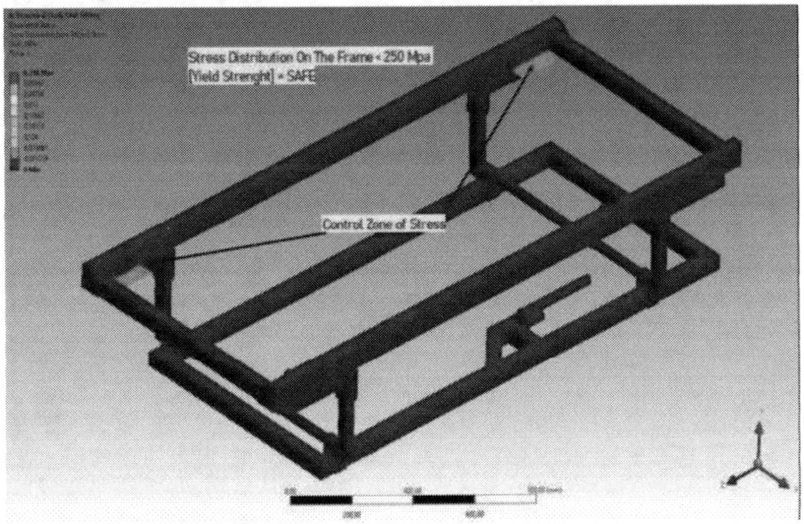

Fig. 5. Structural analysis of the patient bed lifting unit

 https://doi.org/10.1051/e3sconf/202345002008

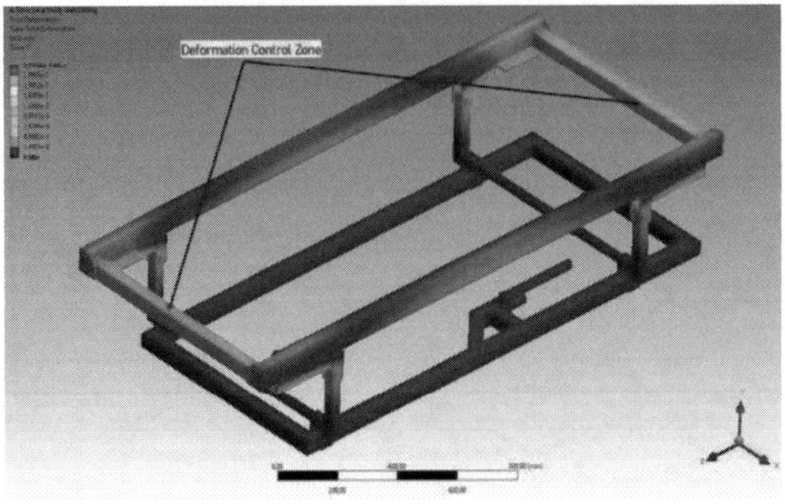

Fig. 6. Structural analysis of deformation control zone

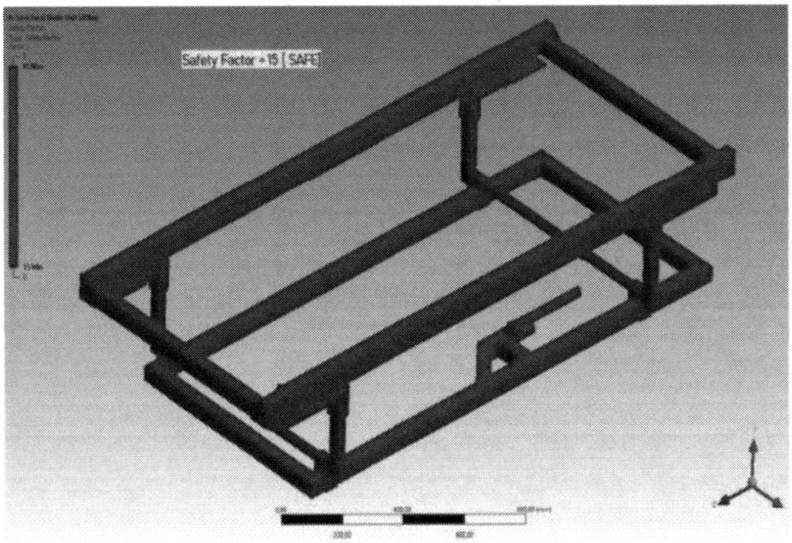

Fig. 7. Analysis of safety factors for patient bed frames

3.3 Structural analysis of patient beds with rapid upper limb assessment (RULA)

Based on the patient bed simulation scenario for RULA assessment as seen in Figure 2, above, thefollowing results were obtained.

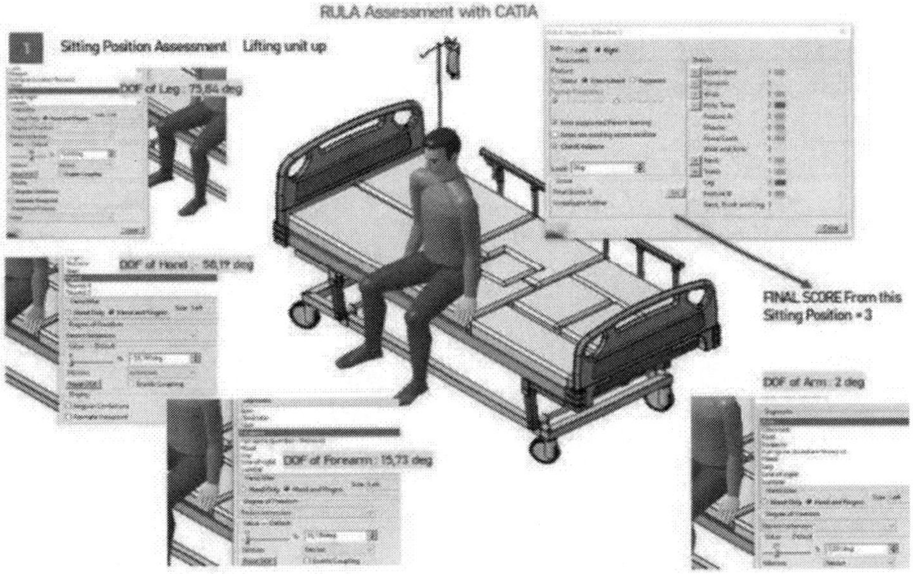

Fig. 8. Sitting Position (lifting unit) RULA Assessment

Referring to the results of structural analysis in the sitting position, namely when lifting the unit, the DOF of the leg or the degree of body posture of a person's legs when sitting with the lifting bed up is 75.84 deg, the DOF of the hand is -58.19 deg, the DOF of the forearm is 15.73 deg, and DOF of the arm is 2 deg, resulting in a final score in the sitting position of 3 or still in the safe category and can be investigated further or can be said to be at level Low risk, change may be needed, if based on MSD risk.

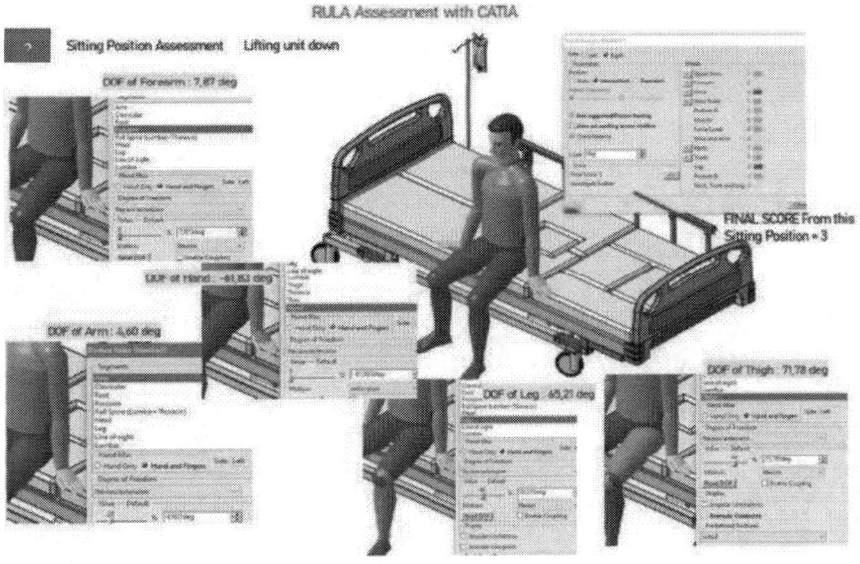

Fig. 9 Sitting Position (lifting unit down) RULA Assessment

Referring to the results of structural analysis in the sitting position, namely when the lifting unit is down, the DOF of legs or the degree of the body posture of a person's legs when sitting with the lifting bed down is 65.21 deg, DOF of hand is -61.83 deg, DOF of forearm is 7.87 deg, and DOF of arm is 4.60 deg, DOF of thigh is 71.78 deg resulting in a final score in the sitting position when lifting down which is 3 or still in the safe category and can be investigated further or can be said in level is low risk, change may be needed,

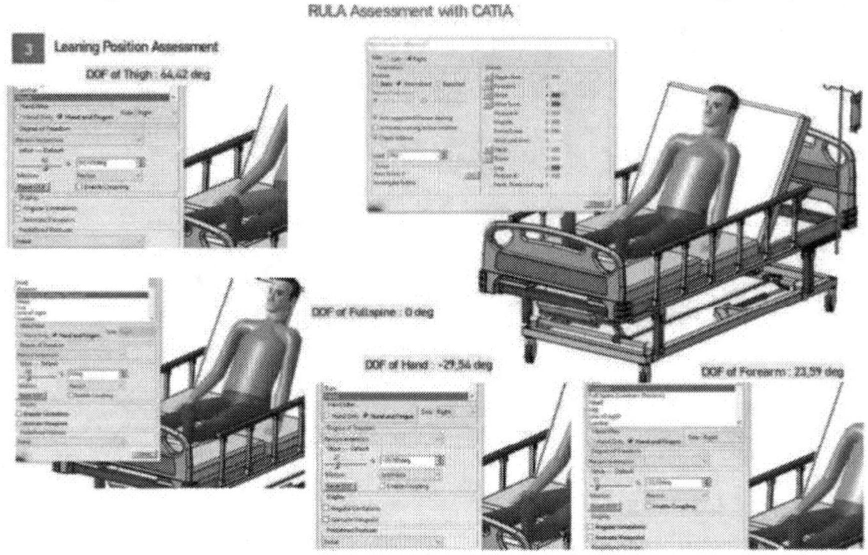

if based on MSD risk.

Fig. 10. Leaning Position RULA Assessment

Referring to the results of structural analysis in the lying position, namely when the DOF of thigh is 64.42 deg, DOF of hand is -29.54 deg, DOF of forearm is 23.59 deg, and DOF of full spine is 0 deg, resulting in a final score in a sitting position when lifting down, namely 3 or still in the safe category and further investigation can be carried out or can be said to be at level low risk, change may be needed, if based on MSD risk.

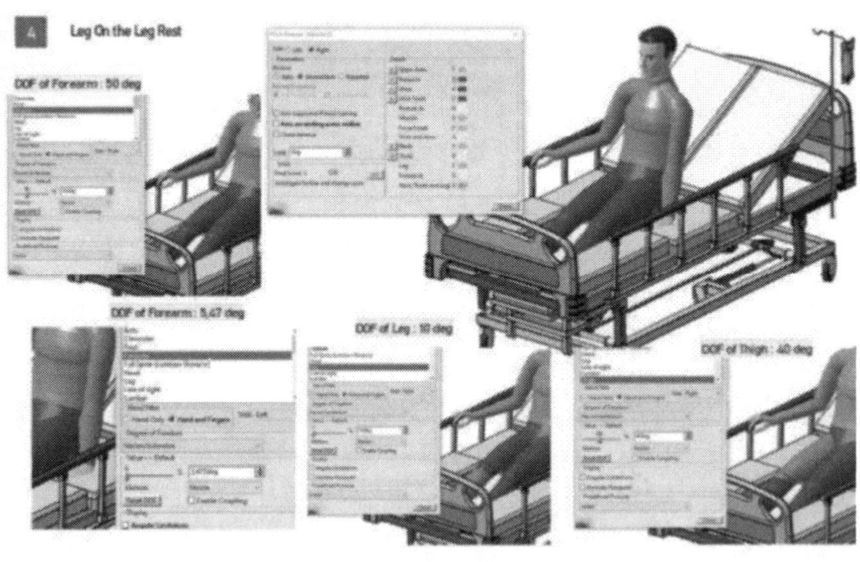

Fig. 11. Leg On the Leg Rest RULA Assessment

Referring to the results of structural analysis in the position of the foot on the footrest or leg on the leg rest, namely when the DOF of the thigh is 40 deg, DOF of right forearm is 50 deg, the DOF of left forearm is 5.47 deg, and DOF of the leg is 10 deg, produces a final score in a sitting position when leg on the leg rest, namely 5 or investigate further change soon or can be said in medium-risk, further investigation, change soon, if based on MSD risk.

Based on these results, it can be seen that several positions that were analyzed when the automatic patient bed was operating, in 3 conditions, namely the sitting position when lifting up, the sitting position when lifting down, and the lying position received a final score of 3 or in the low-risk category which can be said to be still in the safe and comfortable category. However, when the foot is in the final leg rest, the RULA score is 5, which is in the medium risk category, which can be said to require further investigation to avoid serious risks. Analysis using the RULA method was also carried out by M. Massaccesi et.al (2003) in analyzing the position of truck drivers where drivers often experience complaints due to inappropriate work postures[8]. Apart from that, research by Nelfiyanti, et.al (2022) regarding MSD analysis using the RULA approach for chassis assembly workers where there are several groups analyzed and there are groups analyzed that have a high risk so corrective action needs to be taken as soon as possible [5]. Therefore, it can be seen that evaluating or assessing work posture is very important, especially for several products and work positions to avoid injury or problems with body posture [9, 10].

This is also supported by research by Widiyawati, et.al. (2020), regarding the risks of hunched work postures, using maximum reach and the position of certain body parts is uncomfortable, resulting in decreased worker productivity. Therefore, this analysis evaluates employee work postures to identify problems with the work postures used. The RULA method is also used to evaluate musculoskeletal loads in a job where a person has risks to upper body activities. The samples used in this research were workers who worked in a sitting position and used computers in their activities. Based on the analysis, it is known that worker activities received a score indicating that the workplace requires changes to improve ergonomic aspects[11].

4 Conclusion

Based on the structural analysis carried out on the patient's bed, there are several conclusions, namely: (1) The results of the analysis of the comfort structure of the patient's bed automatically obtain an average RULA score of 3 or low risk and are included in the safe category which is carried out in the structural analysis when sitting in a lifting up, lifting down position, then when lying down, and also when in a sitting position feet are on the leg rest; (2) There needs to be further improvement and investigation into the leg on the leg rest condition becauseit is in the medium risk category; (3) The designed patient bed has been evaluated and used as a consideration for the mass production process of automatic patient beds.

Acknowledgements

We would like to thank the Ministry of Education and Culture, with the 2023 Kedaireka Matching Fund Program, we can develop innovations regarding automatic patient beds to facilitate patient care.

References

1. S. Mat, M. A. Abdullah, A. R. Dullah, S. A. Shamsudin, M. F. Hussin, *Car seat design using RULA analysis,* in Proceeding of mechanical engineering research day (2017)
2. Donni, S. Perdana, *Perancangan Ulang (Redesign) Tempat Tidur Untuk Pasien Rumah Sakit Dengan Metode Kansei Engeneering Di Pt Xyz,* in Prosiding Seminar Nasional Universitas Islam Syekh Yusuf, Indonesia (2020)
3. R. Patrisina, S. W. Utami, *Work Posture Analysis at The Spinning Department of Textile Industry using Rapid Upper Limb Assessment (RULA) Method,* In ASEAN/Asian Academic Society International Conference Proceeding Series (2019)
4. B. M. Nuryaningtyas, T. Martiana, Indones. J. Occup. Saf. Heal. **3** (2014)
5. N. Nelfiyanti, Mohamed, M. F. F. A. Rashid, Int. J. Automot. Mech. Eng. **19**, 2 (2022)
6. Z. Abidin, Sugiyanto, Humanit, Manag. Sci. Proc. **02**, 1 (2021)
7. L. Mcatamney, E. N. Corlett, Appl. Ergon. **24**, 2 (1993)
8. M. Massaccesi, A. Pagnotta, A. Soccetti, M. Masali, C. Masiero, F. Greco, Appl. Ergon. **34** (2003)
9. Y. Y. Rohmatin, S. Benedictus, Int. J. Sci. Technol. **2**, 1 (2023)
10. S. Yavuz, B. Gur, A. D. Cakir, Hittire J. Sci. Eng. **8**, 2 (2021)
11. S. Widiyawati, R. P. Lukodono, A. T. Lustyana, I. A. Pradana, Int. J. Hum. Mov. Sport. Sci. **8**, 1 (2020)

Automatic shoe drying oven integrated with Raspberry Pi Cloud system for advanced footwear industry

Yunus[1*], *Ade Priyo* Widhikdho[2], *Ali Hasbi* Ramadhani[1], and *Bima* Anshari[1]

[1]Departement Mechanical Engineering, State University of Surabaya, Surabaya, Indonesia
[2]School of Mechanical and Manufacturing Engineering, The University of New South Wales, Sydney NSW, Australia

Abstract. The drying oven functions to reduce the moisture content of raw materials until it reaches a certain moisture content to slow the rate of product damage due to biological and chemical activity. This research aims to design an automatic shoe-drying oven equipped with an LCD screen, exhaust fan, 8-level shelf, and electric heating temperature controller to provide maximum and even heating of leather shoes when gluing and drying leather shoes. Raspberry PI aims to combine and collect production process data in cloud storage in the form of a database, which will then be sent and displayed directly to the operator. The research methodology includes design, manufacturing, assembly, operational testing, performance testing, and evaluation stages. It can be concluded that the research results and progress reports include: (1) The design of the work system and mechanical system for the shoe drying oven can work well. (2) The Raspberry PI Cloud system can collect and display actual drying time data, (3) glue drying temperature of 60°C with a drying time of ±5 minutes, (4) skin heating temperature (texture creation) of 50°C-55°C requires warm-up time ±40 minutes, (5) This machine requires 850 watts of power.

1 Introduction

Technological developments in the era of the revolution industrial 4.0 are now supporting competitive competition with the intervention of an intelligent and automated system in the industry [1]. The widespread movement of the fourth-generation industrial revolution in recent years has had an impact on changes in work patterns in various sectors, especially in the business sector. Micro, small, and medium enterprises (MSMEs) have an important role, this sector has excellent potential in supporting the country's economy [2]. It is a reality for medium-sized business owners to meet market demand and economic needs.

UD. Candi Emas Jaya is one of the MSMEs in Sidoarjo which is engaged in shoe craftsmen. UD Candi Emas Jaya already has many customers. Every day UD. Candi Emas

* Corresponding author: yunus@unesa.ac.id

Jaya is able to produce 150 pairs of shoes. However, the shoes made by UD. Candi Emas Jaya has a quality and quantity that can be said to be lacking. Based on direct observations and interviews with the owner, the factors that affect the low quality and quantity, namely still using traditional tools, especially in the shoe drying process so that the process of heating and drying shoe glue becomes very long.

Drying is a method that can extend the shelf life of a wide variety of shoe materials [3]. UD. Candi Emas Jaya in the drying process using sunlight and stoves liquefied petroleum gas (LPG). Drying using manual tools usually takes a long time because the drying process using direct sunlight takes one or two days. [4]. Sunlight can also damage the material, fade the color, and loosen the adhesive on the sole [5]. The drying process using an LPG stove can also make the drying process uneven with a capacity of one pair of shoes per process [6]. The results of drying with LPG stoves have uneven drying quality, and the adhesion strength of the sole and upper is less strong. The drying process can be seen in Figure 1.

Fig. 1. Glue drying process with LPG burner

The slow production process makes UD. Candi Emas Jaya cannot fulfill the demand and the low quality of shoe drying makes it unable to penetrate the export market [7]. Support is needed to develop UD. Candi Emas Jaya because it has the potential to absorb a lot of labor and improve quality so that it has competitiveness in the export market. One of the supports that can be provided is the application of appropriate technology [8]. Appropriate technology such as high-capacity production machinery is urgently needed to improve competitiveness and efficiency, resulting in sustainable economic growth for business owners [9].

The appropriate technology that will be implemented is a shoe-drying machine. Compared to conventional drying and drying using an automatic shoe drying oven integrated with a Raspberry Pi cloud drying system. it has advantages in extremely fast drying, high efficiency, low energy consumption, and easy control [10]. Shimazaki & Aisaka [11], studied the effect of shoe soles on the amount of heat generated inside the shoe and showed that the heat generated from compression and expansion of shoe soles has a significant impact on quality.

Designing an automatic shoe drying oven equipped with a Raspberry PI with an LCD screen, fan, multilevel shelves, and electric heater temperature control to provide maximum and even heating of leather shoes when gluing and drying leather shoes. Raspberry PI aims to combine and collect production process data in cloud storage in the form of a database, which will then be sent and displayed directly to the operator [12]. Copetti et al. [13] The two-dimensional model developed can encourage digitalization and make it easier for MSMEs to face the changes that occur, the government has increased the ease of access and transferred technology to MSME players so that they can survive in business competition. [14]. The need for the development of this machine is to assist MSMEs in improving the quality and quantity of their products, so that they can compete in the national market. In addition, this tool can provide a touch of technology to MSMEs.

2 Methods

This research activity is carried out through several stages, which include design, manufacturing, operational testing, performance testing, and evaluation.

2.1 Design

The design stage begins with the design of an automatic shoes-drying oven integrated with the Raspberry Pi cloud system using SolidWorks software. The design stages in this study consist of: 1) making a design starting from parts, components, and the main unit of the machine, 2) assembly of the design, by assembling all the designs of the parts that have been made, 3) the next step is to make a detailed framework as the basis for making the machine.

2.2 Manufacturing

Completed, then proceed to the manufacturing process stage of the automatic shoes drying oven integrated with Raspberry Pi cloud system based on the finished design, manufacturing starts with making the main components and supporting components. After the component is complete, the Raspberry Pi cloud system is installed.

2.3 Performance testing

After the engine is completed from the manufacturing stage, then proceed to the engine function test stage. In testing the work of the machine there are two types of machine tests, namely function tests without using loads and function tests using loads. The indicators used in testing the function of the tool are by conducting a function test on each component of the tool in the machine to determine the success of the function of the machine that has been made, then testing using a load with indicators of the quality of the drying results of the sole and upper glue, leather texture and production quantity.

2.4 Evaluation

Evaluate the activities during the start-to-finish stage by looking at what needs to be improved on future machines.

3 Result and discussion

This research was conducted for five months from May until September 2023 in collaboration with one of the shoe partners in Sidoarjo. Based on the observations found in the field delivered by UD. Candi Emas Jaya, the next step is to develop a tool to help overcome existing problems, namely by making an automatic shoe-drying oven integrated with the Raspberry Pi cloud system. Automatic shoe-drying oven integrated with the Raspberry Pi cloud system is a machine used for the process of drying shoes, drying glue, and forming profiles from shoe leather so as not to wrinkle. The design of the machine design uses 3D Modeling software and the results of the automatic shoes drying oven integrated with the Raspberry Pi cloud system can be seen in Figure 2.

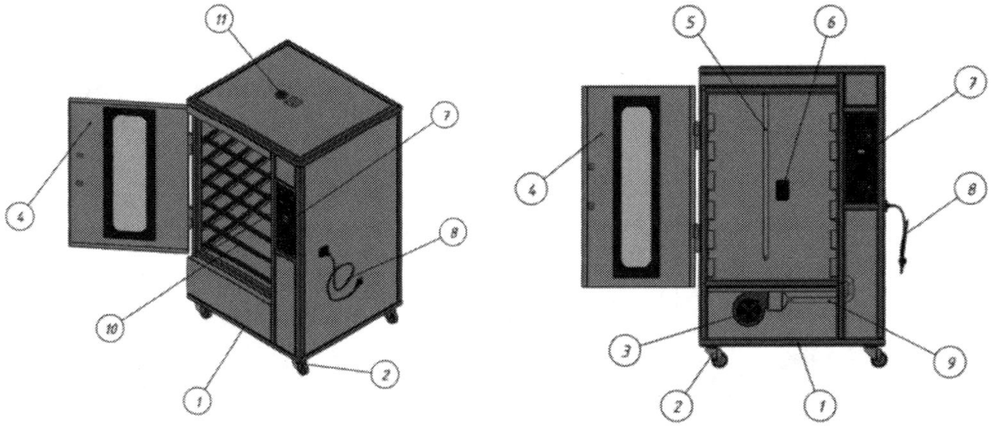

Fig. 2. Design of automatic shoe drying oven machine integrated with Raspberry Pi cloud system

Table 1. Machine description

No	Description	Specification
1	Frame unit	Stainless steel
2	Wheels	Plastic
3	Blower	90 Watt
4	Door unit	Stainless steel
5	Lights	10 Watt
6	Sensor (temperature)	0-150 °C
7	Control panel	1. Temperature control 2. Timer 3. Keypad 4. LCD 5. Bottom ON 6. Bottom OFF
8	Power cable	220 Volt

No	Description	Specification
9	Hot air duct	2"
10	Shoe rack	Stainless steel
11	Exhaust fan	15 Watt, 2500 rpm, 80.4 cmh

Based on Figure 2, it can be seen that the machine has six tiers in the oven designed to accommodate more shoes. Each shelf of the automatic shoe-drying oven integrated with the Raspberry Pi cloud system can be filled with six shoes. After the design is complete, then proceed to the manufacturing stage [15]. In the manufacturing process, the materials used use selected materials that are in accordance with the standards [16]. The material used in manufacturing is stainless 304. The results of the manufacturing stage can be seen in Figure 3.

Fig. 3. Results after manufacturing

The automatic shoe-drying oven integrated with the Raspberry Pi cloud system has been finished then to find out the performance of this automatic shoe-drying oven, a function test is carried out first. The machine performance test is carried out twice, namely the first with a function test by checking all engine components whether they are working properly or if there are still errors [17]. The process of testing the function of the automatic shoe-drying oven integrated with the Raspberry Pi cloud system is done by turning on the machine by pressing the ON button, leaving the machine on for ten minutes then checking the temperature in the oven machine room. The results of the function test obtained all engine components are in good working order and obtain a temperature of 50 °C-110 °C.

The load test is carried out to find out how effective the automatic shoe-drying oven integrated with the Raspberry Pi cloud system is when it is given a load of shoes. The load test process is carried out by turning on the oven by pressing the ON button, then the shoes are put into the machine and waited for three minutes. Based on the results of the load test, the temperature in the machine is around 60 ° C. The temperature inside the machine can also be controlled by automatic thermocontroll and timer. [18]. This is in accordance with the temperature required when heating the skin (texture formation) with a temperature of 50 °C-55 °C [19]. How the machine works can be seen in Figure 4.

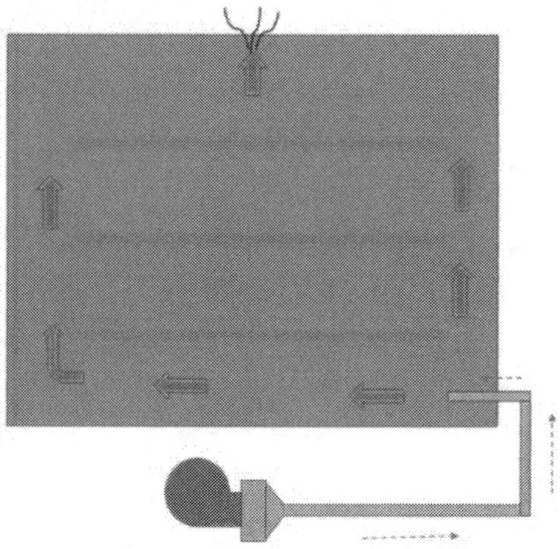

Fig. 4. Diagram of how the machine works

Based on Figure 3 the way the machine works starts from the heat source that comes from the heating element. The heating element temperature can be adjusted as needed with thermocontroll so that it is more evenly distributed [20]. The oven process can be set automatically with a timer so that it can produce good-quality shoes. The design of the working system and mechanical system of the shoe drying oven can work properly. The Raspberry PI Cloud system can collect and display actual drying time data [21]. Raspberry PI makes it easy for us to know how many shoes have dried evenly. In addition to Raspberry PI, the machine is also equipped with a thermocontrol and timer on the machine will make it easy for users to adjust the temperature and time of the oven according to the characteristics of drying shoes. The exhaust fan functions as an air circulation regulator in the oven so that the heat is evenly distributed and maintained constant throughout the oven space so that optimal shoe quality will be obtained, namely evenly dried glue [22]. The required glue drying temperature is 60°C with a drying time of five minutes. This machine requires 850 watts of electricity to operate [23]. The machine specifications can be seen in Table 2.

Table 2. Specifications of automatic shoe drying oven machine integrated with Raspberry Pi Cloud system

No	Description	Specification
1	Dimensions	p=94 cm; l=84 cm; t=1034 cm
2	Heating Element	Heater 1000 watt
3	Material	Stainless steel 304
4	Capacity	18 pairs
5	Power	850 watts

How to operate the automatic shoe-drying oven integrated with Raspberry Pi cloud system, as follows: 1) Plug the machine power cable into the power source, (2) Press Power On/Off, (3) Put the shoes into the rack and close the door, (3) Set the thermocontroller and timer for the shoe oven process, (4) Enter the nominal number of shoes processed in the system, (5) Press the start button to start the curing process, (6)

The machine work process will stop after the curing time has expired (when the time entered has expired, the system will send a signal to the server in the form of a notification that the curing process has been completed and (7) Exit the shoes when the alarm timer has sounded. The machine test results for three days can be seen in Figure 5.

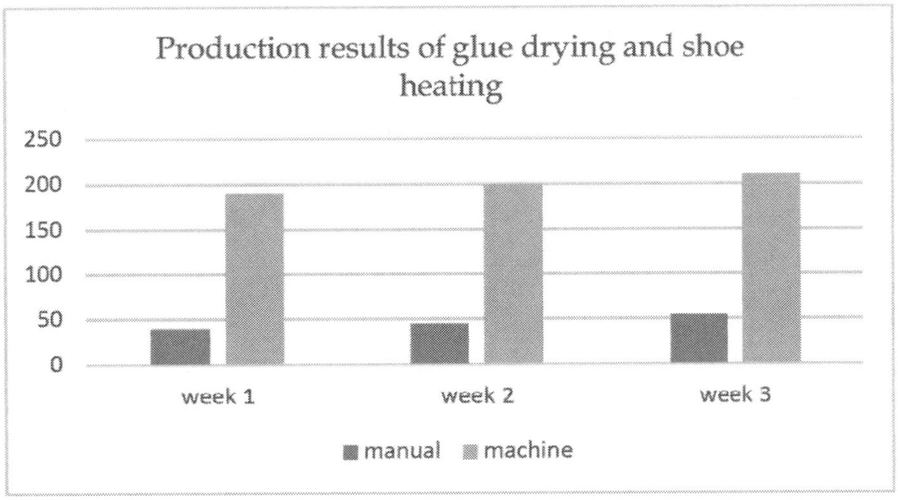

Fig. 5. Graph of shoe production comparison results

Based on Figure 5, it can be seen that the machine comparison test was carried out manually for three days. On the first day, the results obtained were 120 shoes manually while using the machine obtained a result of 450. On the second day, the results obtained were 145 using the manual tools owned while using the machine obtained a result of 420 shoes. The third day obtained a result of 150 using manual tools while using the machine obtained a result of 450 shoes. Based on these results, it can be seen that the use of manual drying equipment only obtained an average of 139 while the use of machines obtained quite stable results with an average of 440 shoes per day.

The oven machine can increase production significantly by almost three times compared to the manual process. The drying results using the tool can dry the glue evenly so that the bonding strength of the sole and upper is very strong. The automatic shoe dryer equipped with timer control and thermos control has a production capacity of 16 pairs of shoes per process. The time required for glue drying is three minutes while heating the leather takes 40 minutes. An automatic shoe-drying oven integrated with the Raspberry Pi cloud system is able to increase production capacity by five times compared to using the LPG gas stove heat dryer. The tool can increase production to meet demand from consumers who also do not forget the quality of the shoes [24]. Drying shoes using an automatic shoe-drying oven integrated with the Raspberry Pi cloud system shows that the drying results of sole and upper glue are very good and evenly distributed throughout the sole and upper parts. The results of drying shoes show that the drying shoe leather is not moist, tight, smooth, and shiny.

4 Conclusion

The design of the work system and mechanical system for the shoe-drying oven can work well. The Raspberry PI Cloud system can collect and display actual drying time data, glue drying temperature of 60°C with a drying time of five minutes, and skin heating temperature (texture creation) of 50°C-55°C requires a warm-up time of 40 minutes, This machine requires 850 watts of power. The automatic shoe-drying oven integrated with the Raspberry Pi cloud system can increase shoe production up to five times, especially in the drying process.

References

1. Ministry of Communication and Informatics, What is Industry 4.0 and how does Indonesia welcome it ? (Kominfo, Indonesia, 2019)
2. Coordinating Ministry for Economic Affairs Micro, small and medium enterprises (MSMEs) are a pillar of the Indonesian economy (Kemenko, indonesia, 2021)
3. G. Suja, S. Sundaresan, k. S. John, J. Sreekumar, R. S. Misra, Agron Sustain Dev **32** (2012)
4. J. M. Cheon, et al., e-Polymers **16** (2016)
5. L. Fransisca, S. Maulana, Kohesi: Jurnal Sains dan Teknologi **1** (2023)
6. D. J. Cho, Y. R. Jeong, *Design of smart shoe box based on IOT*, in 2017 Ninth International Conference on Ubiquitous and Future Networks (ICUFN), IEEE, 2017)
7. W. Lestari, R. Syarief, K. Sumantadinata, Manajemen IKM **8** (2013)
8. I. Mawardi, H. Hanif, Z. Zaini, Z. Abidin, CARADDE: Jurnal Pengabdian Kepada Masyarakat **1** (2019)
9. M. Syahbudi, Indonesia's creative economy: Strategy for the competitiveness of creative industry MSMEs towards going global (a research using the pentahelix model) (CV. Merdeka Kreasi Group, Mojosari, 2021)
10. A. K. S. Chauhan, A. K. Srivastava, Drying Technology **27** (2009)
11. Y. Shimazaki, K. Aisaka, *Novel thermal analysis model of the foot-shoe sole interface during gait motion*, in The 12th Conference of the International Sports Engineering Association 278, MDPI, 2018)
12. A. Joukhadar, L. Chachati, M. Al-Mohammed, O. Albasha, Cogent Eng. **7** (2021)
13. M. I. M. Copetti, J. Durany, J. R. Fernández, L. Poceiro, Appl Math Comput. **305** (2017)
14. R. Slamet, et al., Jurnal Manajemen Indonesia **16**, 136 (2017)
15. D. C. Eddy, S. Krishnamurty, I. R. Grosse, M. Steudel, Journal of Engineering Design **31** (2020)
16. Z. Wu, W. Liu, W. Zheng, W. Nie, Z. Li, Int J Prod Res. **59** (2021)
17. M. S. Gad, K. A. A. Yehia, A. A. Abdelhakeem, Fullerenes, Nanotubes and Carbon Nanostructures **26** (2018)
18. C. S. Thorne, A. Gatt, C. DeRaffaele, A. Bazena, C. Formosa, Gait Posture **105** (2023)
19. I. R. Dewi, H. Herminiwati, Majalah Kulit, Karet, dan Plastik **30**, 61 (2014)
20. A. E. Hassen, L. Pagliano, Results in Engineering **19**, 101373 (2023)
21. K. Žáková, M. Rábek, IFAC-PapersOnLine **51** (2018)

22. L. Zhong, Y. Tong, Z. Zhang, A. Du, Y. Wang, Procedia Eng **205** (2017)
23. I. Iswanjono, D. Widjaya, T. P. A. Setyani, T. Tjendro, P. K. Purwadi, Jurnal Pengabdian Masyarakat Indonesia **3** (2023)
24. M. Font-i-Furnols, L. Guerrero, Meat Sci. **193**, 108941 (2022)

Machine learning on academic education: Bibliometric studies

Hapsari Peni Agustin Tjahyaningtijas[1*], *Nanang* Husin[2], *Hasanuddin* Al Habib[3], *Asmunin* Asmunin[4], *Rindu Puspita* Wibawa[4], *Alim* Sumarno[5], *Jesse R* Paragas[6], and *Endang* Susantini[7]

[1] Department of Electrical Engineering, Universitas Negeri Surabaya, Surabaya, Indonesia.
[2]Department of Economic and Business, Universitas Negeri Surabaya, Surabaya, Indonesia.
[3]Department of Data Sains, Universitas Negeri Surabaya, Surabaya, Indonesia.
[4]Department of Informatics Engineering, Universitas Negeri Surabaya, Surabaya, Indonesia.
[5]Department of Educational Technology, Universitas Negeri Surabaya, Surabaya, Indonesia.
[6] Information Technology Department, College of Engineering, Eastern Visayas State University, Philippines.
[7]Department of Biology, Universitas Negeri Surabaya, Surabaya, Indonesia.

Abstract. The use of Machine Learning exhibits significant promise in facilitating advancements in the field of education. It is vital to conduct a comprehensive review of existing research to ascertain the significance of utilizing Machine Learning as a viable approach to enhance educational advancements. This bibliometric analysis provides a comprehensive overview of the advancements in the application of machine learning techniques within the field of education. This study utilizes publication and citation data from many academic literature sources to elucidate prominent patterns, areas of research emphasis, and scholarly collaborations within this field. The findings of the bibliometric analysis reveal a significant increase in scholarly attention toward the application of machine learning in the field of education during the past several years. The scope of these investigations encompasses a diverse array of subjects, such as personalized learning, predictive analytics, automated evaluation, learning recommendations, and online exam proctoring. The findings of this study also demonstrate a notable rise in the level of collaboration among scholars from many fields, highlighting the significance of interdisciplinary approaches in tackling the intricate challenges associated with the integration of machine learning in education.

1 Introduction

Machine Learning has advanced rapidly in recent years, allowing machines to learn from examples and experiences, establishing a foothold in the educational sector. On the other hand, Deep learning (DL) is a subfield of ML that focuses on the creation of artificial neural networks (ANN). These networks are comprised of algorithms that are modeled after the structure and function of the human brain [1]. In the area of education, machine

* Corresponding author: hapsaripeni@unesa.ac.id

learning is used worldwide. The application of machine learning in education has resulted in a significant shift in the way we teach and learn by analyzing individual student data and customizing materials and teaching methods to their needs, [2]. This not only allows teachers to give more relevant learning experiences, but it also aids in the prediction of student performance, the identification of factors that influence academic success, and the provision of timely interventions. Furthermore, machine learning is employed for automatic assignment and exam evaluation, saving teachers time and enabling deeper analysis of student progress. As a result, machine learning contributes to the creation of educational settings that are more flexible, effective, and responsive to individual needs [3].

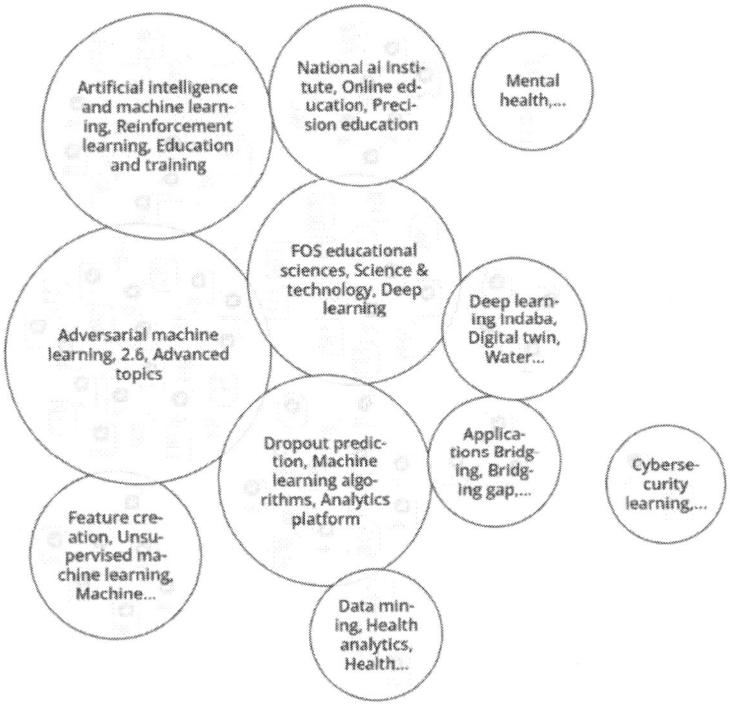

Fig. 1. The area of AI research on education using machine learning

The number of publications, citations, and collaborations between researchers in various disciplines related to education and technology continues to increase. This shows that machine learning has become an integral part of the development of modern education and is the main focus for many researchers who are trying to improve the efficiency and effectiveness of the learning process [4]. It can be seen that the importance of interdisciplinary collaboration in addressing the challenges and opportunities associated with the application of machine learning in educational contexts. For this reason, it is necessary to conduct high research both in quantity and quality and then publish scientifically to encourage discovery and utilization in the field of Machine Learning for Indonesian education. What also needs to be done is to map Artificial Intelligence and Machine Learning research, and then see how far progress related to education is. So, this research answers the following problems: the development of the number of international scientific publications in the field of machine learning from 2015 to 2023; the productivity level of machine learning researchers; and the progress of international publications for machine learning research based on keywords.

The examination of the growth of study in the utilization of Machine Learning holds significant importance for researchers [5]. This analysis is crucial for comprehending the current state of knowledge concerning the art of identifying future research that will be pursued. Through an analysis of the expansion of scholarly inquiry, academics can discern patterns and areas of investigation that necessitate further attention and exploration. In addition to this, researchers can gain insights into the global landscape of Machine Learning research in the education sector, enabling them to assess the relative standing of their research, ascertain its novelty, and identify potential collaborators and available funding opportunities worldwide. In general, monitoring the progression of research concepts is a crucial aspect of remaining informed about the most recent advancements in a certain discipline and propelling the advancement of knowledge through inventive research endeavors.

2 Methods

This study utilized data from international articles on the subject of Machine Learning in education obtained from renowned databases such as ResearchGate, Scopus, and Google Scholar. The data collection process involved utilizing Litmaps and open knowledge maps to search for publications. Machine Learning keywords were then applied to article title categories, abstracts, and keywords within the time frame of 2015 to 2023. However, it was discovered that the research was only documented in 2014. Consequently, data collection was extended from 2014 to 2023, with annual iterations, to ensure greater precision and comprehensiveness in the results. The visualization of development patterns and productivity of authors in the field of machine learning in education can be achieved through the utilization of VOSViewer, a tool commonly employed for such purposes in academic research.

3 Result and discussion

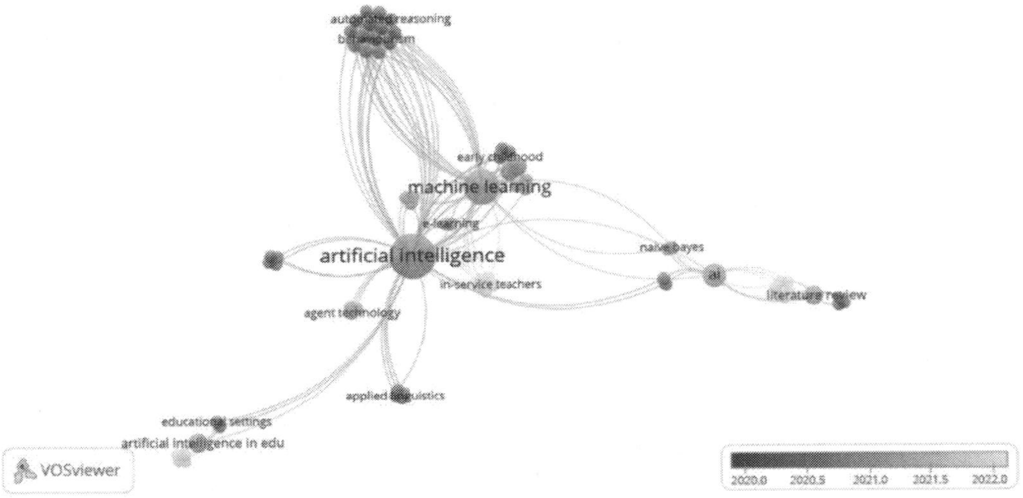

Fig. 2. Overlay visualization map of ML in education

Figure 2 depicts a visual representation of the various study domains where machine learning techniques are being employed within the education field. The visualization was generated by mapping the keywords extracted from the abstracts of 80 articles pertaining to the application of Machine Learning in the field of education, out of a total of 269 articles. The first figure is presented in Figure 2. Based on the analysis of the research landscape spanning from 2020 to 2022, it is evident that the use of Machine Learning exhibits a wide range of applications, encompassing domains such as learning, automated reasoning, applied linguistics, and behavior detection. The visualization reveals a correlation between node brightness and the relative size of the study area associated with each node. Specifically, brighter nodes indicate fewer research areas compared to darker-colored nodes.

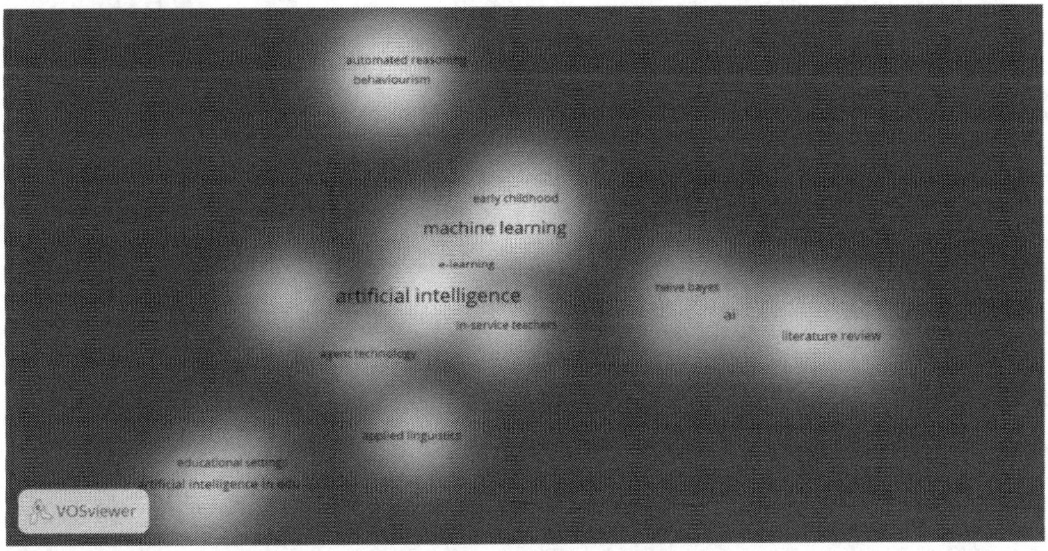

Fig. 3. Density Visualization of the research area of ML in education

Table 1. Number of articles published each year

Year	Number of Articles
2023	66
2022	53
2021	48
2020	23
2019	25
2018	16
2017	15
2016	12
2015	11

Table 2. Number of articles published according to the area of Machine Learning in education

Area of Machine Learning in Education	Number of Articles
Predictive analytic	66
Online exam proctoring	63
Automated evaluation	52
Personalized learning	45
Virtual Assistant	43

Table 1 displays the number of articles that have been acquired from the Scopus and Research Gate databases. The advancement of machine learning (ML) research in the education industry has been progressively growing over time. The 269 papers that were acquired may be classified into four distinct parts, specifically machine learning in the context of personalized learning, predictive analytics, automated evaluation, virtual assistants, and online examinations. According to the data presented in Table 2, it is evident that predictive analytics is the most prevalent area of machine learning research in the education sector, whereas virtual assistants exhibit the lowest level of research activity.

The integration of machine learning into personalized learning has had a profound impact on the educational domain. According to Guilherme (2019) [4], with the utilization of advanced data analysis techniques, machine learning algorithms can comprehend the individual preferences, abilities, and learning styles of each learner. This enables educators to develop customized learning plans that cater to the specific requirements of each student, encompassing the selection of appropriate materials, appropriate levels of challenge, and the implementation of the most efficient instructional strategies. In contemporary educational settings, there has been a shift away from mandating a consistent learning approach for students. Instead, students are now allowed to obtain educational content that is both pertinent and customized to their own level of comprehension. The integration of machine learning into personalized learning systems offers students an enhanced avenue for realizing their maximum educational capabilities. Simultaneously, this integration empowers teachers to deliver improved assistance and concentrate on addressing the individualized requirements of each student.

The utilization of machine learning in the field of predictive analytics has emerged as a highly potent instrument for enhancing operational efficiency and efficacy across many industries, including the realm of education. Machine learning algorithms are employed within the educational domain to derive valuable insights from a wide range of student data. This enables the anticipation of students' forthcoming academic achievements, the identification of elements that impact their scholastic outcomes, and the detection of prospective hazards, such as students who are susceptible to discontinuing their education [6]. Through the utilization of predictive analysis, educational institutions and educators are empowered to enhance their planning of interventions, provision of specialized support, and development of more efficacious educational programs. Machine learning, as evidenced by studies conducted by Tsiakmaki et al. [7] and Zanirato et al. not only enhances our comprehension of historical educational trends but also enables more accurate and efficient predictions of the future of education [8].

It is a significant advancement when machine learning is used for automated evaluation in a learning environment. Machine learning algorithms possess the capability to efficiently and precisely analyze vast quantities of student responses. According to Sanuvala and Fatima [9], this approach facilitates the effective evaluation of examinations and assignments, minimizing the time demands on teachers or tutors. Furthermore, it has been observed that machine learning algorithms can detect and analyze patterns and trends within students' responses, hence aiding educators in gaining insights into certain areas that may necessitate targeted instructional interventions [10]. Automated evaluation further facilitates the maintenance of consistency in scoring, a crucial aspect in upholding fairness and objectivity throughout the scoring procedure. The integration of machine learning into automated assessment systems enhances the efficiency of education, facilitates prompt feedback delivery to students, and enables teachers to dedicate their attention to the pedagogical and innovative dimensions of their instructional practices.

Machine learning algorithms are utilized to examine student data, encompassing their learning history, preferences, level of comprehension, and academic advancement [2]. Based on the provided information, the system possesses the capability to suggest supplementary educational materials, textbooks, online resources, or even courses that align with the student's interests and aptitudes [11]. This not only facilitates pupils in attaining a more pertinent and captivating learning encounter but also enhances the overall caliber of education. Learning recommendations derived from machine learning algorithms can effectively enhance student motivation by offering them suitable tasks that align with their individual ability levels. Therefore, this particular strategy possesses the capacity to revolutionize how pupils obtain and undergo the process of learning, thereby enhancing its level of interactivity and efficacy.

The application of machine learning in the context of virtual assistants has introduced novel approaches within the field of education. According to a study conducted by S.

Iqbal et al. [12], machine learning-based virtual assistants have the capability to engage in direct interactions with students, discern their requirements, and offer suitable recommendations for their learning endeavors. Singh and Jaiswal [13] assert that the utilization of machine learning techniques enables virtual assistants to enhance their recommendations iteratively by leveraging interactions with students and monitoring their learning progress. This not only enhances the individualization of the learning process, but also provides students with supplementary resources, reading materials, exercises, or materials that are pertinent to a specific subject matter. This particular form of virtual assistant can aid students in answering inquiries, offering clarifications, and providing personalized assistance, hence enhancing the accessibility and interactivity of the learning experience. According to Almusaed et al. [14], the utilization of machine learning techniques enhances the cognitive capabilities of this virtual assistant, hence facilitating a more optimized and proficient learning encounter for students.

A crucial method for preserving the integrity of online assessments is the use of machine learning. According to Kaddoura and Gumaei [15], the utilization of machine learning algorithms enables the surveillance of student activities during examinations, facilitating the identification of instances of academic dishonesty. These algorithms are capable of issuing alerts in response to potentially suspect behaviors, including but not limited to screen sharing, accessing unauthorized resources, or engaging in communication with other individuals. The utilization of this technology has the potential to enhance the efficacy and automation of online exam proctoring, hence alleviating the workload of human proctors and offering enhanced safeguards against academic dishonesty [16]. The utilization of machine learning in exam proctoring necessitates ethical and privacy deliberations, as well as the need for transparency to uphold fairness and foster student confidence in the assessment procedure.

4 Conclusion

A bibliometric analysis was conducted to examine the use of Machine Learning in the realm of education, to present a comprehensive perspective to academics of the growing body of research in this domain. The study of the 269 articles revealed that there was a significant focus on the application of machine learning (ML) in the domain of predictive analysis, indicating a high level of interest in this area. Conversely, the research pertaining to virtual assistants exhibited comparatively lower levels of interest among scholars and practitioners.

Acknowledgments

This study is a part of the Innovation Product Development Research Project, which is financially supported by the Directorate of Innovation Rankings and Scientific Publications, administered by the Institute for Research and Community Service Universitas Negeri Surabaya Indonesia.

References

1. J. Kufel, et al., Diagnostics **13**, 15 (2023)
2. K. Seo, J. Tang, I. Roll, S. Fels, D. Yoon, International Journal of Educational Technology in Higher Education **18**, 1 (2021)
3. E. M. Onyema et al., Comput Intell Neurosci. **2022** (2022)
4. A. Guilherme, AI Soc. **34**, 1(2019)
5. J. L. Rastrollo-Guerrero, J. A. Gómez-Pulido, A. Durán-Domínguez, Applied Sciences **10**, 3 (2020)
6. M. Hussain, W. Zhu, W. Zhang, S. M. R. Abidi, Comput Intell Neurosci **2018** (2018)
7. M. Tsiakmaki, G. Kostopoulos, G. Koutsonikos, C. Pierrakeas, S. Kotsiantis, O. Ragos, *Predicting University Students' Grades Based on Previous Academic Achievements*, in 2018 9th International Conference on Information, Intelligence, Systems and Applications (IISA), Zakynthos, Greece: IEEE Xplore (2018)
8. S. Zanirato, Maia, et al., World 2023 **4** 2 (2023)
9. G. Sanuvala, S. S. Fatima, *A Study of Automated Evaluation of Student's Examination Paper using Machine Learning Techniques*, in Proceedings - IEEE 2021 International Conference on Computing, Communication, and Intelligent Systems, ICCCIS, (2021)
10. A. Kashi, S. Shastri, A. R. Deshpande, J. Doreswamy, G. Srinivasa, *A Score Recommendation System Towards Automating Assessment In Professional Courses*, Proceedings - IEEE 8th International Conference on Technology for Education, T4E 2016, (2017)
11. Y. Yamasari, R. A. Ahmad, H. P. A. Tjahyaningtijas, A. Qoiriah, N. Rochmawati, A. Prihanto, *Improving the Performance of Classification via Clustering on the Students' Academic Performance using Stacking Algorithm*, in Proceedings of the 14th International Conference on Soft Computing and Pattern Recognition (SoCPaR) (2023)
12. M. Z. Iqbal, A. G. Campbell, *Touchless Hand Interaction and Machine Learning Agents to Support Kinesthetic Learning in Augmented Reality*, in SIGCSE 2021 - Proceedings of the 52nd ACM Technical Symposium on Computer Science Education (2021)
13. N. Singh, U. C. Jaiswal, *Analysis of Student Study of Virtual Learning Using Machine Learning Techniques,* International Journal of Software Science and Computational Intelligence, IJSSCI (2022)
14. A. Almusaed, A. Almssad, I. Yitmen, R. Z. Homod, Education Sciences **13**, 7(2023)
15. S. Kaddoura, A. Gumaei, Intelligent Systems with Applications **16** (2022)
16. D. R. E. Cotton, P. A. Cotton, J. R. Shipway, Innovations in Education and Teaching International **60** (2023)

Discrimination leading to social intolerance in senior high school "case study in Manado, Indonesia"

Yohanis Rantelangan[1*], *Darsono* Wisadirana[2], *Anif Fatma* Cahwa[2], and *Bambang Dwi* Prasetyo[3]

[1]Doctoral Program Department of Sociology, Faculty of Social and Political Sciences, Brawijaya University, Malang, Indonesia
[2]Department of Sociology, Faculty of Social and Political Sciences, Brawijaya University, Malang, Indonesia
[3]Department of Communication, Faculty of Social and Political Sciences, Brawijaya University, Malang, Indonesia

Abstract. Discrimination behaviours frequently occur in a school setting. The establishment of discriminatory behaviours is due to a variety of factors, including those present in educational settings. Using a descriptive qualitative method, the researcher finds out the discrimination sample as the impact of class differences between special target classes and regular classes at Senior High School Manado. Researchers analysed the psychological and social impact of these differences, especially concentrating on how these differences can lead to discrimination and social intolerance among students. The environment of school, teachers, and the influence of social stratification cannot be separated from the existence of discrimination that breeds social intolerance. Researchers explain how inequality in access to quality education, differences in learning approaches, and social norms applied in the classroom can lead to misunderstanding, stereotypes, and conflicts among students. In addition, this article emphasizes the importance of inclusive education as a solution to overcome differences and reduce the harmful effects of class differences in schools. This research provides a recommendation to the government in education to understand this phenomenon that will enable us to create a more just and inclusive educational environment, and encourage students to be tolerant, which will help achieve for equal education for all.

1 Introduction

One of the most common ways that human rights are infringed and exploited is through discrimination, in all of its forms and manifestations. Discrimination is one of the hardest things to recognize because it affects millions of people every day. The concepts of discrimination and social intolerance are related. Social intolerance is the lack of respect

* Corresponding author: rantelanganj@gmail.com

for customs or views that differ from one's own. It also involves not liking others who are perceived to be like us, such as those who have a different political or sexual orientation from us or who come from a different social or cultural group. Another way social intolerance can show out is through impolite speech, the creation of one's own organizations, and other actions that impede inclusivity.

The government is concerned with preventing social intolerance through education. Education has a significant impact on how people behave and how society functions. However, the reality is that discrimination does occasionally exist in the educational setting. High school is a crucial stage in education since it marks the beginning of students' journey to adulthood and their ability to select more focused academic or career courses. Unfortunately, discrimination is still a problem in the field of education, particularly when it comes to treating students unfairly or discriminating against them based on their social standing, ethnicity, religion, gender, or academic standing.

Discrimination based on distinctions between the usual class and the special target class (Binsus) is a factor that may have an effect on the learning environment in one of the senior high schools in the city of Manado. This may result in inequality, low self-esteem, and decreased academic performance. The phrase "Educating the life of the nation" represents the ideals of the Indonesian nation to educate and disseminate education throughout Indonesia in order to establish an intelligent national life, which is opposite to the action taken. The objectives of multicultural education can also be seen as a learning strategy that makes use of a variety of student backgrounds and personalities [1].

One of the high schools (SMA) in the city of Manado has its own charm by making a class called the Special Assisted Class (Binsus) specifically for students with better academic abilities. Students who are able and have better educational opportunities are often associated with superior classes or special assisted classes (Binsus). Regular classes are often filled with students with diverse backgrounds and abilities and often have more limited resources. On the other hand, they can enjoy a more challenging curriculum, additional support and better facilities. This difference can lead to social differences that are detrimental to academics and have the potential to cause greater social differences.

A just and inclusive society, where everyone has an equal chance to realize their potential, is seen to be best built through education. Policy-makers and researchers urge educators to embrace student diversity and adjust their instruction to the learning needs of the various kids in their classes in order to maximize each student's potential for learning [2]. However, the education gap remains a troubling problem. In the context of education, the phenomenon of class differences, especially between the regular class and the special target class (binsus) or what is often called the superior class, often shows inequality in access to educational resources and social dynamics which can lead to discrimination and even inequality in schools.

Discrepancies in schools not only lead to differences in learning experiences between students in regular classes and students in special target classes; they can also cause conflict, negative stereotypes, and even discrimination among students in contrast to pessimistic perspectives, which anticipate unchanged or decreasing levels of tolerance, optimistic perspectives forecast rising levels of tolerance [3]. Students high-achieving

 https://doi.org/10.1051/e3sconf/202345003001

grades may feel valued more than regular graders. This can damage the school environment, cause disagreements, and prevent good social relations between students.

Therefore, understanding the background of class differences and their consequences is very important. To begin, researchers will discuss the importance of studying this issue, analysing the components that contribute to differences, and studying the social and psychological impacts of these inequalities. In addition, researchers will discuss the importance of inclusive education to overcome differences and reduce the adverse effects of class differences in schools. Having a better understanding of this phenomenon will enable us to create a more just and inclusive educational environment, encourage students to be tolerant, and realize the vision of an equal education for all.

2 Method

In addition to harming universal education, disparities between pupils in classrooms and schools can also increase discrimination and foster social intolerance. There are many social repercussions, such as unfavourable learning conditions for children, poor academic results, a gap between students, and many others, could take place. Case studies with qualitative methods are used by researchers to collect and field the data on the problems that occur and analyse them. Questions about students' experiences, views, and attitudes towards classmates became part of the survey in this study. This will help us understand how well we understand students, their opinions, and possible social intolerance. Researchers conducted interviews with informants to better understand their perspectives on inequality, social interactions, and how these elements affect discriminatory behaviour or attitudes. Researchers can examine the dynamics of student relationships in these two classrooms, find spots of tension or gaps, and examine the effects of these differences with special target classes (Binsus) and regular classes.

3 Result and discussion

3.1 Parental involvement with children

Most of the students place a high priority on their education. School should be a place where they interact and build bonds with their classmates and also a place for giving them opportunities for their cognitive abilities. Students need the sense to get this opportunity and also need stability in their education. The positive relationship between students and their teachers, including their school, which consists of sharing characteristics, will create a strong sense of belonging between one and the other, including the school where students study.

Students' impressions of their teacher's justice in acting are another aspect that was found to be related to good student development in the school environment. The actions taken by the teacher are proven to have an effect on the emotional development of students in the classroom. In addition, school is the most important part that will influence students to determine their choices. Most parents will determine which school based on their observations and school achievements will be the best choice for their children to study. Although the transition of students to high school is not as fast as usual, some children are still under the supervision of their parents. Parents should be aware of the challenges their child may face during the school transition and be prepared to offer additional help and direction while it is needed.

When a child enters adolescence, for the first time, physical development, cognitive capacity, and social expectations align. This ensures the systematization/synthesis of the "acquisitions" that feed into self-identity (childhood identification), an earlier age, and increase to maturity. This is why parents of students should still be involved [4]. Erikson (1968) [5] describes adolescence as a transitional stage between childhood and adulthood. This is also the last and most complex stage of major development [6]. In order to relate to others, identity-building and moral development occur [7]. All youth must begin to develop the emotional skills needed to manage stress and relate to others effectively. "Emotional intelligence" is the term used to describe this ability [8].

3.2 Sentiment and discrimination

An absolutist perspective that quickly results in segregation and dissent is produced by the failure to create a tolerant and reasonable framework. To plant bigotry and social intolerance, like maize, education units require an adequate water supply. This fact has considerably aided in the growth of radicalism, the dissemination of social intolerance, and the rise of prejudice that now uses internet platforms to target the millennial population.

If this condition is linked to multiple bullying events that have recently generated widespread discussion, it is quite concerning. Schools that are meant to instil good morals have changed. A child's school, which serves as a second home, is powerless to stop the feelings that frequently result in discrimination. In other words, schools cannot be used to teach people how to feel secure and at ease. There are still significant difficulties that need to be tackled in order to prevent harassment and discrimination against students, teachers, and school personnel.

The National Education System (Sisdiknas) Nomor 20 Tahun 2003, states that the administration of education in schools must be democratic, fair, and non-discriminatory while maintaining national plurality, religious freedom, and human rights. Therefore, it is the responsibility of educational institutions to promote and educate variety.

There are a number of behaviours that show social intolerance, such as disregarding other people's rights; discrimination or persecution of individuals due to their ethnicity, religion, gender, or sexual orientation; interfering with another person's right to follow a particular group, religion, or political ideology; not wanting to get along with or acting badly around others who hold different opinions; hate and inflict harm on those who hold opposing thoughts or viewpoints; Concentrate on or believe they are a member of their own group [9].

Some examples of Actions that refer to sentiments that end in discrimination such as: (1) Bullying motivated by certain characteristics: Students who identify with a particular group may be subjected to verbal, physical, or psychological abuse as a result of sympathy for that group, such as racial or sexual orientation. (2) Unfair educational treatment: If kids exhibit certain characteristics, teachers or other school personnel may have a bad opinion of them. This may have an impact on how teachers interact with their pupils, the opportunities for learning, and academic evaluations. (3) Inclusion in school activities: If students or staff members believe that having unfavourable attitudes against particular groups is acceptable, this may result in such groups being excluded from social activities, athletics, the arts, and other school events. (4) Unjust penalties or punishments: For the same action, some students may experience harsher sanctions or punishments than other students. (5) Inadequate psychological assistance Students who have unfavourable opinions of particular groups may feel awkward or alone, which could impede their emotional and social development. (6) Uninclusive curriculum: Ignoring diversity or the contributions of particular groups might result in negative attitudes in those communities.

By encouraging equality, inclusivity, and respect for diversity, schools must deal with the feelings that breed prejudice. This can be accomplished with the help of fair school regulations, staff training in cultural awareness, anti-bullying programs, inclusive curricula, and psychological support for kids who might feel prejudiced.

3.3 Discrimination from social stratification

When a society's social and economic structure is represented in the educational environment, it is known as the grievances produced by stratification phenomena. This phenomenon results in unequal educational opportunities, access, and experiences for pupils. This effect is referred to as school stratification-related distraction.

Students from various social and economic backgrounds may have varying educational possibilities as a result of inequality in the distribution of educational opportunities in the form of stratification. Access to necessary educational resources like textbooks, computers, and extra courses may be restricted for students from low-income households. Along with equal opportunities Class Separation Based on Achievement is a type of stratification where students are put into classes based on their academic performance, which can lead to inequity. The chances for students who are enrolled in classes with a less rigorous curriculum to reach their full potential and take advantage of higher educational opportunities may be diminished.

In reality, the school serves as both the coordinator of the curriculum and the most crucial factor in the growth of student accomplishment. Pupils may feel undervalued or underrepresented if the curriculum is not inclusive and does not reflect the cultural diversity and experiences of the pupils. In the context of the school, this may foster prejudice and preconceptions. Social stratification outside of school can result from environmental influences, and this can have a significant impact on the learning environment. Student self-perception and educational goals, for instance, may be influenced by peer pressure and social pressure, but not just the pupils who are picked out for hate and bias because of their identity suffer from a toxic school environment [10].

The theory of social stratification is fundamental to the study of inequality. Social stratification in education is evident in how distinct student groups are handled differently. According to this study, the disparity in treatment between pupils in ordinary classes and those in special target classes is an example of the social stratification that occurs in schools (Binsus). Even though they are in the same environment as other classes, most kids in the superior category who are part of the special target class category (binsus) are constantly the centre of attention.

The distribution of students who are part of special assistance classes (binsus) is fundamentally based on social stratification. This issue cannot be disregarded in schools. In order to protect pupils from harm, educators must actively and proactively fight prejudice and advance equality and inclusiveness. They must take swift and immediate action to address any instances of bias and hatred when they arise, using a paradigm that prioritizes support for people who have been affected, dialogue, empathy, and reconciliation [6].

The attainment of universal education depends on pupils feeling appreciated by their peers. There are solid arguments in favour of the idea that character development depends significantly on one's sense of self. At the very least, having a strong sense of self-worth should prevent kids and teenagers from turning to unconventional, antisocial methods of getting attention. Children who develop a morally-based sense of self-worth are less likely to participate in antisocial activity intended to win the approval of peers who exhibit antisocial motivations and behaviours. Moreover, demonstrated that influences accompanying positive self-evaluations can encourage moral behaviour. Children who feel good about themselves when they act in ways that are consistent with their moral ideals tend to experience increased motivation to act morally.

Special target classes or binsus classes, compared to regular classes can be considered a form of social stratification in schools, which can lead to discrimination and social intolerance among students because students in special assistance classes can be stigmatized and considered inferior to students in regular classes. This can perpetuate social inequality and strengthen existing power structures, because students in Special Assistance classes may have more opportunities and advantages than students in regular or regular classes. Educators cannot be excluded from implementing policies on the separation between special target classes and regular classes. Education-related bias and discrimination also have long-term consequences. "Children who experience teacher discrimination are more likely to have negative attitudes about school, lower cognitive motivation and achievement, and are at higher risk of dropping out of school, especially for students who join regular classes". For educators, social stratification can be seen in the differences in the treatment of various groups of students. In this paper, the social stratification that exists in schools is exemplified by the differences in treatment that exist between regular-class students and special target-class students (Binsus). But schools are not hermetically sealed institutions. They are not immune from the political and socioeconomic forces gripping our nation [11].

4 Conclusions

This essay tries to demonstrate how disparities in educational treatment can have a substantial impact on social, psychological, and educational development as well as open the door for the establishment of social intolerance in the form of prejudice by drawing on the idea of social stratification. To ensure that every student has an equal chance to

 https://doi.org/10.1051/e3sconf/202345003001

develop and succeed, efforts to close these inequalities must be founded on the principles of equality and inclusion. According to this, making a distinction between the special target class and the ordinary class increases discrimination and makes students more intolerant. It is seen that this stratification has negative impacts, such as stigma, social intolerance, and academic discrepancies. The results of this research can provide an overview of the importance of dealing with this difference in treatment and pushing towards a more inclusive and also education system. Solutions based on inclusion and equity emerge as the right way to tackle this problem and create a more equitable and inclusive educational environment. These issues cannot be simply ignored in schools. In order to protect students from harm, educators must act forcefully and proactively to combat prejudice and advance fairness and inclusiveness. When hate and bias incidents occur, they must respond immediately and aggressively using a strategy that prioritizes dialogue, empathetic understanding, atonement, and assistance for those who have been hurt.

References

1. S. Nieto, Affirming Diversity: The Sociopolitical Context of Multicultural Education (Longman, New York, 1991)
2. Unesco. A guide for ensuring inclusion and equity in education (UNESCO Digital Library, 2017)
3. J. G. Janmaat, A. Keating, Are today's youth more tolerant? Trends in tolerance among young people in Britain (UCL Institute of Education, UK, 2019)
4. V. Albani, L. T. Butler, W. B. Traill, O. B. Kennedy, Appetite **120** (2018)
5. E. H. Erikson, Identity: Youth and Crisis (W. W. Norton & Company, US, 1968)
6. S. M. Sawyer, P. S. Azzopardi, D. Wickremarathne, G. C. Patton, The lancet child & adolescent health **2**, 3 (2018)
7. J. V. Jordan, A relational perspective on selfesteem (Wellesley, MA: Stone Center, 1994)
8. L. M. Bru-Luna, M. Martí-Vilar, C. Merino-Soto, J. L. Cervera-Santiago, Healthcare **9**, 12 (2021)
9. B. S. Eko, H. Putranto, Journal of Intercultural Communication Research **48**, 4 (2019)
10. M. Costello, C. Dillard, American Educator **43**, 3 (2019)
11. Mbilishaka, M. Afiya, Apugo, Danielle, Race Ethnicity and Education **23**, 5 (2020)

Trends research of para-cycling: A bibliometric analysis

Fadilah Umar[1*], *Misbah* Misbah[2], *Ulfa Fatahara Laras* Fadian[1], and *Abdul Aziz Purnomo* Shidiq[1]

[1]Faculty of Sport, Universitas Sebelas Maret, 57126 Surakarta, Indonesia
[2]Faculty of Teacher Training and Education, Universitas Lambung Mangkurat, 70123 Banjarmasin, Indonesia

Abstract. This bibliometric analysis research aims to map research on para-cycling. Based on search results in the Scopus database with the keywords "para cycling" OR 'para-cycling', 47 documents were obtained. The visualization of the data using the VOSviewer. The number of publications increased in 2016-2018 and 2019-2021. The largest number of documents was published in 2021. Much research on this topic is published in Q1 journals, while Elsevier publishes many. The Netherlands, as the most productive country, has done a lot of research on this topic. Authors who publish many articles in this field are dominated by writers from Ireland, such as Clifford, E.; Mannion, P.; and Hajdukiewicz, M. Affiliates with the largest number of publications come from Belgium, such as Université de Liège and KU Leuven. Four clusters were obtained on this topic based on the VOS viewer visualization results. Research on this topic in the last two years included cyclist, race, muscle strength, physiology, and athletic performance. This study shows that research on this topic is still scarce and is expected to continue to grow in the coming years. This study can be used as a basis for further research on this topic.

1 Introduction

Para-cycling is a discipline in high-level competitive sports [1]. The Union Cycliste Internationale (UCI) oversees para-cycling and establishes classes depending on the cyclist's capability [2]. The third-largest Paralympic sport is para-cycling, which is managed by the International Cycling Federation. [3]. Para-cycling is the sport of utilizing bicycles, tandems, tricycles, and handbikes to compete the track and on the road by people with disabilities [4]. Competitors in para-cycling are categorized according to functional disability, so riders with neurological and locomotor disabilities compete against one another [5]. Objective Para-cycling classification aims to generate fair competition [6].

* Corresponding author: fadilahumar@staff.uns.ac.id

However, discussion about para-cycling is still limited. Some only research on aspects of epidemiological data regarding the injuries it causes [1], a para-cycling classification system [7], and sports nutrition [8]. No one has yet conducted a comprehensive analytical review of para-cycling.

Therefore, to get a comprehensive picture of para-cycling, a bibliometric analysis was carried out based on research that has been published in the Scopus database. This bibliometric analysis research aims to map research on para-cycling from 2011-2023*. The research questions are as follows: (1) What is the trend in the number of para-cycling publications from 2011-2023*? (2) Which top ten journals publish the most articles about para-cycling?, (3) Which top ten countries do the most? Who published articles on para-cycling? (4) Who are the top ten authors on this topic in terms of the number of citations and the number of documents?, (5) Which top ten affiliates published articles on para-cycling?, and (6) How are the visual network and overlay of this para-cycling topic based on co-occurrence?

2 Methods

This study is a bibliometric analysis [9]. The reference search was conducted on August 21, 2023, and obtained 47 documents in the Scopus database using the keywords "para-cycling" OR "para-cycling". Based on the 47 documents obtained, consisting of 89% or 42 articles and 11% or five reviews. Then, the data is saved in CSV form, and analysis and visualization are carried out using VOSviewer.

3 Results and discussion

The following number of publications each year on the topic of para-cycling is presented in Figure 1.

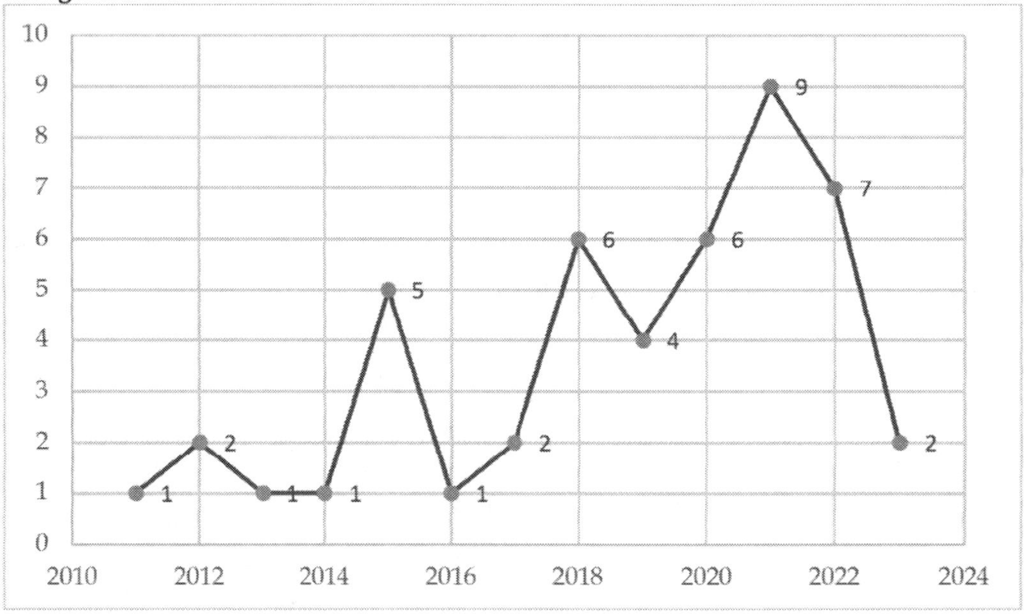

Fig. 1. Number of publications on para-cycling per year.

Figure 1 shows increased publications in 2016-2018 and 2019-2021. The highest number of documents was published in 2021. This is in line with studies in the discipline of Adaptive Physical Education, which also experiences an increase in publications yearly [10]. The number of publications on this topic will likely increase because there is little research discussing para-cycling. This has the opportunity to be researched further and in-depth.

The top ten journals on para-cycling are listed in Figure 2.

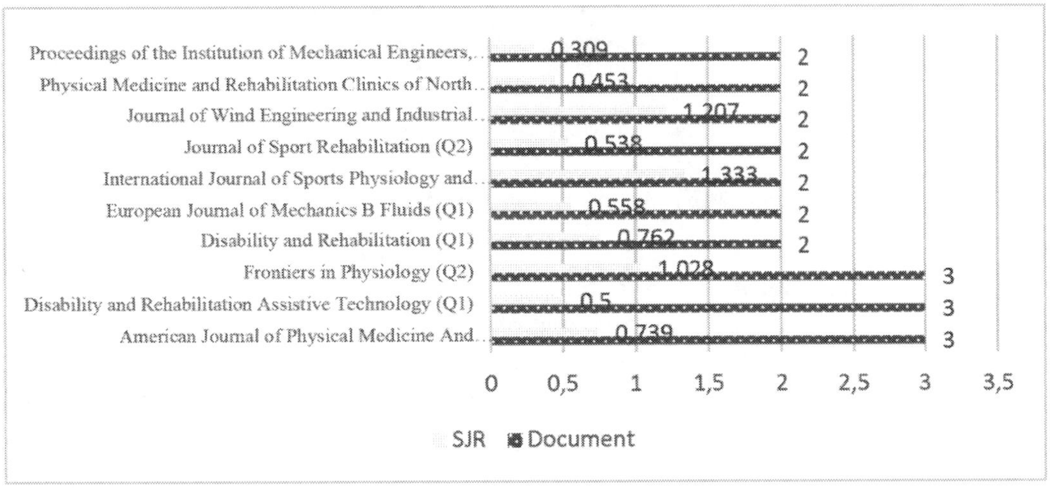

Fig. 2. Top 10 journals that publish research on para-cycling.

Figure 2 shows that research on this topic is mostly published in Q1 journals, while most publishers are Elsevier. In three articles in the American Journal of Physical Medicine and Rehabilitation, para-cycling is discussed. One of the articles maps the effects of lower extremity diseases on para-cycling performance by examining the substantial correlation between isometric and dynamic strength with sprint power and race pace. Potential for para-cycling classification using MMT and its isometric and dynamic metrics [6]. As for the articles published by publisher Elsevier, one of them discusses aerodynamics in cycling science. This is important because aerodynamic clothing and equipment, athlete posture, and racing strategy utilize scientific aerodynamics knowledge [11].

Top ten countries publish considerably on para-cycling, as presented in Figure 3.

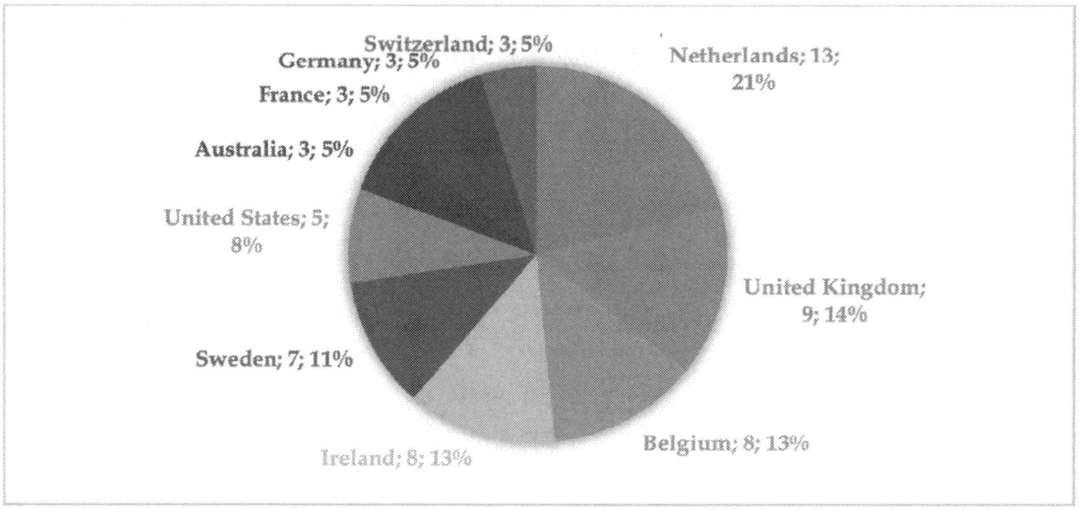

Fig. 3. Top ten countries publish considerable about para-cycling.

Figure 3 demonstrates that, according to numerous studies on this subject, the Netherlands is the most productive nation. Like crosswinds' effects in tandem aerodynamics studied by Mannion et al. [11]. When cyclists ride at a non-zero angle to the wind, crosswinds happen. Aerodynamic features associated with cyclists' leg rotation new light on flow structure and drag trends [11]. Effects of the torso angles of the pilot and stoker on the aerodynamics of tandem para-cycling [12]. Top ten authors in the field of para-cycling based on the numbers of citations are presented in Table 1.

Table 1. Top ten authors in the field of para-cycling based on the numbers of citations.

Authors	Q	Cited by	Publisher
West et al. [13]	Q1	31	Wiley-Blackwell
Mannion et al. [14]	Q2	26	Springer London
Krassioukov et al. [15]	Q1	24	Nature Publishing Group
Mannion et al. [16]	Q1	22	Elsevier B.V.
Mannion et al. [17]	Q1	20	Elsevier B.V.
Flueck [18]	Q2	18	Frontiers Media S.A.
Dyer [19]	Q2	17	SAGE Publications Inc.
Zeller et al. [20]	Q1	15	Rehabilitation Research and Development Service
Mannion et al. [11]	Q1	12	Elsevier Ltd
Stephenson [21]	Q1	12	Blackwell Munksgaard

Based on Table 1, it is found that the authors whose articles are widely cited come from international journals indexed by Scopus with the Q1 quartile [13], [15] and are dominated by the publisher Elsevier [11], [14], [22].

Top ten authors in the field of para-cycling based on the numbers of documents are presented in Figure 4.

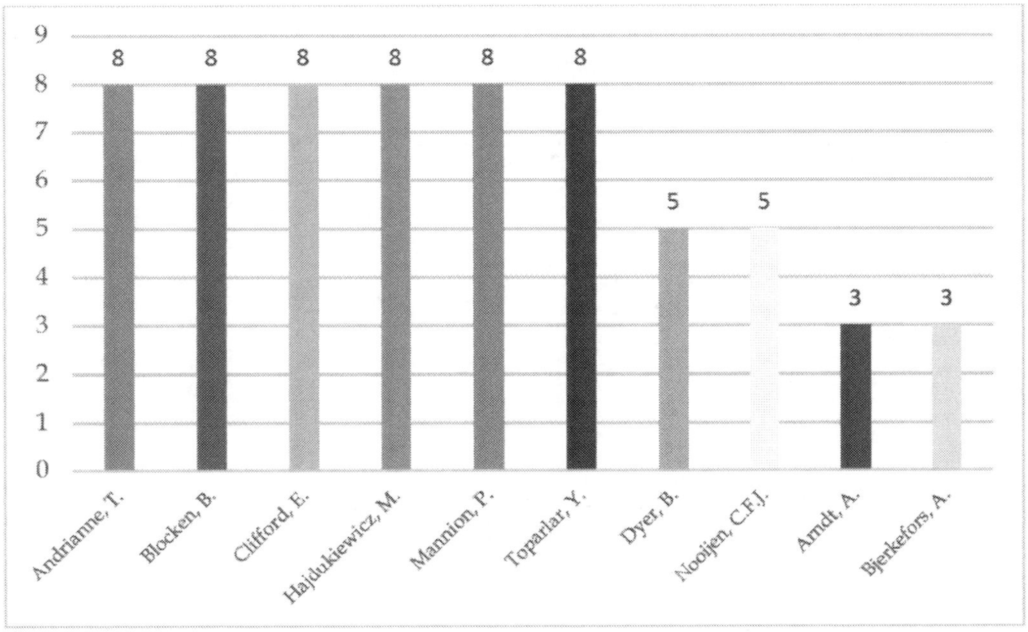

Fig. 4. Top ten authors in the field of para-cycling based on the numbers of documents.

Figure 4 shows the top 6 authors with each 8 (eight documents) majority from Ireland, such as Clifford, E.; Mannion, P.; and Hajdukiewicz, M [14], [16], [17].

The top ten affiliations in this topic are listed in Figure 5.

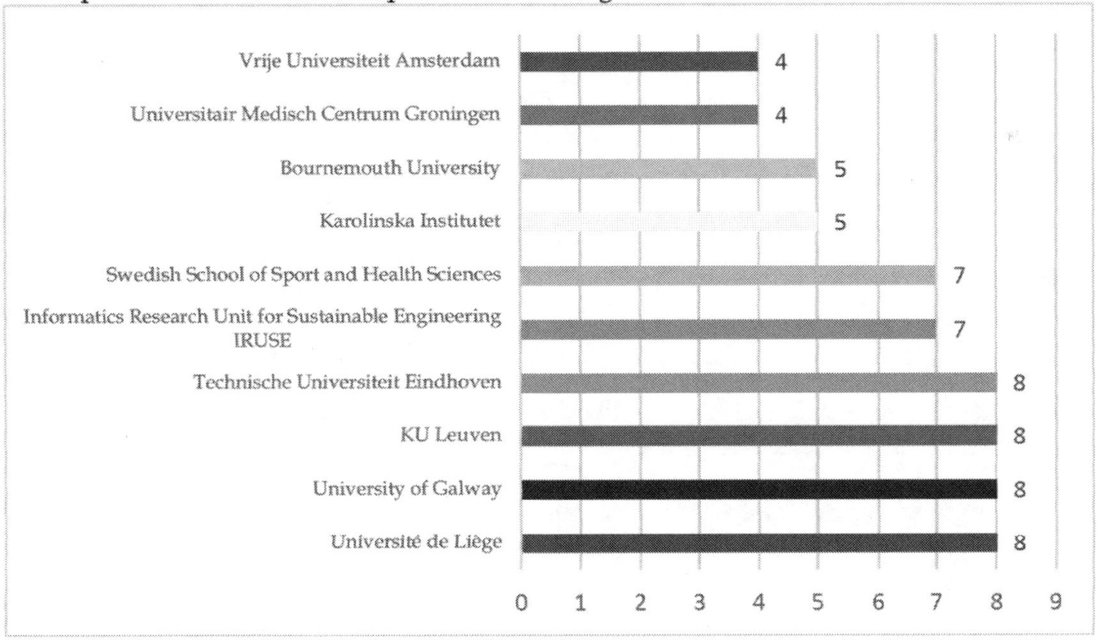

Fig. 5. Top ten affiliation in this topic.

Figure 4 shows Affiliates with the largest number of publications coming from Belgium, such as Université de Liège and KU Leuven. One of the authors from KU Leuven is Mannion et al, who have done a lot of research in para-cycling [14], [16], [17]. The results of the network and overlay on para-cycling are presented in Figures 6 and 7.

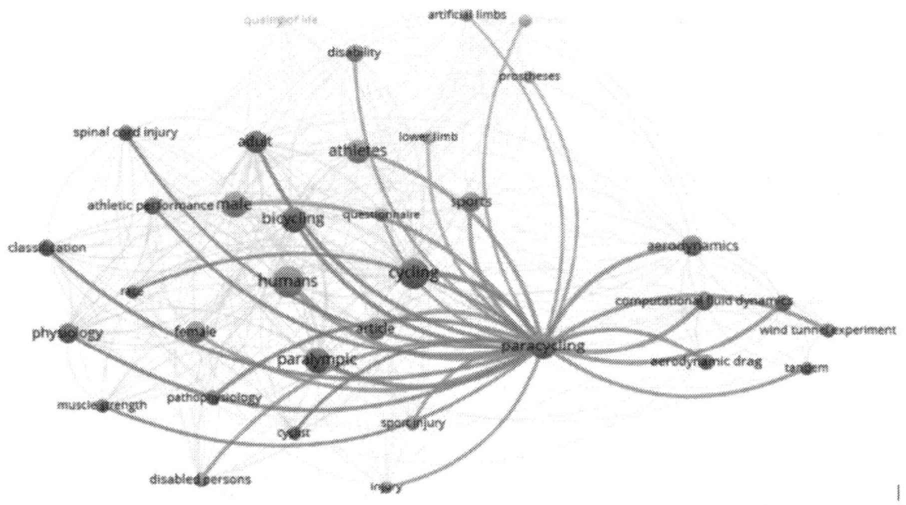

Fig. 6. Network visualization for para-cycling.

Figure 6 shows four clusters characterized by blue, red, green, and yellow. Each cluster shows the para-cycling article numbers shown in Table 2.

Table 2. Research development for each cluster.

No	Cluster	Number of Items	Keywords
1	Red	13	Adult, Athletic performance, bicycling, classification, cycling, cyclist, muscle strength, paralympic, pathophysiology, physiology, race, spinal cord injury
2	Green	12	Article, athletes, disability, disabled persons, female, humans, injury, male, quality of life, questionnaire, sport injury, sports
3	Blue	8	aerodynamic drag, aerodynamics, computational fluid dynamics, para-cycling, tandem, wind tunnel experiment, wind tunnels
4	Yellow	7	artificial limbs, limb prosthesis, lower limb, prostheses

There is cluster 1 discusses a lot about the Paralympic and classification in cycling, such as the identification of performance characteristics associated with the classification of visual impairment in para-cycling [23]. Cluster 2 discusses a lot about athletes, disabled persons, and injury. In recent years, the importance of standardized monitoring systems for injuries and illnesses in elite athletes has been emphasized. Although the training and competition load for athletes with disabilities is increasing, documentation regarding their injuries and illnesses remains scant [24]. Cluster three discusses a lot about the-cycling, tandem, dan aerodynamics. Trajectory measurements, wind tunnel tests, and numerical models using computational fluid dynamics can be used to examine tandem aerodynamics [14]. Cluster 4 discusses a lot about prostheses. Aerodynamic enhancement of cycling prostheses is now possible with the goal of enhancing para-cycling performance [25].

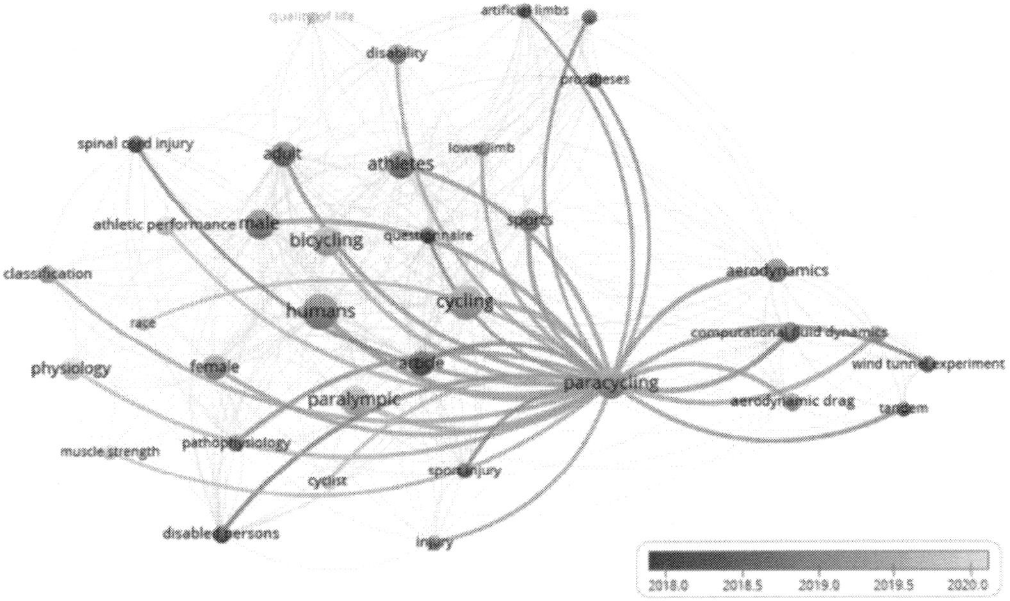

Fig. 7. Overlay visualization for para-cycling.

Figure 7 shows that the topic of para-cycling in the last 3 years is related to cyclists, race, muscle strength, physiology, and athletic performance. There were no differences between sprint power and road time trial performance across cyclists, based on impairment type or bike type. For those competing on a bicycle, tricycle, or hand cycle while lying down or kneeling, the sprint test may be useful for predicting or monitoring time trial performance [26]. Several studies looked at the performance of tricycles (class T) and bicycles (class C) at road race competitions held by the International Cycling Federation between 2011 and 2019 [27], as well as the performance of 427 cyclists who took part in the Championship in the 1 km tandem race. The Paralympics and World Para-cycling [23]. Based on the results of the analysis of the articles reviewed, there is still little research on para-cycling in Indonesia. This could be an opportunity to research para-cycling in Indonesia related to coaching para-cycling athletes, improving athletes' performance, etc.

4 Conclusions

Based on results and discussion, it can be concluded that there are four clusters were obtained on this topic based on the VOS viewer visualization results namely: (1) cluster 1 discussing a lot about Paralympic and classification in cycling; (2) cluster 2 discusses a lot about athletes, disabled persons, and injury; (3) cluster 3 discussing a lot about the-cycling, tandem, dan aerodynamics; and (4) cluster 4 discussing a lot about prostheses. Research on this topic in the last two years included cyclist, race, muscle strength, physiology, and athletic performance. This study shows that research on this topic is still scarce and is expected to continue to grow in the coming years.

References

1. P. Kromer, K. Röcker, A. Sommer, H. Baur, L. Konstantinidis, A. Gollhofer, and A. Hirschmüller, Sport. Sport. Organ Der Gesellschaft Fur Orthopadisch-Traumatologische Sport. **25**, 167 (2011)
2. B. Dyer, Prosthet. Orthot. Int. **40**, 538 (2016)
3. P.M. Leprêtre, T. Weissland, J. Slawinski, and P. Lopes, Front Physiol **3**, 327 (2012).
4. Y.N. Hanief and F. Umar, Adv. Rehabil. **34**, 37 (2020)
5. R.L. Wright, Front. Physiol. **6**, 425 (2016)
6. A. Liljedahl, J. B., Arndt, A., Nooijen, C. F., & Bjerkefors, Am. J. Phys. Med. Rehabil. **102**, 461 (2023)
7. J.B. Liljedahl, Disabil. Rehabil. **43**, 3440 (2021)
8. J.L. Flueck, Sports **9**, 1 (2021)
9. I. Zupic and T. Čater, Organ. Res. Methods **18**, 429 (2015)
10. F. Umar, M. Misbah, F.F. Ekawati, and Y.N. Hanief, J. Phys. Educ. Sport **22**, 2996 (2022)
11. P. Mannion, Y. Toparlar, E. Clifford, M. Hajdukiewicz, T. Andrianne, and B. Blocken, Eur. J. Mech. **74**, 68 (2019)
12. P. Mannion, Y. Toparlar, B. Blocken, M. Hajdukiewicz, T. Andrianne, and E. Clifford, Sport. Eng. **22**, 1 (2019)
13. C.R. West, C.M. Gee, C. Voss, M. Hubli, K.D. Currie, J. Schmid, and A. V Krassioukov, Scand. J. Med. Sci. Sports **25**, 476 (2015)
14. P. Mannion, Y. Toparlar, B. Blocken, M. Hajdukiewicz, T. Andrianne, and E. Clifford, Sport. Eng. **21**, 123 (2018)
15. A. Krassioukov, J.J. Cragg, C. West, C. Voss, and D. Krassioukov-Enns, Spinal Cord **53**, 78 (2015)
16. P. Mannion, Y. Toparlar, B. Blocken, E. Clifford, T. Andrianne, and M. Hajdukiewicz, J. Wind Eng. Ind. Aerodyn. **179**, 92 (2018)
17. P. Mannion, Y. Toparlar, B. Blocken, E. Clifford, T. Andrianne, and M. Hajdukiewicz, J. Wind Eng. Ind. Aerodyn. **180**, 182 (2018)
18. J. Flueck, Front. Nutr. **7**, (2020)
19. B. Dyer, Prosthet. Orthot. Int. **39**, 232 (2015)
20. S. Zeller, T. Abel, P.M. Smith, K. Strueder, Influence of noncircular chainring on male physiological parameters in hand cycling (National Center for Biotechnology, 2022)
21. B.T. Stephenson, Scand. J. Med. Sci. Sport. **31**, 4 (2021)
22. P. Mannion, Y. Toparlar, E. Clifford, M. Hajdukiewicz, T. Andrianne, and B. Blocken, Comput. Methods Biomech. Biomed. Engin. **22**, 386 (2019)
23. G.B. Lima, R.L. Kons, D. Detanico, and G. Fischer, Am. J. Phys. Med. Rehabil. **100**,

1190 (2021)
24. E.J. Kubosch, K. Fassbender, K. Steffen, N. Südkamp, and A. Hirschmüller, Br. J. Sports Med. **51**, 347 (2017)
25. B. Dyer, Disabil. Rehabil. Assist. Technol. **15**, 166 (2020)
26. C.F. Nooijen, R. Muchaxo, J. Liljedahl, A. Bjerkefors, T. Janssen, L. Woude, and S. Groot, J. Sci. Med. Sport **24**, 1193 (2021)
27. C.F. Borg, D. N., Osborne, J. O., Tweedy, S. M., Liljedahl, J. B., & Nooijen, Am. J. Phys. Med. Rehabil. **101**, 384 (2022)

Normalization methods analysis of career pattern using self-organizing map

Purwanto[1], *Hapsari Peni Agustin* Tjahyaningtijas[2*], *Jesse R* Paragas[3] and *I Gusti Putu Asto* Buditjahjanto[4]

[1]Post Graduate of Electrical Engineering Department, State University of Surabaya, Indonesia.
[2]Electrical Engineering Department, State University of Surabaya, Indonesia.
[3]Information Technology Department, College of Engineering, Eastern Visayas State University, Philippine
[4]Electrical Engineering Department, State University of Surabaya, Indonesia.

Abstract. Clustering the distribution of student graduates is an approach used to analyze and understand the success of Vocational High School education programs in preparing graduates to enter the workforce or start their own businesses. The purpose of clustering is to evaluate the effectiveness of educational programs, identify entrepreneurial potential, formulate career planning, and develop entrepreneurial skills, all contributing to the fulfilment of Sustainable Development Goals (SDGs) related to quality education (SDG 4), decent work and economic growth (SDG 8), and industry, innovation, and infrastructure (SDG 9). Through this clustering, schools can evaluate the extent to which quality high school graduates achieve career or entrepreneurial success, supporting the objectives of SDG 4. This information helps in designing educational programs that are more in line with the needs of the job market, providing better career guidance to students, and promoting entrepreneurial skills among high school students, contributing to SDG 4 and SDG 8. Clustering the distribution of vocational high school students by working, continuing, and entrepreneurial status plays an important role in strengthening the link between education and the world of work, aligning with the aims of SDG 4 and SDG 8. Self-Organizing Map (SOM), as an Artificial Neural Network, assists in data clustering or mapping tasks, aiding in the discovery of patterns and trends within the vocational high school graduate population. The result of clustering using Z-Score and Min-Max normalization techniques is 5.31% and 3.98%, respectively, providing insights into the career and entrepreneurship trends and patterns of vocational high school students. This valuable information can be used for the development of educational programs, career guidance initiatives, and improved alignment between education and the needs of the world of work, ultimately contributing to the realization of SDGs 4, 8, and 9.

* Corresponding author: hapsaripeni@unesa.ac.id

1 Introduction

After completing the basic education level, students can continue vocational education at a secondary school called Vocational High School. Vocational High Schools have a more specific focus on practical learning and skills that are directly related to the world of work or industry. The effective transition of vocational high school graduates into the world of work or entrepreneurship is a crucial component of their educational journey. Distribution of graduates by graduate status can provide valuable insights for schools, policymakers, and stakeholders to improve the quality of education and enhance the alignment between education programs and labor market needs. In this context, the use of approaches to cluster and analyze the distribution of high school graduates, to address this issue can use clustering technology as a support for the work of school staff in charge of clustering the distribution of graduates followed by the students. An example of an unsupervised learning technique that creates a low-dimensional representation of the input making use of a framework for an artificial neural network is self-organizing maps [1]. Although a Self-Organizing Map is a kind of simulated artificial neural network, the way it works is slight.

Different from ordinary artificial neural networks. When an error in the Artificial Neural Network system, it performs the correction learning step using the gradient decrease. A self-organizing Map uses competitive learning, each node will compete to be the winner. A Self-organizing Map will maintain the input topological structure by applying non-stop functions.

Clustering is carried out on data on the distribution of 2019-2020 graduates and the results are displayed in the form of topological visualization of self-organizing maps. Previous research on clustering alumni data has been carried out, including research on clustering alumni data of Semarang State Polytechnic using the K-Means method and the results are presented in the form of a web-based digital map. The clustering mechanism is based on four attributes, namely: type of company, office classification, field of work, and study program competency. The test results show that 51 alumni work in accordance with their competence, while 23 alumni with less suitable jobs and 26 alumni are not in accordance with their competence. The results of this study are used to help make decisions whether to make curriculum changes or not [2]. Further research uses fuzzy C-Means to cluster graduates using GPA and length of study variables. Based on clustering, four clusters were obtained. Of the four clusters, it is known that cluster 4 has graduate members with an age range of 5.91 years. This shows that there are still many students majoring in mathematics who spend more than 10 semesters or 5 years studying. The results of this study are expected to be used as a consideration for the department in increasing the cumulative grade point average of students so that they can complete their study time quickly [3]. The Bayesian Network model was used for Hepatitis Diagnosis in [5] and combined with Relief feature selection in [6]. In [7-9], artificial neural network models were used to predict student academic performance from initial semester input grades, and in [10] combined with Linear Regression and Support Vector Regression. Self-organizing map applications have been used in unsupervised learning to group students into scholarship programs [11-12]. The Self-Organizing Map method is also used in education to show students [13-14] cognitive structure models and to track learning activities [15]. Many studies using the SOM method have been carried out, such as in previous research named Self-Organizing Map (SOM) Algorithm-Based Applicant Mapping of Predicting Seriousness in Selecting Private University It was found that the SOM method for predicting the interest of private university registrants in 2017 was 0.0065%, in 2018 it was 0.00067%, and in 2019 it was 0.0047% [16], the next research entitled Student modelling using SOM cluster principal component analysis showed that the SOM method for modeling the learning system used the technique of clustering of the student data set using principal, the results were a map of clustering [17] , the next research is entitled the data from the student higher education census: a source of knowledge It was found that the SOM method was used to visualize profiles based on genre and academic status [18], the next research was entitled Using Self-Organizing Maps and Neural Networks, an analysis of student behavior in online learning environments using users clustering to analyze face-to-face systems, online and mixed learning systems obtained face-to-face and online results connected to a methodological approach [19].

This research will do clustering with attributes of the field of work, and cumulative grade point average. The method that will be used in clustering student distribution data is Self-Organizing Map, which is an unsupervised algorithm using Z-Score and Min-Max normalization techniques. Self-Organizing Map is chosen because in general, data cannot be explicitly separated into groups but has a tendency expressed by a degree of membership that is valued between 0 and 1 to its prevalence. This study's outcomes are anticipated to categorize the distribution of student graduates at vocational high school institutions.

2 Material and method

2.1 Self-Organizing Map (SOM)

Teuvo Kohonen initially presented SOM in 1996. Self-organizing neural networks are used by SOM to map high-dimensional data into a single lower dimension, to lower the dimensions of the data in order to view the element's target input without guidance or unsupervised learning data with the assumption of structured topology into clusters. The SOM algorithm treats each cluster unit's weight vector as an illustration of the cluster-related input pattern. Unit clusters that match the closest input vector pattern in terms of weight are considered winning. In order to match the input other winning unit and its surrounding unit keep updating their weights. Unlike previous JST models, the target neurons in the SOM network are arranged in two dimensions, with tunable shapes, as opposed to being arranged horizontally. Different forms will produce neurons surrounding different neurons that win, hence the ultimate weight may also differ. Weight adjustments are made in SOM for both the line weight to the neighboring neuron and the line weight linked to the winning neuron. Basic SOM implementations come in a variety of forms, from square long/six square, 2D/3D maps, to different distance function possibilities for defining surrounding surroundings. Implementation architecture is the implementation

SOM method, data from a high-dimensional vector space (input) each input layer neuron's connection to all other neurons in the output layer is translated into a two dimensional vector region output at a local location. A class (cluster) of the input is represented by each neuron in the output layer. With its SOM dimensional reduction properties, it is widely used as a dimension reduction tool, such as the principal component analysis (PCA). SOM is seen as a grouping approach, but, since dimensional reduction may also be understood as a decrease in the number of dimensions (or clusters) of the data input. The formula is as follows:

$$d(i,j)=\sqrt{[\sum(x(i)-w(j))^2]} \qquad (1)$$

Where:

•d(i,j) is the distance between the input vector i and the weight of the neuron j
•x(i) input vector of i.
•w(j) weight of j neuron.
•∑ a symbol for counting.

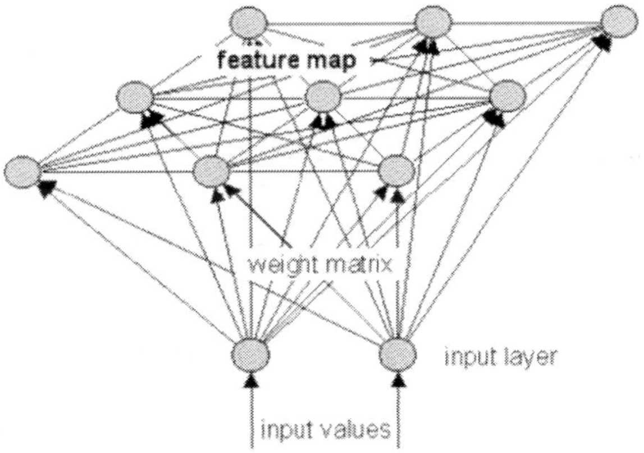

Fig. 1. Architecture Self-Organizing Map

At the training stage, each input vector will be compared to all neuron weights in the SOM. The neuron with the weight closest to the input vector will be selected as the winner (winner neuron) then the weight of the winning neuron and its surroundings will be updated (adjusted) using a calculation formula called the rule of Self-Organizing Map. These are the guidelines for the self-organizing map:

$$\Delta w(i,j) = \eta * h(i,j) * (x(i) - w(j))] \qquad (2)$$

Where:

•Δw (i,j) is the change in the weight of the neuron j for the input vector i.
•η (eta) is the learning rate, which controls how much weight adjustments are made.
•h(i,j) is a responsibility function, which measures how close the j neuron is to the winning neuron.
•x(i) input vector of i.
•w(j) weight of j neuron

2.2 Normalization techniques

A data mining technique called data normalization is used to convert a dataset's values into a common scale. It is crucial to note that numerous machine learning algorithms exhibit sensitivity to the input feature scale and can yield superior outcomes upon normalization of the data. In data mining, a variety of normalization approaches can be applied, like as:

2.2.1 Normalization of min-max:

This method reduces a feature's values to a range of 0 to 1. To do this, deduct the feature's minimum value from each value, then divide the result by the feature's range [20]. The Min-Max normalization formula is as follows:

$$X\_scaled = (X - Xmin) / (Xmax - Xmin)] \quad (3)$$

The dataset's minimal value is denoted by Xmin, its maximum value by Xmax, and the original value is represented by X.

2.2.2 Normalization of Z-Score:

The values of a feature are scaled using this method to have a mean of 0 and a standard deviation of 1. To calculate this, take each number, remove the feature mean from it, and divide the result by the standard deviation. The following is the Z-Score normalization formula:

$$X\_scaled = (X - mean) / Standard\ Deviation \quad (4)$$

The original value is represented by X, the dataset mean is denoted by Mean, and the dataset standard deviation is represented by Standard deviation.

2.3 Geographical Information System (GIS)

Geospatial Information Systems, also known as Geographic Information Systems (GIS), became known in the early 1980s. A GIS is a system for obtaining, storing, analyzing, and managing spatial data along with data-related attributes that are remotely referenced to the Earth. Based on technology and its implementation, geographic information systems can be categorized into three applications:

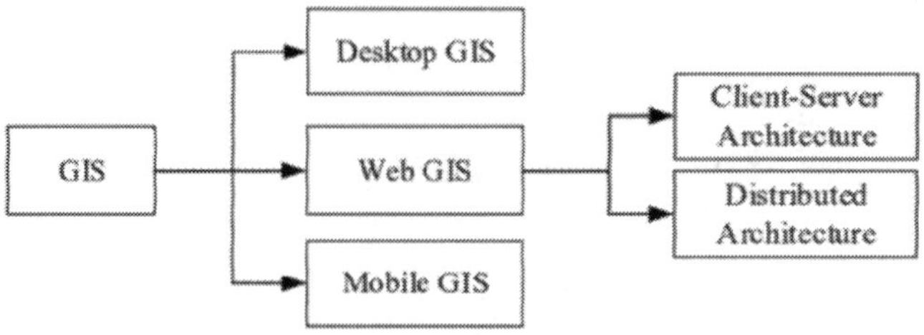

Fig. 2. Category Geographic Information Systems

3 Result and discussion

Figure 3 below provides a summary of the investigation.

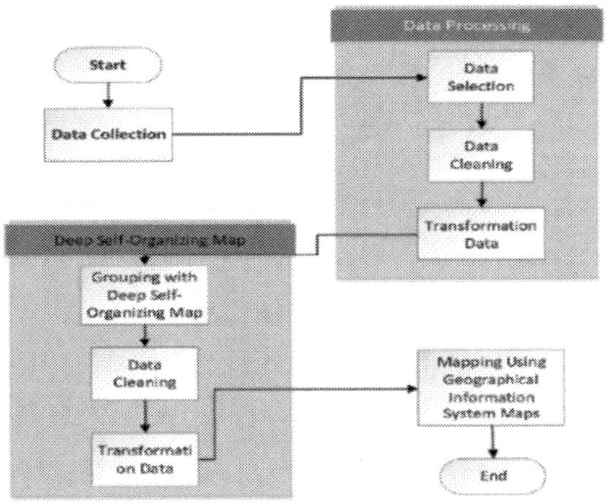

Fig. 3. Overview of the research

3.1 Collection of data

At this stage, it is done to collect the data that will be used in this research. This data is a type of secondary data, namely high school alumni data. The alumni data attributes to be used in this study include Student Number, student name, gender, address, no phone, study program, generation, year of graduation, final grade, place of work, address of workplace, and suitability of field of work.

3.2 Data processing

In this phase, there are three steps: data selection, data cleaning, and data transformation. The goal of this phase is to convert raw data into quality data, such as correcting incomplete (incomplete), noisy (data containing value errors), and inconsistent data.

- Data selection: In the data selection stage, we will select the attributes to be used. Of course, not all attributes are included in the dataset used in the data mining process because only attributes that become identification references will be selected. Of all the attributes mentioned above, the ones that will be used in the data selection process are Student Identification Number, Student Name, Gender, Major, and Final Grade.
- Data Cleaning: Before the data mining process can be implemented, the data cleaning process must be carried out. The purpose of this process is to ensure the quality of the data that has been selected at the data selection stage. Suppose gender writing is still not inconsistent there are still men and women using it.
- Data transformation: At this stage, the data will be transformed so that it is suitable for use in the clustering process. An attribute of the field of work is the categorical

data type Final Grade of the major. This process is carried out using the Self-Organizing Map method.

3.3 Self-Organizing Map

Z-Score and Min-Max are two of the three normalization procedures available in the Self-Organizing Map approach:

Clustering with a Self-Organizing Map using Z-Score and Min-Max normalization methods, aims to know data normalizations before forming SOM map training data. The results of normalization use Z-Score and Min-Max with the following datasets and header components in Table 1. As follows Information header Column:

1. NIS : Student Identification Number
2. TL : Place of birth
3. JK : Gender
4. KK : Skill Competency
5. AGM : Religion
6. PKN : Pancasila and Citizenship Education
7. BIN : Indonesian Language
8. MTK : Mathematics
9. SJI : History of Indonesia
10. BIG : English Language
11. SB : Arts and Culture
12. PJK : Physical Education, Sports and Health
13. BDH : Regional Language
14. UKK : Competency Test

Table. 1. Examples of Student Data

NIS	TL	JK	KK	Status	School Test Scores													UKK
					AGM	PKN	BIN	MTK	SJI	BIG	SB	PJK	BDH	SIMD	EKB	ADM	DP	
7130	1	1	1	1	93.5	94.5	93	87.5	91.5	93.5	93.5	90	91.5	93	91	89.5	92	98
7131	1	1	1	1	92.5	96.5	92	90.5	92	93.5	93.5	89	91	95	93	90	94	98
7132	1	1	1	1	90.5	92.5	85.5	84.5	90	88	88	89	90.5	88.0	85.0	87.5	90.5	90
7133	2	1	1	2	88	96.5	93.5	92.5	93.5	95.5	95.5	90	92.5	95.0	94.0	92.5	94.5	98
7134	3	1	1	1	92.5	95	88.5	84.5	88.5	92.5	92.5	88	91	90	91	91	90	91
7135	4	1	1	1	91.5	93.5	88.5	83.5	88.5	92	92	90	91	89	86	90	90	90
7136	2	1	1	1	92.5	92.5	87.5	83.5	87.5	94.5	94.5	85	91	92	88	88	91	92

NIS	TL	JK	KK	Status	School Test Scores													UKK
					AGM	PKN	BIN	MTK	SJI	BIG	SB	PJK	BDH	SIMD	EKB	ADM	DP	
7137	1	1	1	1	92.5	94	88.5	82.5	88.5	95.5	95.5	94	90	89	85	89	88	92

From an example of the initial dataset in Table 1. Normalization data such as Z-Score and Min-Max are produced, from which data is used to identify clustering data whose map size is specified as 10 x 10 and each cell holds 20 random weight values assigned. Table 2. Shows the initial weight sample for 1 cell:

Table 2. The Early Cell (1000 Train)

Data Weight																			
Min Max Initial Cell Weight	0.2	0.2	0.2	0.7	0.2	0.2	0.1	0.1	0.1	0.2	0.2	0.2	0.2	0.2	0.1	0.1	0.1	0.1	0.1
Z-Score Initial Cell Weight	0.9	0.9	0.9	1.0	0.9	0.9	0.9	0.9	0.6	0.6	0.6	0.7	0.7	1.1	1.1	1.6	1.0	1.0	1.0

On the SOM algorithm for each iteration, the value of the node weight is adjusted using equation (1). In Table 3. List the final weight for 1 cell:

Table 3. End of cell (1000 Train)

Data Weight																			
Min Max Final Cell Weight	0.1	0.1	0.1	0.2	0.1	0.1	0.1	0.4	0.3	0.3	0.2	0.2	0.2	0.2	0.2	0.3	0.3	0.3	0.2
Z-Score Final Cell Weight	0.6	0.7	0.7	1.0	0.9	0.9	0.9	0.9	0.9	1.2	1.0	1.0	0.8	0.8	0.8	1.1	0.6	0.6	0.5

On the SOM algorithm for the initial weight of 1 cell on the train 2000, the value of the weight of the node can be seen in Table 4. Listing the starting weight for 1 cell:

Table 4. The Early Cell (2000 Train)

Data Weight																			
Min Max Initial Cell Weight	0.2	0.2	0.1	0.1	0.1	0.1	0.1	0.3	0.3	0.3	0.3	0.4	0.4	0.4	0.3	0.3	0.1	0.1	0.1
Z-Score Initial Cell Weight	0.6	0.5	0.8	0.9	0.9	1.8	1.3	1.3	1.0	0.9	0.9	1.2	1.2	1.3	1.0	1.0	0.8	0.7	0.9

On the SOM algorithm for each iteration, the value of the node weight is adjusted using equation (2). In Table 5. List the final weight for 1 cell:

Table 5. The end of cell (2000 Train)

Data Weight																			
Min Max Initial Cell Weight	0.1	0.1	0.2	0.4	0.4	0.4	0.3	0.3	0.3	0.3	0.2	0.2	0.2	0.3	0.3	0.5	0.1	0.1	0.1
Z-Score Initial Cell Weight	0.5	0.4	0.6	1.5	1.2	1.2	1.2	1.2	1.2	1.9	1.1	1.1	1.0	1.0	1.0	1.1	1.1	1.1	1.0

After 1000 trains and 2000 trains were performed, the results were then visualized with the U-Matrix of each normalization. Visual representation of Self-Organizing Maps is achieved primarily via the use of a color scale in the integrated distance matrix (U-Matrix). Fig. 4 shows that nearby nodes are represented identically and share the same hue, suggesting that they are part of the same cluster. This is a visualization of the Self-Organizing Matrix-U Map produced from 10x10 with 1000 trains as seen in the Fig. 5. Of the matrix -U produced by 10×10 with 2000 trains.

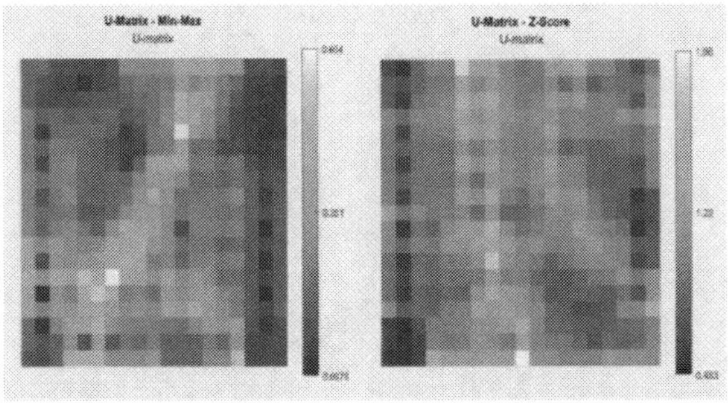

Fig. 4. U-Matrix 1000 Train

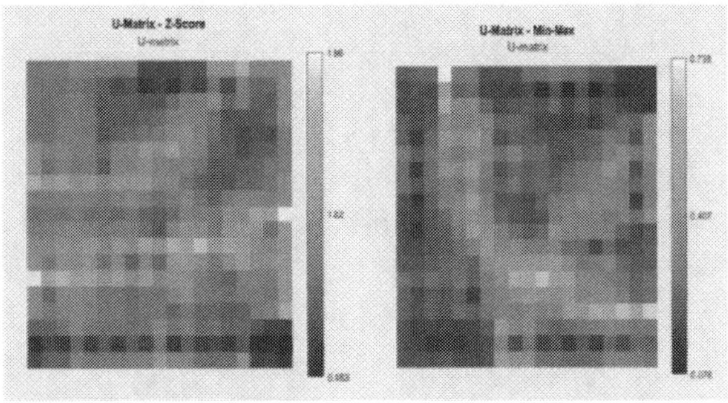

Fig. 5. U-Matrix 2000 Train

Analysis of the results of SOM clustering using Z-Score and Min-Max normalization by giving different train weights has an influence, namely the pattern of clustering when the train value is changed from 1000 trains to 2000 trains, when 1000 trains with two normalizations running shows there are two different clusters in the best cluster position with 5.31% Z-Score and 4.91% Min-Max, while when 2000 trains show different numbers but in the same cluster, with a value of 5.31% Z-Score and 3.98% Min-Max. The visualization results in Fig. 4. And Fig. 5. Which can be seen that some data falls into one cluster: the blue-colored cluster. In general, the map indicates that there are two to three clusters made up of one dominating cluster, while there are other clusters near the blue hue, especially yellow and orange in certain areas. This result is consistent with the conventional clustering of vocational high school students that categorizes status into 3 groups: working, continuing, and entrepreneurship. The major cluster is generally the employment cluster because the main goal of vocational high schools is to generate graduates who are ready to work, but in this analysis, there is a cluster that shows that those who continue to higher education are the most dominant shown in blue, then there is an entrepreneurial cluster that occupies the second dominant position shown in orange, and finally is employment shown in yellow.

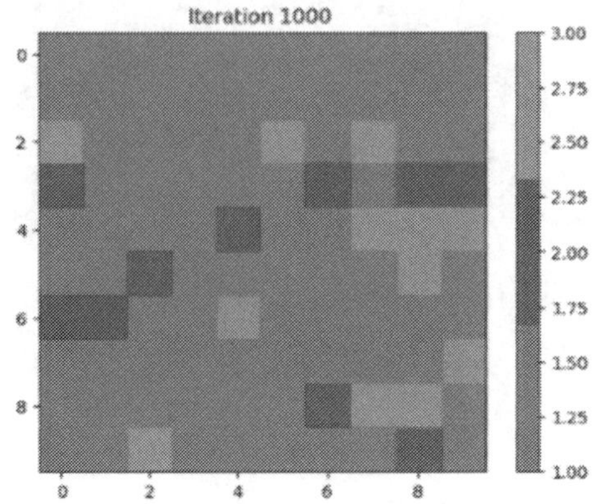

Fig. 6. SOM Label on the trained max iteration

To construct a label map, assign one label to each neuron mapped using random data. Figure 3 shows the label maps created for the first and last iterations. Initially, many neurons have values that are not 0 or 1 and the class labels appear randomly distributed, whereas, in the final iteration, there is a clear separation between classes 0 and 1 even though there are some cells that do not belong to one class in the last iteration, from the separation between class 0 and 1 produces an accuracy of 69%. The low accuracy value in Neural Networks with Self-Organizing Map (SOM) Architecture which uses the Min-Max algorithm is caused by many factors, therefore in the future, it is necessary to optimize parameters, add features, and develop more complex modeling so that system reliability and performance can be achieved improved.

4 Conclusion

This research has generated initial insights into grade data, and graduation status of student participants by identifying clusters of students within specific areas of expertise competencies. The total number of possible clusters is three, which correspond to the conventional categories of Working, Continuing, and Entrepreneurship. The cluster with the largest number of members is the Continuing cluster, and in this research, a self-organizing feature using a neural network is added to the Self-Organizing Map, thus providing a comparison where the normalization results produce an accuracy of 69% for the results from the last iteration. Future research plans to improve accuracy will be developed using Deep Self-Organizing Map.

References

1. T. Kohonen, J. Hynninen, J. Kangas, J. Laaksonen, "SOM PAK: The Self-Organizing Map Program Package SOM PAK: The Self-Organizing Map Program Package," 1996.
2. T. Kohonen, Biol. Cybern. **43**, 1 (1982)
3. S. H. Handoko, E. Sediono, S. Suhartono, J. Sist. Inf. Bisnis **1**, 2 (2014)
4. E. Sehirli, K. Arslan, Expert Syst. Appl. **205** (2022)
5. O. J. Alabi, J. W. Ng'ambi, D. Norris, Asian J. Anim. Vet. Adv. **7**, 2 (2012)

6. S. Lakho, A. Hussain Jalbani, M. S. Vighio, I. A. Memon, S. S. Soomro, Q.-N. Soomro, Sukkur IBA J. Comput. Math. Sci. **1**, 2 (2017)
7. F. T. Anggraeny, I. Y. Purbasari, E. Suryaningsih, *ReliefF Feature Selection and Bayesian Network Model for Hepatitis Diagnosis*, in 3rd Int. Conf. Inf. Technol. Bus. (2017)
8. A. Usman, O. L. Adenubi, J. Sci. Inf. Technol. **1**, 2 (2013)
9. K. L. Wu, M. S. Yang, Pattern Recognit. Lett. **26**, 9 (2005)
10. F. Höppner, F. Klawonn, Mathware & soft computing **11**, 5 (2004)
11. E. Yohannes, S. Ahmed, Int. J. Comput. Appl. **180**, 40 (2018)
12. M. W. Bara, N. B. Ahmad, M. M. Modu, H. A. Ali, *Self-organizing map clustering method for the analysis of e-learning activities*, In Majan international conference (MIC), IEEE (2018)
13. Y. Lee, J. Educ. Comput. Res. **57**, 2 (2019)
14. P. N. E. Nohuddin, Z. Zainol, A. Nordin, Zulfaqar Journal of Defence Science, Engineering & Technology **1**, 1 (2018)
15. D. Ifenthaler, I. Masduki, N. M. Seel, Instructional Science **39** (2011)
16. B. H. Kunaryo, M. Somantri and A. Sofwan, *Applicant Mapping of Predicting Seriousness in Choosing Private University using Self Organizing Map (SOM) Algorithm*, in 2020 7th International Conference on Information Technology, Computer, and Electrical Engineering (ICITACEE), Semarang, Indonesia (2020)
17. L. Chien-Sing, Y. P. Singh, *Student modeling using principal component analysis of SOM clusters*, in IEEE International Conference on Advanced Learning Technologies, Joensuu, Finland (2004)
18. S. R. M. de Campos, R. Henriques, *The knowledge discovery through the student's higher education census data*, in 13th Iberian Conference on Information Systems and Technologies (CISTI), Caceres, Spain (2018)
19. S. Delgado, F. Morán, J. C. S. José and D. Burgos, IEEE Access **9** (2021)
20. S. S. Ramakrishna, Int. J. Adv. Res. Comput. Sci. **9**, 2 (2018)

Forecasting poverty in East Java using vector autoregressive method and vector error correction model

A'yunin Sofro[1*], *Safira Diah Nur* Aidha[2], and *Khusnia Nurul* Khikmah[1,2]

[1]Mathematics Department, Faculty of Mathematics and Natural Sciences, Universitas Negeri Surabaya, Surabaya, East Java, Indonesia 60231
[2]Statistics Department, Faculty of Mathematics and Natural Sciences, IPB University, Bogor, West Java, Indonesia 16680

Abstract. People experiencing poverty are people who are unable to fulfil their basic needs. A region with a dense population is prone to problems overcoming poverty. In this instance, the gross regional domestic product, the human development index, and the open unemployment rate are the variables impacting poverty. Therefore, more study is required to address this issue of poverty. The vector autoregressive and error correction models are two possible approaches. The East Java Central Bureau of Statistics provided the data, which included gross regional domestic product, human development index, open unemployment rate, and percentage of poverty. Forecasting the number of poverty people is obtained using estimates from data that can affect forecasting results. In this article, the best forecasting results are obtained with an RMSE value of 21.51062 using the vector error correction model, namely with a percentage of poverty value of 7.2619.

1 Introduction

The inability to fulfill basic things is called poverty, where the poverty rate in Indonesia in 2020 reached 10.19% [1]. This poverty directly affects the economy, one of which is the growth of household consumption expenditure on gross regional domestic product, where economic growth is a benchmark indicator of the success of development implementation [2,3]. In addition to gross regional domestic product, other influencing factors include the human development index and the open unemployment rate [4]. The increase in poverty in Indonesia can be predicted by using previous data on the factors that cause an increase in poverty. In this case, the author uses the factors of gross regional domestic product, human development index, and open unemployment rate. This forecasting can be used to prevent or overcome so that the influencing factors can be overcome first so that they do not experience a decline.

* Corresponding author: ayuninsofro@unesa.ac.id

Forecasting is an activity of guessing or estimating a future situation based on past and present circumstances that are needed as a decision support system to determine when an event will occur so that appropriate action can be taken [5]. This research uses data with four variables, namely the consumption rate of gross regional domestic product, human development index, open unemployment rate, and percentage of poverty. This variable was chosen because it refers to an existing article, where it is mentioned that the regression analysis of the variables of gross regional domestic product consumption rate, human development index, and open unemployment rate shows that the three variables influence the poverty rate.

Regression analysis with multivariate data can be predicted using the vector autoregressive method. This method was chosen because it moves from regression analysis, where in prediction, there are two methods, namely the ARIMA method for univariate data and the vector autoregressive method for multivariate data. Vector autoregressive is used because the behavior of a variable is expected to be affected by other variables, but the influence does not have an immediate impact but requires an interval or can be called a time lag. the advantages of the vector autoregressive model, include a simple model form where all variables in this model can be considered as endogenous variables. The vector autoregressive model uses ordinary least square (OLS) parameter estimation and better forecasting accuracy between other more complex simultaneous equation models [6, 7].

Other research [8, 9] shows that the vector error correction model has good forecasting results. The Vector Error Correction Model is a derivative of the Vector Autoregressive model where the vector error correction model must be stationary in the first difference. By using the Vector Error Correction Model, data can be analyzed for short-term and long-term relationships between variables. Referring to previous research conducted by [10, 11], Vector Autoregressive and Vector Error Correction Models provide good forecasting results. Both methods will be used in this study.

2 Methods

2.1 Vector autoregressive (VAR)

The vector autoregressive method is a combination of several autoregressive models whose variables will affect each other. The vector autoregressive model is based on the statistical properties of the data. In vector autoregressive, each endogenous variable in the system is considered as the lag value of all endogenous variables in the system [12]. Thus, the univariate autoregressive model is generalized to a vector autoregressive model consisting of multivariate time series variables. The vector autoregressive model is one of the dynamic linear models (MLD) that is widely used in the application of predicting economic variables both in the long term and in the medium to long term. In addition, this model can also be used to determine the causal relationship. [13] wrote the vector autoregressive model equation with k variables and order p or VAR(p) as follows:

$$Y_t = \Phi_0 + \Phi_1 Y_{t-1} + \Phi_2 Y_{t-2} + \cdots + \Phi_p Y_{t-p} + a_t \qquad (1)$$

Where $Y_t = (Y_{1,t}, Y_{2,t}, \dots, Y_{k,t})$ denotes the vector Y_t of size $(k \times 1)$ of the time series variable, Φ_i denotes the matrix of size $(k \times k)$, $\Phi_0 = (\Phi_{10}, \Phi_{20}, \dots, \Phi_{k0})$ is a vector of dimension k and $a_t = \left(a_{1,t}, a_{2,t}, \dots, a_{k,t}\right)^{\mathrm{T}}$ is the error of size $(k \times 1)$ which is assumed to be multivariate normal with $E(\mu_t) = 0$, $E(\mu_t\mu_t^{\mathrm{T}}) = \sum\mu$ and $E\,E(\mu_t\mu_s^{\mathrm{T}}) = 0$ for $s \neq t$. The covariance matrix $\sum\mu$ must be positive definite.

The vector autoregressive model in equation (1) will be stable if:

$$\det\left(l_k - A_1 z - A_2 Z^2 - \dots - A_p Z^p\right) \neq 0, |z| \leq 1 \quad (2)$$

Equation (2) is the reverse characteristic polynomial of the VAR(p) model which has no roots and is on the unit circle.

2.2 Vector error correction models

Place the figure as close as possible after the point where it is first referenced in the text. If there is a large number of figures and tables, it might be necessary to place some before their text citation. The vector error correction model is a vector autoregressive model with cointegration constraints. Since there is a cointegration relationship in the vector error correction model, when there is an extensive range of short-term dynamic fluctuations, the vector error correction model expression can be restricted to the long-term behavior of the endogenous variables and converge to the cointegration relationship [9]. Assuming $y_t = (y_{1t}, y_{2t}, \dots, y_{kt})$ as a k-dimensional stochastic time series, $t = 1,2,\dots,T$ and $y_t \sim I(1)$ where $y_{it} \sim I(1)$, $i = 1,2,\dots,k$ is affected by the d-dimensional exogenous time series $x_t = (x_{1t}, x_{2t}, \dots, x_{dt})'$; then the vector error correction model can be formed as follows:

$$y_t = A_1 y_{t-1} + A_2 y_{t-2} + \dots + A_p y_{t-p} + Bx_t + \mu_t \ , t = 1,2,\dots,T \quad (3)$$

If y_t is not influenced by a d-dimensional exogenous time series, then the vector autoregressive model from equation (1) can be written as follows:

$$y_t = A_1 y_{t-1} + A_2 y_{t-2} + \dots + A_p y_{t-p} + \mu_t \ , t = 1,2,\dots,T \quad (4)$$

With the cointegrating transformation of equation (2), we can obtain

$$\Delta y_t = \prod y_{t-1} + \sum_{i=1}^{p=1} \Gamma_i \Delta y_{t-1} + \mu_t \quad (5)$$

where

$$\prod = \sum_{i=1}^{p} A_i - I \quad (6)$$

and

$$\Gamma_i = \sum_{i=1}^{p} A_j \quad (7)$$

If yt has a cointegrating relationship, then $\prod y_{t-1} \sim I(0)$ and equation can be written as

$$\Delta y_t = \alpha\beta' y_{t-1} + \sum_{i=1}^{p=1} \Gamma_i \Delta y_{t-1} + \mu_t \quad (8)$$

Where $\beta' y_{t-1} = ecm_{t-1}$ is the error correction term, which describes the long-run equilibrium relationship between variables, so it can be written with the following formula:

$$\Delta y_t = \alpha ecm_{t-1} + \sum_{i=1}^{p=1} \Gamma_i \Delta y_{t-1} + \mu_t \quad (9)$$

Equation (3) is a vector error correction model where each equation is an error correction model.

2.3 Root Mean Square Error (RMSE)

Root mean square error (RMSE) is a calculation method that is often used to obtain the minor error value between two different methods or equation models [14]. The formula is as follows:

$$\text{RMSE} = \sqrt{\overline{(f - o)^2}} \quad (10)$$

Where, f is the forecast and o is the observed value.

2.4 Data Sources

In this study, the data used are data on gross regional domestic product, human development index, open unemployment rate, and percentage of poverty in all provinces in Indonesia. Where gross regional domestic product, human development index, and open unemployment rate are independent variables, and the percentage of poverty is a dependent variable. Each variable is taken from 2004 to 2020 for East Java Province. The data is taken from the Central Statistics Agency (BPS). The variables used in this study are divided into 4, namely the gross regional domestic product variable ($Y_{1,t}$), the human development index variable ($Y_{2,t}$), the open unemployment rate variable ($Y_{3,t}$), and the percentage of poverty variable ($Y_{4,t}$). The dataset used is secondary data taken from the Central Bureau of Statistics.

3 Result and discussion

This study begins with data stationarity testing. Testing is done using the Augmented Dickey-Fuller method where the hypothesis used is $H_0 = |statistic\ t| < |critical\ value\ t|$ and $H_1 = |statistic\ t| > |critical\ value\ t|$. The test criteria will reject H_0 if the test statistic value of ADF has a value smaller than the critical region value or

$p - value < \alpha$. The test results show that the data is not yet stationary, so it is necessary to do differencing. The first difference results in the data and the following Table 1.

Table 1. ADF first difference test results.

	Gross regional domestic product	Human development index	Open unemployment rate	Percentage of poverty
Statistic t	-7.668	-4.519	-2.878	-4.936
Critical Value t	-1.95	-1.95	-1.95	-1.95

From Table 1, it can be seen that all data has been stationary because of the results of $|statistic\ t| > |critical\ value\ of\ t|$. So, the data becomes stationary in the first difference. Then, the requirements for the vector error correction model are also met because the data has been stationary in the first difference. Furthermore, the differencing results are modeled using vector autoregressive. The first step in this modeling is to determine the optimal lag based on the optimal AIC value, which is 1.260790 at lag-1 and -0.90666 at lag-2 so that the optimal lag of the data used is two or it can be said to be a VAR(2) model. Furthermore, parameter estimation is carried out, and parameter estimates for the VAR(2) model with four variables can be obtained in Table 2.

Table 2. Parameter estimation results of vector autoregressive.

	Y			
	1	2	3	4
Gross regional domestic product.l1	0.1328	−0.1552	0.0364	−0.2625
Human development index.l1	0.0105	0.0680	−0.0419	−0.0365
Open unemployment rate.l1	−0.0006	0.7176	0.5498	0.4432
Percentage of poverty.l1	−0.3816	−3.1447	0.1361	0.9063
Gross regional domestic product.l2	−0.1484	−0.9799	0.0981	0.3362
Human development index.l2	−0.1716	−2.1576	−0.0542	0.031
Open unemployment rate.l2	0.3015	2.3788	−0.0166	−0.5474
Percentage of poverty.l2	0.0114	−0.9414	−0.1232	0.1349

Const.	22.138	278.844	8.0933	−0.9967

The results of parameter estimation are tested with Granger Causality, which aims to determine the relationship between related variables. If the variables are not interconnected, then the vector autoregressive model cannot be formed, and the results are obtained in Table 3.

Table 3. Granger causality test results.

	$p - value$
Gross regional domestic product	4.99×10^{-10}
Human development index	9.83×10^{-1}
Open unemployment rate	8.04×10^{-4}
Percentage of poverty	1.37×10^{-2}

Based on the test results obtained, gross regional domestic product affects the human development index, open unemployment rate, and percentage of poverty. human development index does not affect gross regional domestic product, the open unemployment rate, and percentage of poverty; open unemployment rate affects gross regional domestic product, human development index, and percentage of poverty. percentage of poverty does not affect gross regional domestic product, human development index, and open unemployment rate. So the model that is formed is:

$$Y_{1.t} = 0.1328\text{Gross regional domestic product.}l1 + 0.0105\text{Human development index.}l1 - 0.0006\text{Open unemployment rate.}l1 - 0.3816\text{Percentage of poverty.}l1 - 0.1484\text{Gross regional domestic product.}l2 - 0.1716\text{Human development index.}l2 + 0.3015\text{Open unemployment rate.}l2 + 0.0114\text{Percentage of poverty.}l2 + 22.138$$

$$Y_{2.t} = -0.1552\text{Gross regional domestic product.}l1 + 0.0680\text{Human development index.}l1 + 0.7176\text{Open unemployment rate.}l1 - 3.1447\text{Percentage of poverty.}l1 - 0.9799\text{Gross regional domestic product.}l2 - 2.1576\text{Human development index.}l2 + 2.1576\text{Open unemployment rate.}l2 - 0.9414\text{Percentage of poverty.}l2 + 278.844$$

$$Y_{3.t} = 0.0364\text{Gross regional domestic product.}l1 - 0.0419\text{Human development index.}l1 + 0.5498\text{Open unemployment rate.}l1 + 0.1361\text{Percentage of poverty.}l1 + 0.0981\text{Gross regional domestic product.}l2 - 0.0542\text{Human development index.}l2 - 0.0166\text{Open unemployment rate.}l2 - 0.1232\text{Percentage of poverty.}l2 + 8.0933$$

$$Y_{4.t} = -0.2625\text{Gross regional domestic product. l1} - 0.0365\text{Human development index. l1} + 0.4432\text{Open unemployment rate. l1} + 0.9063\text{Percentage of poverty. l1} + 0.3362\text{Gross regional domestic product. l2} + 0.031\text{Human development index. l2} - 0.5474\text{Open unemployment rate. l2} + 0.1349\text{Percentage of poverty. l2} - 0.9967$$

The second model to be modeled is the vector error correction model, starting with the determination of the Optimal Lag. The optimal lag obtained the optimal AIC value of -0.90666 at lag-2 so that the optimal lag of the data used is 2. Furthermore, stability testing is carried out. The vector error correction model is said to be stable if the modulus value is at a radius < 1. If the most significant modulus value is less than one and is located at the optimal point, then the vector autoregressive model is stable, and the following results are obtained in Table 4.

Table 4. Stability test results.

	Modulus		Modulus
[1]	1.4314739	[5]	0.4453617
[2]	1.4314739	[6]	0.4453617
[3]	1.0153875	[7]	0.4031710
[4]	0.5575469	[8]	0.1289675

Based on Table 4, modulus values 1 to 3 show more than 1, but after that, the modulus gradually decreases. It is concluded that the vector error correction model used is relatively stable. After the stability test is carried out, cointegration testing will be carried out using the Johansen cointegration test. The test results are in Table 5.

Table 5. Johansen cointegration test results.

	Statistic trace	Critical value
$r \leq 3$	4.9043	9.24
$r \leq 2$	16.7792	19.96
$r \leq 1$	57.1662	34.91
$r = 0$	114.8108	53.12

Based on the results in Table 5, at $r = 0$ and $r \leq 1$, the significance value of the trace is more significant than its value (with a significance level of 5%). Then, it can be interpreted that the data has a long-run equilibrium. Then, the model can be estimated parameters. The vector error correction model parameter estimates obtained are as follows.

Table 6. Vector error correction model parameter estimation results.

	Gross regional domestic product	Human development index	Open unemployment rate	Percentage of poverty
Gross regional domestic product.l1	0.177	−0.39	−0.095	−0.168
Human development index.l1	−0.003	0.095	0.0013	−0.045
Open unemployment rate.l1	−0.034	0.514	−0.102	0.353
Percentage of poverty.l1	−0.39	−0.33	0.305	0.095
Gross regional domestic product.l2	−0.125	−0.14	0.048	0.16
Human development index.l2	−0.069	−1.91	−0.031	0.017
Open unemployment rate.l2	0.141	0.176	−0.078	−0.026
Percentage of poverty.l2	−0.07	0.313	0.106	−0.197
Const.	−0.155	1.259	0.222	−0.325

After obtaining parameter estimates, the vector error correction model can be formed as follows:

$$\begin{bmatrix} \text{Gross regional domestic product}_t \\ \text{Human development index}_t \\ \text{Open unemployment rate}_t \\ \text{Percentage of poverty}_t \end{bmatrix} = \begin{bmatrix} -0.155 \\ 1.259 \\ -0.108 \\ 0.155 \end{bmatrix} + \begin{bmatrix} 0.177 & -0.003 & -0.034 & -0.39 & -0.125 & -0.069 & 0.141 & -0.07 \\ -0.39 & 0.095 & 0.514 & -0.33 & -0.14 & -1.91 & 0.176 & 0.313 \\ -0.095 & 0.0013 & -0.102 & 0.305 & 0.048 & -0.031 & -0.078 & 0.106 \\ -0.168 & -0.045 & 0.353 & 0.095 & 0.16 & 0.017 & -0.026 & -0.197 \end{bmatrix} \begin{bmatrix} \text{Gross regional domestic product. l1} \\ \text{Human development index. l1} \\ \text{Open unemployment rate. l1} \\ \text{Percentage of poverty. l1} \\ \text{Gross regional domestic product. l2} \\ \text{Human development index. l2} \\ \text{Open unemployment rate. l2} \\ \text{Percentage of poverty. l2} \end{bmatrix}$$

The model obtained based on the analysis carried out is then carried out forecasting. Vector autoregressive and vector error correction model for forecasting, forecasting is carried out gross regional domestic product, human development index, open unemployment rate, and percentage of poor people in East Java for the next six years. Then, the forecasting results with the VAR(2) model are obtained in Table 7.

Table 7. Forecasting results using vector autoregressive and vector error correction model parameter estimation results.

Year	Gross regional domestic product		Human development index		Open unemployment rate		Percentage of poverty	
	VAR	VECM	VAR	VECM	VAR	VECM	VAR	VECM
2021	7.840	6.409	93.229	90.842	−0.095	4.871	−0.168	11.550
2022	8.587	6.572	96.615	92.954	0.001	4.700	−0.045	10.272
2023	5.055	5.486	48.608	50.238	−0.102	3.692	0.353	10.300
2024	3.692	4.965	41.072	44.835	−0.078	4.241	−0.026	11.542
2025	11.215	6.986	142.001	129.258	0.106	6.435	−0.197	10.620
2026	14.987	8.027	168.217	146.104	0.222	5.711	−0.325	7.2619

Based on Table 7, the forecasting results show the results of forecasting for the next five years, where each method has different results. Then, we will check the RMSE value to find out the best model. By using equation (25), the RMSE value for the vector autoregressive method is 26.91299, while the RMSE value for the vector error correction model method is 21.51062. It can be concluded from this study that the most appropriate forecasting method is to use the vector error correction model with an optimal lag of 2.

This study provides the results of modeling and forecasting poverty data in East Java, Indonesia, based on the vector autoregressive method and vector error correction model. Based on the forecasting results, this study found that in 2021, the level of gross regional domestic product and the percentage of poor people decreased, while the community development index and the open unemployment rate increased. Gross regional domestic product and open unemployment rate significantly affect the poverty rate in East Java. However, when viewed from long-term forecasting, namely until 2026, the community development index decreased in 2022 but gradually improved in subsequent years. This results in the percentage of poor people also reducing. This study also shows that the percentage of poor people in East Java is influenced by gross regional domestic product, community growth index, and open unemployment rate.

In addition, we found that joint modeling, based on the analysis results, provides more complicated information. The data analysis based on the statistical model also shows that the vector error correction model shows more accurate forecasting results than the vector autoregressive model. In addition, the forecasting results can suggest considerations for policymakers in determining steps and attitudes toward the percentage of poor people in East Java.

4 Conclusion

Two different methods can be used, namely vector autoregressive and vector error correction models. Where both methods have standard multivariate variables, two different forecasting models can be obtained. From the different forecasting results, the best method can be selected. In this study, the best model was obtained using the VAR(2) model with an RMSE of 1.1738. The smallest RMSE value indicates that the model has a small error value so that the results obtained are expected to be close to future data. In future research, forecasting using the vector autoregressive and vector error correction model methods can be done using more variables and a more comprehensive range of coverage. This is done to determine the best forecasting and the effect of each variable on the results of forecasting other variables.

References

[1] R. Izzati Al, Situasi Kemiskinan Selama Pandemi (The SMERU Research Institute, 2021)

[2] N. Nursini, Development Studies Research **7**, 1 (2020)

[3] E.-J. De Bruijn, G. Antonides, Theory Decis. **92**, 1 (2022)

[4] D. Dahliah, A. N. Nur, Golden Ratio of Social Science and Education **1**, 2 (2021)

[5] J. Dupuy et al., Ann For Sci, **77**, 2 (2020)

[6] A. Ibrahim, R. Kashef, M. Li, E. Valencia, E. Huang, Journal of Risk and Financial Management **13**, 9 (2020)

[7] S. Guefano, J. G. Tamba, T. E. W. Azong, L. Monkam, Energy **214** (2021)

[8] M. Ding, H. Zhou, H. Xie, M. Wu, Y. Nakanishi, R. Yokoyama, Neurocomputing **365** (2019)

[9] P. Giudici, P. Pagnottoni, Appl Stoch Models Bus Ind. **36**, 1 (2020)

[10] E. N. Ni'mah, URECOL **12** (2017)

[11] A. P. Desvina, Jurnal Sains Teknologi dan Insdustri **13**, 2 (2016)

[12] W. B. Nicholson, I. Wilms, J. Bien, D. S. Matteson, The Journal of Machine Learning Research **21**, 1 (2020)

[13] H. Lütkepohl, New introduction to multiple time series analysis (Springer Science & Business Media, 2005)

[14] M. Ćalasan, S. H. E. A. Aleem, A. F. Zobaa, Energy Convers Manag, **210** (2020)

 https://doi.org/10.1051/e3sconf/202345003005

The influence of disability-friendly corporate branding on company brand equity in East Java

Reynaldi Dwi Junianta[1*], *Sri Setyo* Iriani[1], *Dwiarko* Nugrohoseno[1], and *Dian Anita* Nuswantara[2]

[1] Department of Management, Universitas Negeri Surabaya, Surabaya, Indonesia
[2] Department of Accounting, Universitas Negeri Surabaya, Surabaya, Indonesia

Abstract. This research analyzed the influence of disability-friendly corporate branding on corporate brand equity. Corporate branding is built through corporate values, corporate image, and corporate culture. This research uses a quantitative approach, with a population of corporate employees and customers in East Java. A purposive sampling technique was used to take samples from the population with the following criteria: (1) located in the East Java region; (2) an area with Special Needs High School with more than 50 students; (3) data recorded in the industrial directory in East Java Province region. A total of 180 respondents were obtained through distributing questionnaires online. The research data was then processed using SEM-AMOS to carry out analysis. The research results show that there is a significant influence of disability-friendly corporate branding on corporate brand equity. This research conducted research on company efforts to create disability-friendly branding for their companies using a corporate branding approach. Companies need to carry out corporate branding by through implementing the values of inclusivity, a corporate image that cares for people with disabilities, and a culture of equality. This will help companies to increase corporate brand equity which can provide company benefits that contribute to the implementation of SDGs 4, 8, and 10.

1 Introduction

Every individual in Indonesia is an important resource for Indonesia's development, including people with disabilities. People with disabilities are individuals who have a variety of potential, skills, and talents who can contribute to the development of Indonesia [1, 2]. Creation of inclusive and supportive workspaces, giving people with disabilities equal opportunities to participate in economic activities in Indonesia [1]. When every human resource owned has been properly optimized, it will have an impact on the progress and welfare of Indonesian citizens.

* Corresponding author: reynaldi.19008@mhs.unesa.ac.id

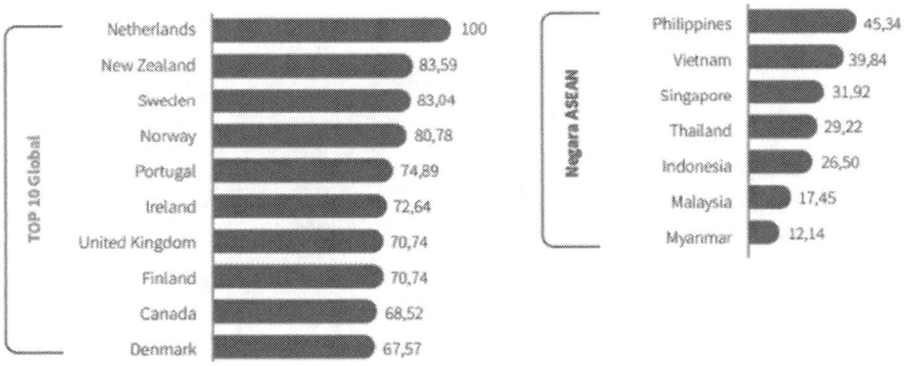

Fig. 1. Global Inclusivity Index 2020 [3]

An environment that upholds the value of inclusivity is important for equality for people with disabilities [4]. To see the level of inclusiveness in Indonesia, the global inclusiveness index can be used as a reference to assess the level of inclusiveness. As can be seen in Figure 1, Indonesia is ranked 125th in the global inclusiveness index. This ranking shows that Indonesia is still far below the ranking when compared to other countries in terms of inclusiveness.

As awareness of the importance of social and economic inclusion for people with disabilities increases, various policies have been created in Indonesia to support and protect their rights. One important step is the issuance of UU Number 8, year 2016 concerning Persons with Disabilities, which provides a legal basis for the protection and fulfillment of the rights of persons with disabilities. Through this law, the government has recognized the right of people with disabilities to have equal access to various aspects of life, such as education, health, work, and public services. Not only laws but policies regarding the acceptance of workers with disabilities in employment providers, so that they get equal employment opportunities. Slowly, Indonesia has shown support for people with disabilities through efforts to create a more inclusive and justice society for people with disabilities, so that they can participate fully in the social and economic life of this country [5]. Of course, this is linear with the achievements of SDGs 4 (quality education), 8 (decent work and economic growth), and 10 (reduced inequalities).

The various supports provided to people with disabilities do not make them free from difficulties in finding work. People with disabilities are still faced with problems that prevent them from getting equal employment opportunities [6]. There are some businessmen who see people with disabilities as a negative thing. Insufficient physical and technological accessibility also becomes an obstacle for them to access the expected work environment. Not many employers understand the benefits of employing workers with disabilities [7].

The problem of absorbing disabled workers is also faced by job providers. These problems create obstacles for them to provide employment opportunities for people with disabilities. In Figure 2, you can see several things that are problems faced by the company. Concerns about decreased productivity, high costs, negative perceptions, and lack of support from the government are some of the problems they face. The biggest problem that is most often faced is the lack of insight in recruiting people with disabilities and the absence of people with disabilities registering for the job vacancies that have been provided.

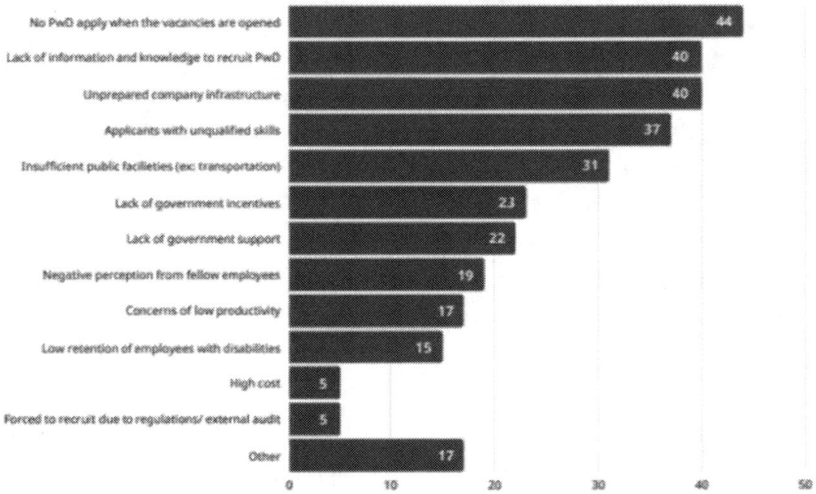

Fig. 2. Barriers to Companies in Recruiting Persons with Disabilities [8]

Various challenges faced by companies in absorbing workers with disabilities, several companies "dare" to appear as disability-friendly companies and employ them. These companies open special recruitment for people with disabilities and employ them within the company environment. This will have an impact on the company, such as the consequences of providing disability-friendly facilities, an equal work culture, and an inclusive work environment. Companies are also faced with the possibility of negative perceptions, such as inefficiency which can have an impact on increasing product costs. This interesting phenomenon of companies with disability-friendly branding is the focus of this research. The company's decision to become a disability-friendly company, which has not been done by many other companies, is an interesting thing in this research. The use of the corporate branding approach as an approach in analyzing this phenomenon was carried out to answer this research. Corporate branding can provide benefits to the company by increasing the company's brand equity. So, can disability-friendly corporate branding affect a company's brand equity? This is the question that wants to be answered through this research.

This research aims to analyze the influence of disability-friendly corporate branding in companies on the company's brand equity. So that the results of the analysis can provide benefits for the company, including determining a strategy to increase its brand equity. Simultaneously, when companies begin to strive to become disability-friendly companies, people with disabilities get the opportunity for equal service as customers and employees, and support the achievement of SDGs from the aspect of education and reducing disparities.

2 Literature review

2.1 Corporate branding

A company with a strong and positive image starts from the company's steps in carrying out corporate branding [8]. Corporate branding is known as a process of forming a company's image and identity to be known and identified by stakeholders [8, 9]. The stakeholders in question are customers, employees, the general public, and other stakeholders. Efforts to implement corporate branding companies aim to make the company appear different from other companies, creating a positive and strong perception as desired by the company. The company's success in carrying out corporate branding will open up various new and interesting opportunities for the company now and in the future [10].

Companies need to pay attention to several things in their efforts to implement corporate branding. Companies need to ensure they are able to build trust [11, 12]. Through this trust, stakeholders will assess the company as a company that is reliable, has social responsibility, and has a positive impact on them. Companies also need to carry out consistent communication in their efforts to carry out corporate branding [9, 13]. The communication carried out aims to strengthen the brand image and build the desired impression in the perception of stakeholders.

In this research, the company's corporate branding is a disability-friendly company. The company establishes itself as a company that supports justice for people with disabilities. Disability-friendly values, culture and image will be built by the company based on the perceptions of its stakeholders. This process will have an impact on decision-making within the company, such as providing disability-friendly facilities, accepting employees with disabilities, as well as company programs and policies related to disability-friendly inclusivity.

2.2 Brand equity

Brand equity is a valuable asset for a company for long-term success [14]. As a company brand value, brand equity can be seen in the form of positive or negative perceptions of the company brand among stakeholders [15]. A company's brand equity can increase the company's value in the market, differentiating the brand from other companies [16]. This supports the company to be known by its stakeholders, so that the company is able to compete in the long term.

A company's brand equity is influenced by the company's corporate branding as a disability-friendly company. Through corporate branding, companies can be assessed as companies that support the values of inclusiveness and strengthen the company's image as a socially responsible company. Through this process, an emotional connection will be created between the company and its stakeholders, supporting the company, and ultimately increasing the company's brand equity [15]. Through disability-friendly corporate branding, companies can generate a positive reputation. The company will appear as a company that actively promotes inclusivity, especially for people with disabilities. This will affect the company's brand awareness. In the context of people with disabilities, corporate branding will contribute to the development of strong brand equity, which can provide benefits for companies in the future.

3 Methods

This research uses a quantitative approach to answer research questions. The research population is East Java Province which has 5,578 large and medium industries based on the 2019 BPS East Java Statistics. Sample selection was carried out using a purposive technique with the following criteria: (1) located in the East Java region; (2) is an area that has senior high school for special needs with more than 50 students; (3) data recorded in the industrial directory in the East Java Province region with complete data so that questionnaires can be sent either by email, telephone or postal mail.

Tabel 1. Research Sample

Details	Total
1. Regencies and cities in East Java	38 regency & city
2. Have less than 50 Senior High Schools for special needs or are not recorded in the directory of schools with special needs in East Java province	26 regency & city
3. Number of industries in 8 regencies and cities	3.089
4. Number of industries that have complete data in the directory of the Regency/City Regional Government Industry and Trade Service	1.012
5. Samples per region	
5.1 Pasuruan Regency	39
5.2 Kediri City	204
5.3 Surabaya City	482
5.4 Malang Regency	238
5.5 Bojonegoro Regency	18
5.6 Sidoarjo Regency	630
5.7 Mojokerto Regency	228
5.8 Gresik Regency	312
6. Total sample sent questionnaire	2.151
7. Completed and returned questionnaires	212
8. Questionnaire that can be processed (completely filled out)	180

The sample selection took into account the number of Senior High Schools for special needs in a region because this indicates a large number of potential disabled workers. Meanwhile, the selection of large and medium industrial areas is adjusted to the existence of the Senior High School for special needs. There were 3 regions excluded from the sample, namely Blitar, Trenggalek and Madiun Regencies, because even though there were more than 50 students at Senior High School for special needs, the number of large and medium industries was not significant. On the other hand, there are areas with less than 50 students of Senior High School for special needs but they are industrial areas so they are still included in the sample, namely Gresik and Sidoarjo districts. Thus, the final sample was 180 companies that filled out the questionnaire completely. The questionnaire was sent via post, email and telephone contact which was followed up by filling it out directly. The weakness of the questionnaire data distributed is that more than 50% of respondents did not fill in data on industrial location areas and industrial classifications so they could not have complete demographic data.

A research questionnaire was created to obtain research respondent data. The questionnaire was distributed online using Google Forms to make it easier for respondents to complete it. A Likert scale of 1-5 was used in the study with a scale of 1 representing a response of "strongly disagree" and 5 representing "strongly agree". Corporate branding was measured using 10 items adopted from previous research [17]. Brand equity was measured using 9 items adopted from previous research questionnaires [18]. The results of the respondent data will then be processed using SPSS and AMOS to carry out hypothesis testing, answering this research question.

Table 2. Research Questionnaire Items

Corporate Branding (Adapted from Chiang et al. 2013) [17]	Our company values employee contributions to inclusiveness and disability-friendly values
	Our company conveys the values and image of a disability-friendly company through official company channels
	Our company conveys the values and image of a disability-friendly company to stakeholders
	Every part of the company strives to support activities to increase its disability-friendly image
	Every part of the company discusses how employees implement a culture of equality
	Our company receives input from stakeholders regarding the implementation of a disability-friendly environment
	Our company has disability-friendly facilities
	Our company ensures that every new employee in the company understands its values and disability-friendly culture
	Our company conveys the company's hopes of becoming a disability-friendly company through various information channels
	Our company conveys disability-friendly values, image and culture to every employee in the company through various channels
Brand Equity (Adapted from Chang et al. 2015) [18]	I consider myself loyal to disability-friendly companies
	Disability-friendly companies are my first choice
	I would buy a disability-friendly company's product over another product option
	The quality of disability-friendly company products is better
	The company's disability-friendly products are useful to me
	I was able to recognize a disability-friendly company brand
	I am aware that disability-friendly companies exist
	The company's disability-friendly characteristics became a part of me
	I easily remember disability-friendly companies

4 Result and discussion

4.1 Respondent characteristics

Research data was obtained through questionnaires distributed to target respondents according to established criteria. From the data collection carried out, information was obtained on the characteristics of the research respondents. The recapitulation of respondent characteristics can be seen in the following table.

Table 3. Respondent Characteristics

No	Characteristic		Total	Freq
1	Gender	Male	94	52%
		Female	86	48%
2	Age	20-30 y.o.	44	24%
		30-40 y.o.	93	52%
		40-50 y.o.	31	17%
		Above 50 y.o.	12	7%
3	How Long Have You Worked?	Less than 1 year	21	12%
		1-2 years	56	31%
		3-4 years	75	42%
		More than 4 years	28	15%
4	Working Location Area	Pasuruan Regency	2	1%
		Kediri City	17	10%
		Surabaya City	43	24%
		Malang Regency	20	11%
		Bojonegoro Regency	1	1%
		Sidoarjo Regency	54	30%
		Mojokerto Regency	17	9%
		Gresik Regency	26	14%

Based on Table 3, the characteristics of the respondents in this study can be seen. For gender, there is an almost even distribution between men and women. For age, most respondents were in the 30 to 40 years range (52%). Regarding how long they have worked, most have worked for a long time in the range of 3 to 4 years (42%). For regional distribution, the majority of respondents were in the Sidoarjo district area, 30%. It is interesting that it was found that most of them were at a time when their careers were already starting to become established, judging from the age at which they had entered the old adult stage and had worked for quite a long time. Some respondents were familiar with the office situation they occupied

4.2 Validity and reliability test

The research data obtained was tested to see whether it met the requirements for validity and reliability. Testing was carried out using AMOS, by looking at the loading factor, CR and AVE values of each variable. The test results can be seen in the Table below.

Table 4. Validity and Reliability Test

Variable	Item	Loading Factor	CR	AVE
Corporate Branding	CB1	0.704	0.926189	0.557091
	CB2	0.74		

	CB3	0.752		
	CB4	0.807		
	CB5	0.74		
	CB6	0.764		
	CB7	0.802		
	CB8	0.708		
	CB9	0.744		
	CB10	0.694		
Brand Equity	BE1	0.764	0.913833	0.541136
	BE2	0.734		
	BE3	0.742		
	BE4	0.731		
	BE5	0.773		
	BE6	0.721		
	BE7	0.73		
	BE8	0.704		
	BE9	0.719		

It can be seen in Table 4 that there are 10 questionnaire items for corporate branding and 9 questionnaire items for brand equity which have been tested for validity and reliability. Validity can be seen from the loading factor value of each question item. A question item is declared valid when the value is more than 0.5. From Table 4, there are no loading factor values below 0.5, indicating that each questionnaire item is valid. For reliability, it can be seen from the CR and AVE values of each variable. It is declared reliable when the CR value is more than 0.7 and the AVE value is above 0.5. Looking at table 3 above, each variable meets the CR and AVE value requirements and is declared reliable. After that, it can be continued at the next analysis

4.3 Normality test

The research data that has been obtained needs to be tested for the normality of the data before proceeding to the hypothesis analysis stage. A normality test was carried out using AMOS. The test results can be seen in the following table.

Table 5. Normality Test

Variable	min	max	skew	c.r.	kurtosis	c.r.
BE9	2	5	-0.537	-2.943	-0.527	-1.443
BE8	1	5	-0.698	-3.825	-0.232	-0.635
BE7	2	5	-0.584	-3.199	-0.425	-1.164
BE6	2	5	-0.696	-3.814	-0.463	-1.268
BE5	2	5	-0.708	-3.88	-0.261	-0.714
BE4	2	5	-0.518	-2.838	-0.541	-1.483
BE3	1	5	-0.669	-3.667	-0.345	-0.945
BE2	2	5	-0.528	-2.892	-0.267	-0.732
BE1	2	5	-0.568	-3.11	-0.419	-1.148

CB1	2	5	-0.51	-2.794	-0.708	-1.94
CB2	2	5	-0.621	-3.404	-0.391	-1.072
CB3	2	5	-0.727	-3.98	-0.352	-0.964
CB4	2	5	-0.791	-4.334	-0.111	-0.304
CB5	2	5	-0.396	-2.169	-0.757	-2.072
CB6	2	5	-0.526	-2.883	-0.604	-1.654
CB7	2	5	-0.667	-3.651	-0.667	-1.826
CB8	2	5	-0.464	-2.539	-0.726	-1.987
CB9	2	5	-0.745	-4.082	-0.356	-0.975
CB10	2	5	-0.674	-3.69	-0.362	-0.992
Multivariate					5.875	**1.395**

In table 5, you can see the results of the normality test using AMOS. Research data is considered normal when the value is smaller than 2.58. From the multivariate values in the table, it is known that the value is 1.395. These results state that the research data is normal. Research data can then be used to test research hypotheses

4.4 Model fit test

Research model fit testing was carried out by looking at several model fit indicators through AMOS. The results of model fit testing can be seen in the following Table 6.

Table 6. Model Fit

Item	Accepted Fit	Results	Conclusion
CMIN/DF	< 3=acceptable fit	1.645	Model Fit
GFI	1 = perfect fit	0.874	Model Fit
AGFI	> 0.9=acceptable fit	0.842	Model Fit
TLI	1 = perfect fit	0.950	Model Fit
RMSEA	< 0.05 = reasonable fit	0.060	Model Fit

The fit model of this research model can be seen from the results of Table 6. The CMIN/DF value indicates an acceptable fit when the value is below or equal to 3. With a research CMIN/DF value of 1,645, it can be assessed as meeting the fit model. The Goodness of fit (GFI) and Adjusted Goodness of fit (AGFI) values are said to be a perfect fit when the value is 1. With a research GFI value of 0.874 and AGFI of 0.842, it can be assessed as meeting the fit model. The TLI value is considered to be closer to 1, the more it represents a very good fit. With a research TLI value of 0.950, it can be assessed as meeting the model fit. The Root Mean Square Error of Approximation (RMSEA) value has a good model fit when the value is below or equal to 0.05. With a research RMSEA value of 0.060, it can be assessed as meeting the model fit. Overall, this research model meets the criteria for being a fit model

4.5 Hypothesis testing

Research answers will be found by testing research hypotheses. In this research, hypothesis testing was carried out using AMOS with the results can be seen in the image below.

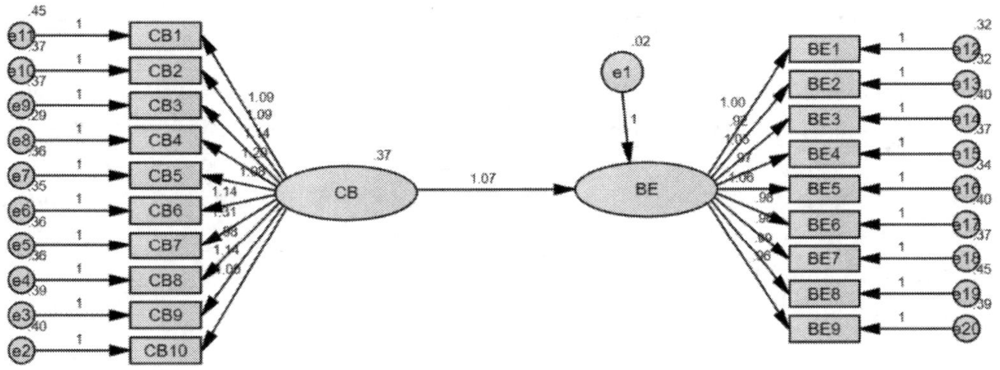

Fig. 3. AMOS Test Results

The AMOS test results can be seen in Figure 3. To answer the hypothesis, the estimated value can be seen in the AMOS test results. The relationship value can be considered significant if the value is below 0.05. The results of this study show that the relationship between corporate branding and brand equity is 0.000. So H1 is accepted. There is a significant relationship between corporate branding variables and brand equity.

Through this research, corporate branding is proven to have a significant influence on brand equity, as in previous research [18, 19]. Strong corporate branding has become an important factor in building sustainable brand equity and winning over stakeholders in the interests of the company. When a company consistently shows its commitment to supporting and understanding the needs of people with disabilities, it creates a positive perception among stakeholders, including customers, employees and the general public. These values form a strong foundation for building sustainable brand equity.

The brand image built through disability-friendly corporate branding is key in influencing consumer perceptions. The brand image reflects how the company is seen by society [20, 21]. In this context, brand image includes the impression of the extent to which a company respects people with disabilities, whether through products, services or internal policies. A positive image of support for people with disabilities can influence consumer loyalty and brand preferences, ultimately contributing to strong brand equity.

Disability-friendly corporate branding has a significant impact on companies. Values that reflect inclusion, a positive brand image, and an inclusive corporate culture play a role in shaping consumer and stakeholder perceptions of the brand. When a company consistently supports people with disabilities in all aspects of its operations, it creates a strong foundation for building positive and sustainable brand equity.

The results of this research have several things that companies can implement to become disability-friendly companies. As is known, companies with a disability image have various benefits for the long-term sustainability of the company. This is because stakeholders can see the company's commitment to creating an inclusive and justice environment, especially for people with disabilities. Especially for stakeholders who care about the values of equality and welfare for all parties.

Through efforts to implement disability-friendly corporate branding, companies must begin to include the values of inclusion, social responsibility and support for people with disabilities as the core of their brand identity. This is reflected in an inclusive company culture, which supports the acceptance and employment of employees with disabilities, and in the design of products and services that take accessibility into account. A positive brand image develops through appropriate communications that highlight the company's commitment to disability issues. As the company's brand equity increases, the company will gain various benefits from this action, which also has the impact of encouraging other companies to do the same. This will also support the achievement of SDGs targets 4 (quality education), 8 (decent work and economic growth), and 10 (reduced inequalities).

5 Conclusion

This research aims to analyze the influence of disability-friendly corporate branding on company brand equity. Disability-friendly company branding is built on the values of inclusivity, the image of a company that cares for people with disabilities, and a culture of equality. Through this research, empirical evidence was obtained that disability-friendly corporate branding has an effect on company brand equity. The research was carried out only in the East Java area which has industrial centers and senior high schools for special needs, which is a research limitation in generalizing the research results. Further research can add variables that can develop research on corporate branding, especially in the context of people with disabilities

References

1. Dugarova, Esuna, Social inclusion, poverty eradication and the 2030 agenda for sustainable development (UNRISD Working Paper, 2015)
2. Pujianto, Agung, E. Tjahjono, *Economic Empowerment Model of People with Disability in the Creative Industries*, 5th Annual International Conference on Management Research, Atlantis Press, (2019)
3. Soekatri, Y. E. Moesijanti, S. Sandjaja, A. Syauqy, Int. J. Environ. Res. Public Health. **17**, 17 (2020)
4. Garcia, Crystal E., et al., J Coll Stud Dev. **62**, 2 (2021)
5. S. Nurhayati, *Prophetic Law Review*, **2**, 1 (2022)
6. T. L. Ta, K. S. Leng, Asia Pac. Disabil. Rehabil. J. **24**, 1 (2013)
7. H. P. Hartnett, H. Stuart, H. Thurman, B. Loy, L. C. Batiste, J. Vocat. Rehabil. **34**, 1 (2011)
8. T. Gunawan, J. F. Rezki, Mapping Workers with Disabilities in Indonesia Policy Suggestions and Recommendations (International Labour Organization, 2022)
9. M. Juntunen, S. Saraniemi, M. Halttu, J. Tähtinen, J. Brand Manag. **18**, 2 (2010)
10. K. De Roeck, F. Maon, C. Lejeune, Eur. Manag. Rev. **10**, 3 (2013)
11. S. L. Burt, L. Sparks, Corp. Reput. Rev. **5**, 3 (2002)
12. B. Merrilees, M. L. Fry, Corp. Reput. Rev. **5**, 2 (2002)

13. R. Abratt, N. Kleyn, Eur. J. Mark. **46**, 7 (2012)
14. N. Z. Ökten, et al., Eur. Res. Manag. Bus. Econ. **25**, 1 (2019)
15. O. Iglesias, N. Ind, M. Alfaro, J. Brand Manag. **20** (2013)
16. J. Kim, S. Sharma, K. Setzekorn, JMM Int. J. Media Manag. **4**, 2 (2002)
17. H. H. Chiang, A. Chang, T. S. Han, D. McConville, J. Gen. Manag. **39**, 1 (2013)
18. A. Chang, H. H. Chiang, T. S. Han, Asia Pac. Manag. Rev. **20**, 3 (2015)
19. M. A. Sallam, Int. j. marketing stud. **8**, 1 (2016)
20. M. Longo, M. Mura, A. Bonoli, Corporate Governance 5, 4 (2005)
21. P. Popoli, Marketing Theory **11**, 4 (2011)

https://doi.org/10.1051/e3sconf/202345003006

How to business environment affect green innovation in small medium enterprises: Systematic literature review

Hujjatullah Fazlurrahman[1*], *Udisubakti* Ciptomulyono[2], and *Irhamah*[3]

[1]Doctor of Management Technology, Sepuluh Nopember Institute of Technology, Digital Business, State University of Surabaya, Surabaya, Indonesia
[2]Dept. Industrial and System Engineering, Sepuluh Nopember Institute of Technology, Surabaya, Indonesia
[3]Dept. of Statistics, Sepuluh Nopember Institute of Technology, Surabaya, Indonesia

Abstract. This paper discovers the concept of green innovation in small and medium-sized enterprises (SMEs). The study deliberates on reputation of green innovation in promoting sustainable development and the need for SMEs to encirculate this practice. The paper offers an overview of the different types of green innovation and their significance in promoting environmental sustainability. It examines the challenges faced by SMEs in adopting green innovation practices, with limited resources and lack of awareness. Furthermore, the paper highlights the benefits of green innovation for SMEs, including increased competitiveness and improved brand image. The study achieves this by providing recommendations for SMEs on how to effectively instrument green innovation practices and strategies. Overall, this paper emphasizes the importance of SMEs role in promoting green innovation and sustainable development. This paper will also discuss the classification of internal and external influences of organizations in green innovation research. Internal organizational influences are classified into human resources, corporate culture, technical capabilities, organizational structure, vision, mission and objectives. The internal influence of the organization consists of political, economic, technology and regulation.

1 Introduction

Green innovation or what is often called Green Innovation is now increasingly popular, especially among small and medium businesses (SMEs) who are increasingly aware of the importance of protecting the environment. Green Innovation in SMEs relates to the development and use of environmentally friendly and sustainable technologies, green innovation is one of the business strategies that have an impact on business and the environment. Not only seeking profit but also reducing negative environmental impacts. This is the reason for the spread of green innovation in SMEs [1]. This is where the role

* Corresponding author: hujjatullahfazlurrahman@unesa.ac.id

of SMEs in efforts to reduce negative impacts on the environment and society. Green innovation is the right solution as a form of environmental and social responsibility around SME locations, this has an impact on improving business sustainability.

The process of developing green innovation is not easy, because it is faced with several obstacles and challenges. One of them is resources and funds for the development of innovations. Government support is needed in the development of green innovation in SMEs [2]. In addition, stakeholders and the community also have the same role. The government and industry need to provide incentives to SMEs that have used green innovation as a form of support for business sustainability.

This green innovation is one of the global issues in the sustainable development goals (SDGs). This program aims to improve community welfare and environmental sustainability. In this case, green innovation in SMEs can help achieve these goals by improving the quality of life of the community and reducing the negative impact of business activities on the environment.

2 Literature review

This author [1], explained that there is a European project that promotes eco-innovation called ECOLABNET. The project aims to develop and implement an innovative eco-service knowledge management system for SMEs. This service is very beneficial for SMEs because it can share knowledge, service management, and best practices with SME partners. The authors examined the implementation of the system in Navarra, Spain, and evaluated the level of effectiveness of promoting environmental innovation in SMEs. The results of this research show that this system has succeeded in transferring knowledge and collaboration among SME partners. In addition, it has contributed to the development of eco-innovation in SMEs in the area.

The role of knowledge management in the development and distribution of eco-innovative service packages for small and medium-sized businesses (SMEs) is the subject of the research presented in [2], which may be found here. The authors contend that effective knowledge management is essential for the development and delivery of eco-innovative services, which can assist small and medium-sized enterprises (SMEs) in improving their environmental performance and competitiveness. The article presents a case study of a Polish consulting company that has built a range of eco-innovative service packages for small and medium-sized enterprises (SMEs). The conclusion of the paper emphasizes the significance of knowledge management for the effective provision of eco-innovative services and makes the recommendation that additional study is required to gain a deeper comprehension of the connection between knowledge management and eco-innovation in small and medium-sized enterprises (SMEs). The purpose of this research [3] is to investigate the function of eco-management in the processes of environmental management in Latvia. The authors argue that eco-management has the potential to play a significant part in the advancement of sustainable development and the enhancement of environmental performance in Latvia, particularly within the context of the transition that the country is undergoing toward a green economy.

This study [4] investigates the challenge of internationalization that is presented to small and medium-sized firms (SMEs) who specialize in eco-innovations with the intention of increasing energy efficiency in other industries. The authors believe that small and medium-sized enterprises (SMEs) confront a unique set of obstacles due to the complicated nature of the eco-innovation value proposition. This value proposition frequently requires close collaboration and customization in order to fulfill the specific requirements of individual customers. The paper presents a case study of a Spanish small and medium-sized enterprise (SME) that has created a range of eco-innovative solutions for energy efficiency in the hospitality sector. In its conclusion, the report highlights the relevance of conducting additional research into the internationalization issues that small and medium-sized enterprises (SMEs) involved in eco-innovation face, as well as the significance of addressing these difficulties in order to advance sustainable development and safeguard the environment [5].

In order to investigate the connections between green innovation, green human resource management, green transformational leadership, and green human resource management, the author [6] suggests a mediation-modernization model. The authors contend that good green human resource management and green transformational leadership can encourage green innovation, which, in turn, can stimulate environmental performance inside firms. In this research, a quantitative study is presented that is based on a survey of employees from a variety of firms in Pakistan, and an analysis of the proposed mediation-modernization model is performed. According to the findings of the study, green human resource management and green transformational leadership have a favorable impact on green innovation, which in turn has a beneficial impact on environmental performance. In addition, the study reveals that the positive association between green human resource management and green innovation is moderated by green transformational leadership, which strengthens the influence of green human resource management on green innovation. In its conclusion, the paper highlights the importance of green human resource management, green transformational leadership, and green innovation in promoting environmental performance in organizations. Additionally, the paper suggests that future research should concentrate on the circumstantial factors that influence the association between these variables [7].

The practices of eco-innovation among small and medium-sized businesses (SMEs) in the food processing and manufacturing industries are investigated by researchers [8]. The authors contend that eco-innovation can give small and medium-sized enterprises (SMEs) a competitive advantage by increasing resource efficiency, lowering waste levels, and catering to the growing demand among consumers for ecologically sustainable products. The paper discusses a qualitative study that was conducted in Pakistan and was based on interviews with small and medium-sized businesses. The authors also identify a number of obstacles that stand in the way of the widespread implementation of eco-innovation, such as a dearth of monetary resources, a scarcity of understanding regarding eco-innovation techniques, and an aversion to change.

The author [9] explores the process by which businesses implement environmentally friendly innovations by means of the growth of environmentally friendly dynamic capabilities, as well as the shrinking importance of environmental vitality and the capacity for big data analytics. The authors contend that green dynamic capabilities can assist businesses in adjusting to changes in the surrounding environment and in adopting green innovation techniques, which can ultimately lead to increased performance and sustainability. The paper presents a quantitative analysis that was based on a survey of companies in Malaysia.

The author [10] investigates the ways in which small and medium-sized businesses (SMEs) have adopted and are putting into practice environmentally friendly, innovative strategies. The authors contend that small and medium-sized enterprises (SMEs) have the potential to profit from green innovation methods by strengthening their environmental performance, lowering their expenses, and becoming more competitive. The paper summarizes the findings of a qualitative study that was conducted in Portugal and was based on interviews with small and medium-sized businesses. In its conclusion, the study highlights how vital it is to encourage the adoption of environmentally friendly innovations by small and medium-sized enterprises (SMEs), and it proposes that supportive policies and programs can assist in creating sustainable development and economic progress [11].

This research [12] examines the possible benefits of green innovation for small and medium-sized enterprises (SMEs) that are technology-based. These potential benefits include developing a green image, taking technological initiatives, and improving market competitiveness. This study investigates the connection between low-carbon services and the performance of businesses. According to the findings of the study, low-carbon service innovation has the potential to boost the performance of technology-based SMEs by more than 20% under specific conditions. These findings show the potential value of green innovation for companies of this sort [13].

Within the context of the countries that make up the European Union (EU), the topic of this article [14] is the participation of small and medium-sized businesses (SMEs) in eco-innovation activities. According to the findings of the research, small and medium-sized enterprises (SMEs) in the EU participate in activities related to environmental innovation a significantly lower percentage of the time than they do in activities related to innovation, and the level of participation in environmental innovation can range from that of innovation. The research also categorizes European Union member states into three distinct groups according to the extent to which they participate in eco-innovation and innovation-related endeavors.

A study that identifies factors affecting the environmental, financial, and social sustainability of manufacturing SMEs in Malaysia is discussed in this article [15]. The study was conducted in Malaysia. For the purpose of evaluating hypotheses and estimating the level of importance they have in terms of sustainable performance, the research makes use of a combination of structural equation modeling and artificial neural networks.

This study [16] addresses the relationship between green innovation, resource efficiency, and sustainable economic growth in developing countries, particularly in the E7 countries (China, India, Mexico, Russia, Turkey, Brazil, and India). The E7 countries are referred to as the BRIC countries. The authors examine the data from 2010 to 2021 using a technique called the Generalized Method of Moments (GMM), and they come to the conclusion that green innovation and resource efficiency both have a beneficial impact on green economic growth. They advocate for the implementation of pragmatic policies such as enhancing the creation of environmentally friendly jobs through SMEs, supporting environmentally friendly financing methods, and fostering environmentally friendly foreign direct investment (FDI).

This research [17] illustrates the environmental issues faced by the world as a result of increasing carbon emissions, waste of energy, pollution, wastewater, and a lack of water. Small and medium-sized businesses (also known as SMEs) are responsible for a major portion of the world's pollution and need to find solutions to environmental problems. This article analyzes the ways in which initiatives for corporate social responsibility (CSR), particularly green innovation, can improve environmental performance in small and medium-sized enterprises (SMEs) in developing countries such as the Maldives and Morocco [18].

According to the findings of this study [19], social innovation can also have a beneficial effect on environmental performance. The term "social innovation" refers to the generation and application of novel ideas, principles, or practices with the goals of satisfying societal requirements and enhancing societal outcomes. This study conducts a survey of small and medium-sized enterprises (SMEs) in Semarang, Central Java, and finds that acceptance of SBMI is influenced by a variety of elements, including external factors, dynamic performance expectations, facilitating conditions, social influence, effort expectations, hedonic motivation, and internal speed considerations [20].

This study helps to understand the relationship between green entrepreneurship, innovation and financial performance in the context of Saudi small businesses. It emphasizes the importance of promoting green business self-efficacy and business orientation to promote sustainable business practices and outcomes [21]. Social and political contributions are most important in promoting sustainable product innovation, followed by reducing pollution, waste and emissions.

In addition, the study found that pollution and waste are the most influential factors in promoting sustainable process innovation, followed by anti-competitive behavior and emission reduction [22]. the impact of knowledge transfer activities on SMEs in the food sector. The KITE project facilitated knowledge transfer between universities and industry, resulting in added value and increased competitiveness for the SMEs. The financial investment in research and technology support activities led to developments in information technology and systems, supplier sourcing, and waste reduction. The university's technical and food science/safety support enabled service activity resulting in certifications, new supply routes, and increased sales for the SMEs. These findings contribute to the understanding of the importance of knowledge transfer in driving growth and competitiveness in the food sector [23]. This study involved conducting a survey among SMEs in Portugal and the UK. The survey was shared with SMEs registered with the Small Business Support Institute in both countries, and a total of 249 responses were received after removing incomplete responses [24] the understanding of the factors influencing green innovation and economic performance in small enterprises and provides insights for policymakers and organizations to support and promote green entrepreneurship in Saudi Arabia [25] eco-innovation plays a crucial role in transitioning towards a more resource-efficient economy. The eco-innovation index is an important indicator of environmental innovation, research, and development, and is one of the thematic indicators in the European Union's Resource Efficiency Scoreboard [26].

Providing insights and recommendations for the management of SMEs in reorganizing their activities and redefining their strategies based on eco-innovation.

The study suggests that future research should focus on expanding cooperation between institutions at the industry and government levels, as well as measuring other innovation indicators using individual clusters to better understand differences [26].

The study focuses on Thai SMEs and may not be generalizable to other contexts. Future research could explore the impact of intellectual capital on open sustainability innovation in different industries or countries. Additionally, the study could consider the inclusion of human capital in the intellectual capital construct and further investigate the consistency of open sustainability innovation with sustainable development goals [27]. The literature on eco-innovation has identified key drivers of eco-innovation at the macro, meso, and micro levels [28]. These drivers include the importance of connecting various sources of knowledge, such as universities and suppliers, in supporting environmentally friendly business efforts. Customer demand is also a critical driver, as more environmentally sensitive customers raise their expectations for the environmental quality of goods and services. Additionally, financial performance has been found to be positively correlated with environmental innovation activities [29].

3 Methods

Since the development of data mining tools, the process known as systematic literature review (SLR) has seen a surge in popularity. Following the completion of a search strategy and the use of bibliometric analysis (BA), the SLR researches the inclusion of data collection. For the purpose of this study, "Biblioshiny" is a bibliometric instrument that is used for the examination of science mapping. The author then uses the string " sustainability innovation" OR "sustainable innovation" OR "environment innovation" OR "environmental innovation" OR "green innovation" OR "eco-innovation" AND "SME" OR "SMEs" OR "small and medium-sized" OR "small-medium businesses.". We found 512 document to generate the data.

4 Result and discussion

In Fig. 1 analysis of Biblioshiny, there are various groupings that can be used to understand and evaluate research results. One of them is a thematic map that consists of four quadrants, namely Niche Themes, Emerging or Declining Themes, Motor Themes, and Basic Themes.

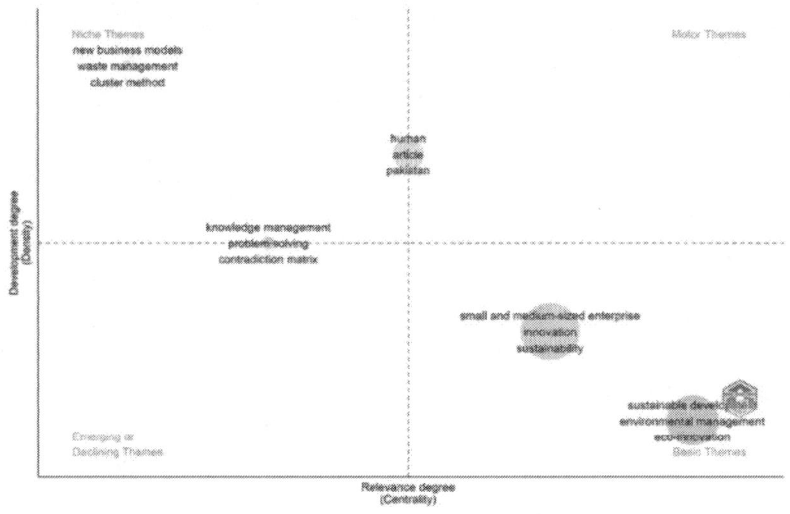

Fig. 1. Bibliometric thematic map.

In the Basic Themes section, there are two groups that are relevant to the paper title "Green Innovation in SME", namely groups that discuss Small and Medium-Sized Enterprise Innovation Sustainability and Sustainable Environmental Management Eco-Innovation. However, even though both are in the same quadrant, their positions are different and have different focuses.

The Small and Medium-Sized Enterprise Innovation Sustainability group tends to focus on things related to technology and trends, as well as things that are rising or falling. Meanwhile, the Sustainable Environmental Management Eco Innovation group tends to focus more on the basic things that form the basis of research.

Even though both have almost the same size and are widely discussed by others, it is still important to understand the difference in focus between the two in the context of green innovation in SMEs.

Fig. 2. Bibliometric thematic evolution

In Fig. 2 it can be seen that Thematic Evolution is divided into 2 time slice periods, namely the first time slice with the 2002-2020 period, initially, there were 8 Discussion

Topics which over time as seen in the second time slice in 2021-2023, the topics was split and conical again with fragments of other discussion topics to become 5 clusters.

If you pay attention, in the 2002-2020 period the topic of discussion was eco-innovation transformed into the baltic sea with a total of 2 papers in the 2021-2023 period which shows that there has been a change in the focus of the discussion from the main topics discussed, namely eco-innovation become baltic sea. they help in understanding the changing focus of trends over time

Then next, if you look closely, at the topic of discussion regarding innovation in the 2002-2020 period which is divided into 4 parts, namely 28 papers that discuss sustainable development, 1 paper that discusses energy efficiency, with 47 papers that discuss small and medium-sized enterprise and 8 papers that discuss human. This shows that there has been a development in terms of research, where researchers are trying to understand and identify more deeply sustainable development, energy efficiency, small and medium-sized enterprises, and humans by using the basic or main topic of innovation.

Table 1. Internal factors of green innovation

No	Classification	Total variable	Percentages
1	Human resource	20	19.8%
2	Corporate culture	11	10.89%
3	Technological capabilities	35	34.65%
4	Organizational structure	22	21.78%
5	Vision, mission, and objective	13	12.87%

Table 1 explains that green innovation is influenced by internal factors and external factors. In internal factors, there are 5 classifications consisting of human resources, corporate culture, technological capabilities, organizational culture and vision, mission and objective. In the classification of human resources, there are 20 variables with a percentage of 19.8% of 100% and variables dominated by green human resource management and transformational leadership. Then in corporate culture, there are 11 variables with a percentage of 10.89% of 100% and corporate social responsibility as the variable that dominates this classification. Furthermore, technological capabilities have a variable number of 35 variables with a percentage of 34.65% and variable frequencies that often appear, namely corporate dynamic capability, and green absorptive capacity. Organizational structure has a number of variables as many as 22 variables with a percentage of 21.78% where the organizational support variable is the dominating variable in this classification. The last one is vision, mission and objective has 13 variables with a percentage of 12.87% which are dominated by knowledge management variables.

In this classification, the highest number of variables is found in the classification of technological capabilities, which amounts to 35 variables or 34.65%.

Table 2. External factors of green innovation

No	Classification	Total variable	Percentages
1	Political	8	17.7%
2	Economic	22	48.88%
3	Technology	3	6.66%
4	Regulation	7	15.5%
5	Natural	22	11.11%

Table 2. In external factors, there are 5 classifications, namely political, economic, technology, regulation, and natural. In the political classification, there are 8 variables with a percentage of 17.7% of 100%. The for economics, there are 22 variables with 48.88%. Then technology has a variable number of 3 with a percentage of 6.66%. Regulation with the number of variables is 7 variables with 15.5%. And the last one is natural which has a total variable of 5 with 11.11%. The economic classification is the classification with the highest number of variables among other classifications, amounting to 22 variables or 48.88%.

5 Conclusion

Based on the results of the analysis conducted on thematic maps and thematic evolution, we can draw the conclusion that these two analytical tools can help provide a comprehensive view of trends and patterns in research results, and can help us understand the focus and developments that occur on the topic of discussion from time to time. Thematic Map on Biblioshiny can help us to group related topics of discussion and provide information about how much the topic of discussion is covered in the research literature. Besides that, Thematic Evolution on Biblioshiny helps us understand changes in focus and shifts in topic of discussion from time to time, so that we can evaluate the success of research and sustainable research strategies.

References

1. R. Kuceba, *System for Transfer and Management of Knowledge on Eco-Innovative Services for Smes In The Network Of Service Partners Ecolabnet: A Case Study*, in Proceedings of the 22nd European Conference on Knowledge Management: A Virtual Conference hosted by Coventry University, UK (2021)
2. G. Chmielarz, *Knowledge Management and Eco-Innovative Service Packages for Smes*, in Proceedings of the 22nd European Conference on Knowledge Management: A Virtual Conference hosted by Coventry University, UK (2021)

3. U. Libkovska, I. Lemss, *The Role of Eco Management in Environmental Management Processes in Latvia,* in 22nd SGEM International Multidisciplinary Scientific GeoConference 2022, Albena (2022)
4. N. Calvo, A. Monje-Amor, O. Villarreal, Technological Forecasting and Social Change **185** (2022)
5. M. Wasiq, M. Kamal, and N. Ali, Sustainability **15**, 3 (2023)
6. F. H. Awan, L. Dunnan, K. Jamil, and R. F. Gul, Environ Sci Pollut Res. **30**, 2 (2023)
7. R. Al-Hanakta, M. B. Hossain, L. Pataki, and A. Dunay, PLoS ONE **18**, 2 (2023)
8. A. Riaz, F. Ali, K. Ashfaq, A. Bhatti, and S. U. Rehman, British Food Journal **21** (2023)
9. D. Yu, S. Tao, A. Hanan, T. S. Ong, B. Latif, and M. Ali, IJERPH **19**, 16 (2022)
10. M. Rodrigues and M. Franco, Sustainability **15**, 5 (2023)
11. C. Phonthanukitithaworn, W. A. Srisathan, C. Ketkaew, and P. Naruetharadhol, Sustainability **15**, 3 (2023)
12. X. Huang, P. Huang, and T. Huang, IJETM **25**, 6 (2022)
13. E. D. Vaisman, M. V. Podshivalova, and A. A. Alola, Bus Strat Env. **31**, 3 (2022)
14. E. Sobczak and D. G. luszczuk, Sustainability **14**, 4 (2022)
15. E. Yadegaridehkordi, B. Foroughi, M. Iranmanesh, M. Nilashi, and M. Ghobakhloo, Sustainable Production and Consumption **35** (2023)
16. Y. Sun, P. Gao, W. Tian, and W. Guan, Resources Policy **81** (2023)
17. W. Simmou, K. Govindan, I. Sameer, K. Hussainey, and S. Simmou, Journal of Cleaner Production **384** (2023)
18. W. Omri, J.-M. Courrent, and A. Nemeh, IEEE Trans. Eng. Manage. **21** (2023)
19. A. Khattak, International Journal of Innovation Science, **15** (2023)
20. A. Utaminingsih, S. Y. Widowati, and E. H. Witjaksono, International Journal of Innovation Science, **15** (2023)
21. A. S. Alshebami, Sustainability **15** (2023)
22. B. Mukhtar, M. K. Shad, L. F. Woon, Predicting the Effect of Environment, Social and Governance Practices on Green Innovation: An Artificial Neural Network Approach. (Universiti Teknologi PETRONAS, Malaysia, 2023)
23. S. Mayho, et al., International Journal of Food System Dynamics **14**, 1 (2023)
24. F. Almeida, J. Santos, Sustainability **14**, 4 (2022)
25. S. Borghesi, G. Cainelli, M. Mazzanti, Res. Policy **44**, 3 (2015)
26. M. Taušová, L. Domaracká, K. Čulková, S. Matušková, N. Pena, M. Mikita, Acta Montanistica Slovaca **26**, 4 (2021)
27. F. Alvino, A. Di Vaio, R. Hassan, R. Palladino, Journal of Intellectual Capital **22**, 1 (2020)
28. J. Guo, M. Chen, X. Sun, Z. Wang, J. Xue, Plos One **15**, 12 (2020)
29. M. D. Parrilli, D. Radicic, A. Vargas-Sánchez, Technovation **119** (2021)

The importance of moral norms in constructing extended theory of planned behaviour: A Study of food-based SMES in urban agricultural community

Jun Surjanti[1*], *Rahayu Dewi* Suyono[2], *Tony Seno* Aji[3], *Heny* Musfidah[4], and *Rosa Prafitri* Juniarti[1]

[1]Economic Education, Faculty of Economic and Business, Universitas Negeri Surabaya, Indonesia
[2]Nutrion Sciences, Faculty of Mathematics and Science, Universitas Negeri Surabaya, Indonesia
[3]Nutrion Sciences, Faculty of Sport Science and Health, Universitas Negeri Surabaya, Indonesia
[4]PhD Management, Postgraduate, Universiti Sains Malaysia, Malaysia

Abstract. Growth and development in big cities often cannot be separated from the phenomenon of environmental quality degradation such as flooding and the availability of clean water, so a sustainable development program is needed that involves community participation, including SMEs. Their role in running a sustainable business, especially in the case of urban farm communities in relation to the moral norms between them, is important to study. Previous research has indeed studied moral norms in forming pro-environmental behavior, but this research has not tried to study them in the context of a society that was deliberately formed to foster environmental values, such as an urban agricultural community. This research intended to test the effect of moral norms, along with attitudes, subjective norms, and perceived behavioral control on sustainable business behavior. This research was a quantitative approach toward the food MSME population at the Urban Agricultural Society of Surabaya (UASS). Sampling used a purposive sampling technique. Research data was obtained from questionnaires distributed to 177 participants. The findings revealed that attitude, moral norm, and subjective norm all had a direct impact on intention but not perceived behavioral control.

1 Introduction

As a result of its economic growth, Surabaya, Indonesia's second-biggest metropolitan area, has a significant impact on environmental deterioration [1]. Waste is one of its many causes, and it is the most damaging to the environment. Flooding and solid waste generated by businesses and households are recurring challenges in the city [2]. Surabaya is prone to floods due to its high population, which is caused by subsidence of the land [3], and its location adjacent to the Brantas River [4]. This hypothesis is in line with Miller and Hutchins [5], who found the urban phenomena of population development in large cities as a result of increased floods and clean water availability. A sustainable development program is required to address

* Corresponding author: junsurjanti@unesa.ac.id

all existing and potential challenges. Sustainability is a development concept that aims to accomplish economic, social, and environmental goals [6, 7]. Attempts to establish a decent or sustainable city, according to Margolin [8], can occur at three levels: micro (individual activities), group, and organization (encompassing government, international, and corporate). Furthermore, the notion of sustainability may be found in a wide range of studies, such as visions of a sustainable future, local implementation of the transition to sustainability, public perception, and participatory method development to aid the transition [9]. As a consequence, it is evident that people contribute to initiatives toward sustainable development, either individually or collectively. This study emphasized the third role of sustainable development: SMEs' goals to run a sustainable business.

This study was conducted on SMEs in the Surabaya Urban Agricultural Society (UASS). The UASS is similar to the Urban Agriculture Community, which functions to integrate big city food supply with more environmentally friendly modes and is capable of "feeding the city" by integrating multiple resources, networks, and opportunities while contributing to the enhancement of community welfare [10]. Urban agriculture can also be used to mitigate flooding and ensure clean water security [11]. Surabaya is currently constructing a Big Biopore Infiltration Hole (BBIH), which was initially designed to tackle flooding and later functioned as the catfish cultivation area aside from its original function. This means that the UASS started to establish an urban farm as a means of developing UA [12]. Furthermore, SMEs were chosen as subjects for this study due to their significant environmental impact, which cannot be overstated [13]. Although it has a significant influence on environmental studies, this issue is not new and should not be regarded as novel [14].

This study employed Caldera et al. [15], natural resource-based view (NRBV) as a theoretical basis as well as Aboelmaged SMEs [16] and Andersen [17] to examine green manufacturing practices. Furthermore, in order to examine factors of sustainable practice in SMEs, this study referred to Ghazilla et al. [18] as well as other similar studies as the theoretical frameworks. This decision was in line with Caldera et al. [15] who state that quantitative sustainable practices analysis is more often carried out through the Analytical hierarchy process approach. Moreover, despite its rarity to implement the Theory of Planned Behavior (TPB) as a framework in SMEs this implementation was commonly found in other research based on environmentally friendly behavior [19]. In their research, Singh et al. [19] include religiosity as a predictor of ethical attitudes and judgments in the development of TPB to examine sustainable intentions in family businesses. Investigating certain cases related to ethics such as cheating, shoplifting, and lying should consider the issue of norms [20]. Hare et al. (2018) mention that conserving ethical actions (moral beliefs, attitudes, intuitions, and norms regarding other species) can be adapted to support the cooperation between human beings and non-human beings [21]. Further, Ghasemi & Kyle [22] also mention that conserving wildlife is a moral obligation for human beings. Similarly, Latombe et al. [23] detail the values in conservation behavior from various ethical perspectives, including Anthropocentrism which views humans as the only species that holds a moral position. They also mention the perspective of Ecocentrism which views that a value system that considers species, their collections, and functions, as well as the wider ecosystem, is a direct moral obligation concern [23].

Sustainable efforts are also part of the ethical behavior of human conservation towards the environment and the species in it. This responsibility (moral aspect) is often neglected in pro-environmental behavior research/study [24]. Thus, Culiberg et al. behavior [25] include the moral foundation in research on pro-environmental consumption. Similarly, Chen [26] also includes moral obligation as an antecedent to the study of coffee consumption behavior labeled as sustainability. Personal norm or moral norm affects attitude [27], which was explained by Lülfs & Hahn [28]. They agree that moral norms can influence the intention of sustainable behavior within the framework of the Theory of Planned Behavior [28]. Meanwhile, the subjective norm variable in the original construction of the TPB only concerned the dimensions of the injunctive norm and descriptive norm [29] while the descriptive norm and individual moral norm were not included [30], so the consistency of the influence of subjective norm on intentions was lower in an extended period of times [31]. However, so far it seems important to conduct research on the role of moral norms on pro-environmental behavior in the context of urban farm communities, such as UASS which was deliberately initiated by the government to develop community participation in sustainable development.

In subsequent developments, Liu & Liu [32] and Karimi & Saghaleini [33] positioned moral norms as mediating between subjective norms and attitudes, which was actually the opposite idea of Ajzen [20]. In addition, they highlight other mediating roles that are formed from the construction of the Extended Theory of Planned Behavior (ETPB) that was found [32, 33]. Although there were similarities in ETPB construction, there were different mediating effects used in the research. This research discussed the differences between subjective norms and moral norms. In line with this, sustainable business research among SMEs should include moral aspects in the construction of ETPB.

1.1 Urban Agricultural Society of Surabaya (UASS)

There are 3 main functions that can be activated in the development/procurement of a UA, namely: sustainability, resilience, and multifunctionality [35]. Of these three, the third function multifunctionally allows UA not only to serve as food security and natural sustainability but also for sociocultural functions, economics, and configurations [36]. UA development can also be carried out in an integrated manner that connects the functions of production, distribution, and processing of products but it requires cooperation between a) vacant land owners; b) farmers, or those who are willing to cultivate vacant land; and c) other communities [10]. The research conducted by Newell et al. [37] shows that there were obstacles to developing UA in Detroit (US), including land tenure uncertainty, environmental pollution, misunderstanding of possible environmental harm and a lack of government backing and financial investment as the biggest challenges. In relation to the UASS, from the data obtained, Surabaya is recorded to have only 1179.60 ha of agricultural land and 5,055 ha of non-rice field land [38]. However, these limited areas were developed multifunctionally. As in Jambangan, one of the sub-districts in Surabaya, UA has been developed not only to serve as food security but also for other functions. UA was developed in vacant areas such as government-owned land, public facilities (roadsides), and vacant land. The citizens used these areas to plant and develop urban farms in order to create environmental sustainability, social sustainability, and economic functions. For example, using the Big Biopori Infiltration Hole (which is supposedly used for flood prevention) for catfish farms adds economic function and social function because it is used as a recreational site for the surrounding people.

1.2 Extended theory of planned behavior (ETPB)

The Theory of Planned Behavior is a theory from Ajzen and an extension of the Theory of Reasoned Action (TRA) [39]. The main construction of TPB explains that behavioral intentions are influenced by three main variables, namely attitudes, subjective norms, and PBC where subjective norms are only limited to injunctive norms [30]. TPB is a theory that considers social and psychological variables in the individual decision-making process [40]. Zhang [30] introduced the origins of TPB to be applied in many fields such as social psychology, health communication, marketing, management, and clinical medicine. This theory is most often cited to predict human social behavior [41].

Many studies have added other variables to the TPB model to improve the predictive ability of the model because the TPB only focuses on three variables [42]. The existence of ETPB indicates that scientists do not simply use the original version of TPB to provide an acceptable explanation for human behavior [43]. Other researchers Gao et al. [44], Savari & Gharechaee [45], Shalender & Sharma [46], López-Mosquera et al. [47], and Hoeksma et al. [48] use extended TPB in sustained behavior with moral norm variables. Chen & Tung [49] not only include moral but also environmental concerns in the TPB model to predict consumers' intention to visit green hotels. In contrast to these studies, Singh et al. [50] expand TPB with variables other than the norm in circular economy readiness research in manufacturing MSMEs (Micro, Small, and Medium Enterprises) about nature but not moral norms. Moreover, Zhao et al. [51] use ETPB by including non-norm variables in the driving intention for the connected vehicle environment. Various research and studies explain that ETPB can be obtained by combining TPB with other theoretical models, such as integrating TPB with TTM (The Transtheoretical Model) [52, 53]; with TAM (Technology Acceptance Model) [54 - 56]; with the norm activation model (NAM) [57 - 60]; with Theory of Value-Belief-Norm [61]; with the Stimulus-Organism-Response (M-SOR) Model [62]; with consumer satisfaction theory [63]. Considering the combination of theories used research that includes moral norm variables to predict behavior includes the results of integration between TPB and NAM. The integration of TPB and NAM is almost entirely used to predict sustainable or pro-environmental behavior. NAM has been studied independently and separately from TPB in predicting behavior [64].

1.3 Variables in extended theory of planned behavior (ETPB)

The degree to which a person has a favorable or negative opinion or appraisal of the action in issue is referred to as attitude [65]. Attitude towards behavior is a function of one's belief in the consequences of behavior that is carried out where it is about a person's perceived probability that engaging in an appealing action would result in particular outcomes or deliver specific experiences [29]. Attitudes can be categorized into cognition, affect, and connection [66]. The attitude toward sustainable business in this research is associated with the evaluation of SMEs on sustainable business practices as referred to by Aboelmaged [16].

Subjective norms are social pressures from reference group members to enforce behavior [27] based on normative beliefs or individual perceptions of others being considered. This factor is called subjective norm because it relates to perceived normative references [66]. Yet, Ajzen states that subjective norms are related to injunctive norms and descriptive norms [29]. Injunctive norm is the subjective expectation that a particular individual or reference group agrees or opposes engaging in the action under consideration, whereas the descriptive norm is the belief about whether another important person himself or herself performed the behavior. Perceived Behavioral Control (PBC) relates to the perceived ease or difficulty in carrying out the activity, and it is thought to reflect previous experiences as well as predicted hurdles and barriers [27]. PBC shows how much the behavior is under one's control [29]. PBC indicates a person's understanding of the target behavior's restrictions and capabilities [67]. A moral norm is the perception of the level of moral correctness of behavior [27]. Similarly, Schwartz (1977) defines moral norms as a person's perception of right or wrong and whether he feels morally obligated to take certain actions. Personal moral norms become obligation-based intrinsic motivation [69]. In pro-environmental behavior, individuals engage in public activism because they feel compelled to do so [70]. For this reason, the moral norm in this study leads to a feeling of obligation to take part in sustainable business actions as referred to by Aboelmaged [16]. Besides the variables mentioned, Hati et al. [71] found other factors that influence the intention in education are moral obligations in ethical theory. According to Zhang et al. [72] individual norms are the strongest direct determinant of intention. Similarly, Liu & Liu [73] agree that the extending theory of planned behavior (TPB) to environmentally friendly products can be done by including moral norms that influence purchase intention. Further, they mentioned that moral norms can also serve as mediators in the TPB model. Meanwhile, Chen & Lee [74] found that moral norms were significant predictors of attitudes which were supported by Shi, Wang, et al. [75] who found the positive effect of moral norms on every intention.

1.4 Research Hypotheses

Kumar et al. [80], Cahigas et al. [77], Ru, et al. [78], and Tommasetti et al. [79] stated that attitude has a significant effect on intention . Even Liu & Liu [73] and B. Chen & Lee [74] mention that extending the theory of planned behavior (TPB) to environmentally friendly products which includes attitudes was found to have the most significant role in predicting intentions. Furthermore, it is possible to research attitudes as a mediating moral norm towards intentions as was done by Ataei et al. [82], in addition to being extended to the prediction of feelings of moral regret as an additional predictor of intention.

H1: Significant influence of Attitude toward Sustainable Business (ATT-SB) to Intention toward Sustainable Business (Int-SB).

Kumar et al. [80], Shi, Wang, et al. [80], and Govindharaj et al. [83] prove that morality is a component that affects the variance of intentions. Esfandiar et al. [82] and Ataei et al. [84]

found the construct of moral norms on intentions. The results of the structural analysis by Zhang et al. [72] reveal that the proposed model of personal norms is the strongest direct determinant of intention. Liu & Liu (2019) expanded TPB to environmentally friendly products and included moral norms [73]. Similarly, Nayum & Th [86] used TPB in the conception of pro-environmental behavior showing that personal norms are the strongest predictor of pro-environmental behavior. This research will analyze the moral norm in sustainable business.

H2: Significant influence of Moral Norm (MN) to Intention toward Sustainable Business (Int-SB).

Cahigas et al. [77]; Liu & Liu [73]; Govindharaj et al. [83] proved that subjective norms had a significant positive effect on the intention component in the (TPB) framework. Subjective norms were also found to have a significant effect on the intention to behave pro-environmentally. Shi, Wang, et al. [82] found subjective norms have a positive effect on each vehicle reduction intention to reduce pollution. Furthermore, Wan et al. [87] show that subjective norms have a positive effect on people's intentions to use urban green areas. Tommasetti et al. [79] found that subjective norms influence consumer behavioral intentions to choose green restaurants. The results of this research are different from Shi, Fan, et al. [88] who state that subjective norms do not affect city emission control intentions.

H3: Significant influence of Subjective Norm (SN) to Intention toward Sustainable Business (Int-SB).

Cahigas et al. [77] and Shi, Fan, et al. [88] found evidence of planned behavior theory (TPB) intention to significantly and positively support perceived behavioral control. This finding is similar to Zhou et al. [90] who found that there is a relationship between perceived behavioral control and purchasing behavior. Liu & Liu [73] also found that perceived behavioral control affects purchase intentions. Perceived behavioral control was also found to have a significant effect on ethical behaviors. In addition, Wan et al. [87] show that behavioral control has a positive effect on people's intentions to use urban green areas. Tommasetti et al. [79] found that perceived behavioral control influences consumers' behavioral intentions to choose green restaurants. Lastly, Kumar et al. [80] found that perceived behavioral control influences intentions on unethical intentions. However, Sniehotta et al. [43] in a previous study found that the belief control intervention (PBC) did not affect intention. In contrast to the basic theory of TPB [20], Ru, Wang, & Yan [91] found that perceived behavioral control and subjective descriptive norms were negatively related to energy-saving intentions although both had a significant effect. This study will further analyze the perceived behavioral control towards sustainable business intentions.

H4: Significant influence of Perceived Behavioral Control (PBC) to Intention toward Sustainable Business (Int-SB).

Moral norms play a variety of roles, including as attitude predictors [92]. Environmental attitudes can be influenced by important moral norms for pro-environmental studies [93]. Moral norms influence intentions through attitudes. According to the results of the meta-analysis from Klöckner [94], some of the influence of moral norms on behavioral intentions will be conveyed through attitudes. People will weigh both the pros and cons of the behavior [95]. Liu et al. [32] and Karimi & Saghaleini [33] explained and hypothesized how moral norms were influenced by subjective norms, moral norms affect attitudes, and how moral norms serve as a bridge between subjective standards and intents. This study serves as a further discussion of the position of the moral norm as intended by the two researchers.

H5: Significant Influence of Moral Norm (MN) to Attitude toward Sustainable Bussiness (ATT-SB).

As previously stated, moral norms appear to have a major influence on pro-environmental behavior. Apart from attitude, PBC, and subjective norms, the moral norm is the fourth variable in the TPB model [92]. Nonetheless, researchers rarely explore how it correlates with subjective norms. Subjective well-being does not necessarily have an immediate impact on intentions, but it can be used to determine whether certain behavioral choices are simple or effective [32]. Moral norms reflect internalized societal and subjective norms, which can be explained by the Norm-Activation model and the Value-Belief-Norm theory [33]. This theory focuses on how the reference group's beliefs on what is right and wrong are absorbed and altered into one's particular moral norms [96]. According to Onwezen et al. [64], through the integration of the Norm-Activation model with TPB, subjective norms can influence moral norms. This finding is consistent with the findings of Liu et al. [97] and Savari et al. [98], who both hypothesize that TPB modulates moral norms.

H6: Significant Influence of Subjective Norm (SN) to Moral Norm (MN).

2 Methods

2.1 Participants and data procedure

Respondents of this study include food-based SMEs in the UA area, Surabaya. These food-based SMEs were: 1) someone who owns and manages a food/beverage business, 2) who owns but does not manage a food/beverage business, or 3) who manages but does not own a food/beverage business. They were selected because of their roles in the sustainable development chain of the UA area in Surabaya. They are expected to be able to further cultivate UA's crops. The questionnaires were distributed to 222 respondents, but 193 were accepted and only 177 could be used as data for this. Although not all the questionnaires can be used, the number that were used as data already met the minimum requirement of 5 constructs according to [99]. Data was collected through a paper-based questionnaire because as stated by Wang et al. (2013), the instrument was suitable to collect where the respondents were located in the same public area and asked to complete it at the same time [100]. The questionnaire implemented a 5-scale Linkert scale with the criteria: Strongly Agree (SS),

Agree (S), Moderately Agree (CS), Disagree (TS) and Strongly Disagree (STS). Paper-based questionnaires were also selected because of the characteristics of the respondents who were mostly from X generation (51.10%), 13.74% were baby boomers, 28.57% were Y generation, 3.85% were Z generation and 2.75% of respondents did not provide this information. Gravili & Fait (2016) state that the X generation is the first generation for "literate technology", unlike the following generations, the Y generation is considered the "technology controls" generation [101]. Therefore, using a paper-based questionnaire was to anticipate the respondents who had difficulties in using technology.

2.2 Measurement

Instrument measurement was adapted from Kim (2014), Liao & Fang (2019) and Shalender & Sharma (2020) [104][103][46]. The instrument developed by adapting Shalender & Sharma (2020) has the advantage of adding a moral norm variable, the attitude is more directed to a negative statement to measure belief in an opinion, while the instrument developed by Kim (2014) was suitable for food-based business respondents. Meanwhile, the instrument developed by Liao & Fang (2019) emphasized more on consumers. Considering the aims of this study, the variable employed was attitudes toward the intention to use Sustainable Bussiness (ATT-SB). Subjective norms (SN), perceived behavioral control (PBB), moral norms (MN), and intention toward Sustainable Bussiness (INT-SB). The analysis technique used was a regression in order to determine the causal relationship of the application of TPB on Sustainable Bussiness to food business actors in the UASS program area.

2.3 Data analysis

The data analysis technique used was Structural Equation Modeling (SEM) using smartPLS3. The steps involved in SEM analysis are as follows.

(1) Describing the research construct or variable. Researchers define each variable to find indicators that build the construct/variable in the research.
(2) Developing and determining a measurement model. Researchers establish indicators to measure the constructs/variables being studied, including determining the relationships between variables in the research.
(3) Designing the research design. Researchers determine the type of data collection techniques, including anticipatory measures to deal with missing data.
(4) Evaluating the measurement model. Researchers evaluate the validity and reliability of research instruments.
(5) Determining the structural model. Researchers test the significance of paths in research designs.
(6) Evaluating the structural model. Researchers only evaluated the R^2 level.

The tests used include the measurement model test and the structural model test with the criteria as mentioned by Hair et al. [99]. Hypothesis testing became the main part of the discussion and part of the structural model test.

3 Result and discussion

3.1 Result

The results of data processing in general show the results of the measurement model test and structural model test. The measurement model test consists of convergent, divergent, and reliability validity. Furthermore, the structural model test consists of hypothesis testing, and R^2.

3.1.1 Convergent validity

Convergent validity is seen from the outer loading and AVE values. Figure 1 shows that the value of the outer loading of all indicator variables/constructs was > 0.5. It shows that the construct has met convergent validity from the outer loading side. Furthermore, the AVE value can be seen in table A (AVE value of all variables was greater than 0.5). The table shows that the AVE value of the ATT-SB variable was 0.776, the intention towards SMP was 0.904, the moral norm (MN) was 0.585, the subjective norm (SN) was 0.531, and the PBC was 0.712. Thus, it can be concluded that the existing instrument has met convergent validity.

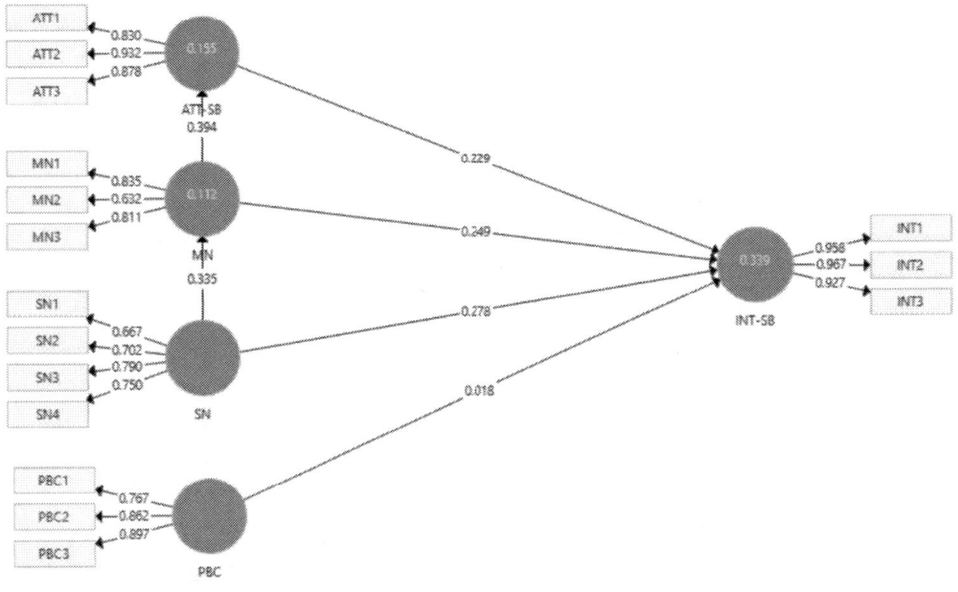

Fig. 1. Figure 1 Outer Loading

Table 1. AVE and Square Root AVE Scores

Variable/Construct	AVE	Cronbach's Alpha	$\sqrt{AVE}$				
			ATT-SB	INT-SB	MN	PBC	SN
ATT-SB	0.776	0.855	0.881				
INT-SB	0.904	0.947	0.430	0.951			
MN	0.585	0.647	0.394	0.441	0.765		

PBC	0.712	0.800	0.540	0.370	0.468	0.844	
SN	0.531	0.713	0.338	0.446	0.335	0.403	0.729

3.1.2 Divergent validity.

The AVE root value for attitude toward SMP was 0.881 and it was greater than the correlation coefficient with other constructs of 0.430 (correlation with intention toward SB), 0.394 (correlation with moral norm), 0.540 (correlation with PBC), and 0.338 (correlation with subjective norm). The AVE value for intention was 0.951 and this value is also greater than its correlation with other constructs (see Table 1). Likewise, the AVE root value for moral norm, PBC, and subjective norm which has a value greater than the correlation of these variables with other constructs. Thus, the outer model of this research has met discriminant validity. Another measurement of discriminant validity was seen from the value of outer loading compared to cross loading with other constructs (see Table 2).

Table 2. Cross Loading

Variable/Construct	ATT-SB	INT-SB	MN	PBC	SB
ATT1	0.830	0.382	0.323	0.398	0.296
ATT2	0.932	0.400	0.383	0.548	0.325
ATT3	0.878	0.354	0.333	0.476	0.269
INT1	0.447	0.958	0.430	0.385	0.460
INT2	0.383	0.967	0.423	0.351	0.443
INT3	0.395	0.927	0.403	0.315	0.363
MN1	0.389	0.394	0.835	0.414	0.249
MN2	0.211	0.203	0.632	0.223	0.226
MN3	0.280	0.378	0.811	0.403	0.296
PBC1	0.417	0.227	0.334	0.767	0.327
PBC2	0.444	0.331	0.417	0.862	0.375
PBC3	0.503	0.357	0.423	0.897	0.323
SN1	0.244	0.222	0.227	0.300	0.667
SN2	0.206	0.258	0.167	0.345	0.702
SN3	0.211	0.388	0.302	0.242	0.790
SN4	0.321	0.384	0.255	0.321	0.750

The value of outer loading on attitude toward SB, namely 0.830, 0.932, and 0.878 is greater than the cross loading with other constructs. The value of the outer loading variable intention toward SB, moral norm, PBC, and subjective norm is also greater than the cross loading with other constructs. From these results, the model in this study has good discriminant validity.

3.1.3 Reliability testing

The results of processing reliability data show that the Cronbach Alpha value (see Table 1) for all variables > 0.6 meaning that the construct was reliable [99]

3.1.4 Hypothetical testing

The results of hypothesis testing are seen from the p values or t statistics. A relationship between variables is said to have an effect if the significance level is: P-value < 0.05 or T-value > 1.96. The findings of the hypothesis testing are showed in Table 3.

Table 3. Result of Path Coefficient and Hypothetical Testing

Path	Path Coefficient	T Statistics	P Values	Conclusion
H1: ATT-SB -> INT-SB	0.229	2.506	0.013	H1 Accepted
H2: MN -> INT-SB	0.249	2.623	0.009	H2 Accepted
H3: SN -> INT-SB	0.278	3.455	0.001	H3 Accepted
H4: PBC -> INT-SB	0.018	0.184	0.854	H4 Rejected
H5: MN -> ATT-SB	0.394	4.823	0.000	H5 Accepted
H6: SN -> MN	0.335	4.212	0.000	H6 Accepted

3.1.5 Attitude toward Sustainable Business (ATT-SB) to Intention toward Sustainable Business (Int-SB)

The results of this research prove that ATT-SB has a significant impact toward Int-SB as indicated by the results of hypothesis testing which states that P-Value (0.009) is smaller than (0.05) and T-Value (2.607) is greater than 1.96, which means that H1 was accepted. This study supports the results of previous research which suggests there was a relevance and relationship between behavior towards sustainable business and the intention to carry out these activities (Kumar et al., [80], Cahigas et al. [77], Ru, Wang, Chen, et al. [78], and Tommasetti et al. [79]. Besides, this research also supports the research conducted by Liu & Liu [73] and Chen & Lee [74] that explain the environmental behavior through ETPB.

3.1.6 Moral Norm (MN) to Intention toward Sustainable Business (Int-SB)

This research proves that MN had significant influence on Int-SB (H2 was accepted). The findings were consistent with research conducted by Kumar et al. [80] , Shi et al. [82] , Esfandiar et al. [83] , Ataei et al. [84] who stated that moral norms were components that can affect the variance of intentions. In addition, the results of this study strengthen the results of research conducted by Liu & Liu [73] who discussed pro-environmental behavior. Therefore, MN can be calculated as a variable that can predict intentions as evidenced by the results of the path coefficient test which states a high value (0.249) in relation to Int-SB.

3.1.7 Subjective Norm (SN) to Intention toward Sustainable Business (Int-SB)

This research supports results of previous research Cahigas et al. [77], Liu & Liu [73], Govindharaj et al. [83] who found a important effect of subjective norms on intentions in the TPB framework. This can be seen from the results of hypothesis testing and path coeficient which shows the P value (0.000) is smaller than Sig. and the value of T (3,612) is greater than 1.96 which means that the hypothesis is accepted (H3). The significant impact of subjective norms on sustainable business intentions in this study also supports the results of previous pro-environmental research Wan et al. [79] and Tommasetti et al. [87]. However, this research did not support Shi, Fan, et al. [88] who mention the absence of SN influence on intention.

3.1.8 Perceived Behavioral Control (PBC) to Intention toward Sustainable Business (Int-SB)

The results of the analysis prove that in this study PBC was found to have no effect on sustainable business intentions in SMEs in MPP Surabaya, which means that H4 of this study cannot be proven. This is evidenced by the P value (0.860) which is greater than Sig. (0.05) and T value (0.177) which is smaller than 1.96. The results of this research are different from the findings of previous studies Cahigas et al. [77], Shi, Fan, et al. [88], Zhou et al. [90]; Liu & Liu [73] who stated that there was a significant influence of PBC on sustainable business intentions. In addition, this result also refutes the findings of Wan et al. [87] who agree on the influence of PBC toward intentions to use green areas in their research on environmental ethics, Tommasetti et al. [79] on intentions to choose green restaurants, and Kumar et al. [80] on unethical behavioral intentions. However, this research supports the findings in the study carried out by Liu et al. [107] which mentions the non-significance of PBC to ethical fashion consumption.

3.1.9 MN to Attitude toward Sustainable Business (ATT-SB) and SN to MN

However, in this study, the researcher emphasizes on two roles of MN, namely as a predictor of ATT-SB (H5) and the relationship between SN and MN (H6). From the results of the analysis and hypothesis testing, it was found that both hypotheses (H5 and H6) were accepted. This can be seen from the P value which is smaller than Sig. (0.05) and T value higher than 1.96. The results indicate that MN had a supportive role in Attitude toward sustainable business (ATT-SB) which means supporting the results of Liu & Liu [32] and Karimi & Saghaleini [33] who agreed on the positive roles for MN as predictors for sustainable business attitude. However, the results shown by H6 in this study were different from the opinion of Liu & Liu (2019), because they found that SN and MN did not have a significant influence to one another as it (SN) directly influenced on ATT-SB and did not require MN as mediator. In addition to proving the above hypothesis, this study discovered an indirect or mediating influence that could potentially be utilized to encourage discussion on the function of moral norms in the development of intentions.

3.1.10 R Square (R^2)

The amount of exogenous factors' influence on endogenous variables is explained by the value of R^2 Adjusted. The previous table explains that moral norms are influenced by subjective norms. The subjective norms can explain the exogenous moral norm variable of 10.7%. Yet the nature of the influence of them was weak because $0 < R^2 < 0.25$ (see Table 5). Similarly, it also explained the attitude was influenced by moral norms. Furthermore, considering the R^2 value, the R^2 attitude toward SMP was 0.150, which means that this variable can be explained by the moral norm of 15.0% with the nature of its influence was weak. Finally, the R^2 value for intention toward SB was 0.323 meaning that this variable was explained by attitude toward SB, moral norm, subjective norm, and PBC by 32.3% with the nature of its influence was moderate. Moreover, overall, the int-SB can be explained by its exogeneous variables (ATT-SB, MN, SN, and PBC) as 32.2 % which indicated that the influence of exogenous and endogenous variables was low (see Table 4).

Table 4. R2 Adjusted

	R^2 Adjusted
Attitude toward SB	0.150
Intention toward SB	0.323
Moral Norm	0.107

3.2 Discussion

Among of four predictors, PBC becomes the only predictor which did not have role in establishing Int-SB. The existence of this difference is possible because of external elements, such as culture in the community and the community's point of view. This can be seen from the results of observational analysis from MPP Surabaya which has formed a positive culture and perspective on eco-green so that this allows for different forms of culture and people's perspectives on sustainable business behavior that can affect their intentions indirectly. This is promoted by research Langemeyer et al. [35] and Hassan et al. [36] which stated that other functions of urban agriculture are as a form of business for natural, social, economic sustainability, and others. However, this assumption has not been proven because in this study cultural factors and people's perspectives were not included in the variables studied.

This study indicated that moral norms were the second-most significant predictor of int-SB after subjective standards, emphasizing the importance of moral norms. This demonstrates that moral norms must be incorporated into the TPB model in the case of environmental conservation behavior [93]. Furthermore, moral norms are predictors of attitudes toward sustainable business, which supports the findings of Klöckner [94], who suggest that some of the impact of moral standards toward behavioral motives is conveyed through behavior. Table 2 illustrates that attitude acts as a bridge between moral norms and intentions (MN -> ATT-SB -> INT-SB). This study, however, does not go into detail about the function of mediating attitudes in the connection between intentions and moral norms. In the case of conducting a sustainable business, humans have a moral imperative to safeguard the environment by decreasing pollution from their operations [57]. According to prior studies, SMEs account for 60–70% of industrial pollution in Europe [108, 109]. Accordingly, the role of SMEs in safeguarding the environment Corresponding with the various evidence, SMEs at UASS demonstrated that they have fulfilled their moral commitment to safeguarding the environment, such as through the installation of a Waste Water Treatment Plant (WWTP), a waste bank, and community 3R activities.

Subjective Norms affect moral norms. It was proven by the result of Hypothesis 6 (H6), which is accepted, meaning that the UASS group believes that business promoting cleanliness has been internalized and changed into personal or individual moral values for community food SMEs. Views from reference groups can be internalized and assimilated into personal moral norms [96]. UASS is a rural location where locals are more enthusiastic about waste management than residents in residential areas [110]. UASS, which has also expanded its function as a recreational facility, enables the development of a positive attitude toward the necessity of economic activities while safeguarding the environment. Thus, the subjective norms that play a role in the creation of moral standards in SMEs are inextricably linked to the research area's character, namely UASS. The issue of moral norms in SMEs cannot be separated from UASS outreach operations. Ryder (1965) defined socialization as an environmental process that involves mechanisms [111] such as observation, imitation, and internalization [112] to acquire skills, knowledge, and attitudes [113, 114] that influence people's behavior. Moral norms can be developed through socialization [68].

Morality is developed through the socialization process, which includes learning from experience, observation of others, and instructive stories (Moskalenko, 2008). Internalization of moral norms occurs through identification, discourse, and interaction with those who are referred to or embraced [115]. The openness trait shared by the people of Surabaya helps the socialization process for SMEs flow smoothly, resulting in the internalization of moral norms. When other variables are ineffective, attitude becomes a crucial variable for intention to fluctuate across activities and settings [20]. Table 2 further illustrates that attitude acts as a mediator between moral rules and intention. Attitudes are generated through the process of learning about an item through experience, which is a response to the evaluation of attitudes so that information about the object of the attitude improves [116]. In the example of food SMEs in UASS, their knowledge of sustainable business will determine whether their views toward sustainable business are positive or negative, which will affect their intention to execute sustainable business. The discussion that follows is about the influence of moral norms and other variables on SMEs' long-term business intentions. Moral norms with ATT-SB, PBC, and SN can only explain intention variables to a limited extent. This suggests that there may be additional variables that were not considered in this study but have a greater impact on intention formation. Wang and Zhang [117] Ahmed [118], Kabir et al. [119] and Moghavvemi et al. [120] present research on extending TPB with internal variables like self-efficacy, prior behavior, and locus control, as well as external variables like education and external assistance.

4 Conclusion

Attitude, moral norm, and subjective norm proved to have a direct influence on intention, but not with PBC. From these results indicate the need for an appropriate control to form appropriate behavior. For this reason, the public policy that is designed needs to include elements of pro-environmental education to improve attitudes and morals. Moreover, the government policy is necessary to be directed in establishing community awareness to optimize the role of subjective norms on sustainable business intentions and moral norm establishment. In connection with PBC which is not significant to sustainable business intentions, future researchers need to consider the construction of intentions by using other theoretical approaches. Liu et al. [32] suspected that in the case of pro-environmental behavior, the Theory of Reasoned Action (TRA) is more suitable than the TPB model. This study recommends the need for further analysis of the TRA model which is expanded with moral norm variables for future pro-environment studies. Finally, moral norms remain the key to consider in the expansion of the TRA model. Because moral norms are part of the Norm-Activation model in behavior theory [33], integrating TRA with NAM can be a recommended option for examining pro-environmental behavior in people with compliance issues. This study did not reveal the complete relationship of ETPB, without SN to ATT as the original TPB construct. In addition, it also does not discuss SN to PBC, the model is only studied to photograph field conditions at MPP, by proving moral norms as the core policy in the pro environmental area as a real manifestation of ethics in sustainable business.

Acknowledgments

The research can be carried out well due to the support of all related parties, for that the researcher sends his thanks to all those who support the implementation of this research, such as UNESA, the Ministry of Education, Culture, Research, and Technology of Republic of Indonesia, and Directorate of Research, Technology, and Community Service for their supports and fundings. In addition, our thankfulness was given to all parties who have collaborated and provided data and support.

References

1. H. Santosa, Environ. Urban. **12**, 2 (2000)
2. I. D. A. A. Warmadewanthi, et al., Waste Manag. Res. **39**, 8 (2021)
3. A. B. Rimba, P. E. Yastika, E3S Web Conf. **148** (2020)
4. World Bank, *Surabaya, Indonesia*, in Energizing Green Cities in Southeast Asia (2019)
5. J. D. Miller, M. Hutchins, J. Hydrol. Reg. Stud. **12** (2017)
6. B. Cristina, A. Diana, Procedia - Soc. Behav. Sci. **116**, 1999(2014)
7. W. E. Mabee, M. J. Blair, J. T. Carlson, C. N. M. Deloyde, Sustainability **11**, (2020)
8. V. Margolin, J. Des. Econ. Innov. **1**, 1 (2015)
9. S. Huttunen, M. Ojanen, A. Ott, H. Saarikoski, Energy Res. Soc. Sci. **91** (2022)
10. S. Miccoli, F. Finucci, and R. Murro, Agric. Sci. Procedia **8** (2016)
11. G. Ebissa, H. Desta, City Environ. Interact. **14** (2021)
12. N. Hosseinpour, F. Kazemi, H. Mahdizadeh, Land use policy **112** (2022)
13. Yuliani, B. E. Soetjipto, Green Practices **101** (2018)
14. H. Yanto, A. Susanti, Kiswanto, N. Baroroh, A. Wibowo, Int. J. Innov. Creat. Chang. **7**, 11 (2019)
15. H. T. S. Caldera, C. Desha, L. Dawes, J. Clean. Prod. **177** (2018)
16. M. Aboelmaged, J. Clean. Prod. **175** (2018)

17. J. Andersén, Technovation. **104** (2021)
18. R. A. R. Ghazilla, N. Sakundarini, S. H. Abdul-Rashid, N. S. Ayub, E. U. Olugu, S. N. Musa, Procedia CIRP **26** (2015)
19. G. Singh, S. Sharma, R. Sharma, Y. K. Dwivedi, Technol. Forecast. Soc. Change **173** (2021)
20. I. Ajzen, Organ. Behav. Hum. Decis. Process **50** (1991)
21. D. Hare, B. Blossey, H. K. Reeve, R. Soc. Open Sci. **5**, 11 (2018)
22. B. Ghasemi, G. T. Kyle, Biol. Conserv. **259** (2020)
23. G. Latombe et al., NeoBiota. **72** (2022)
24. V. K. Wells, C. A. Ponting, K. Peattie, J. Mark. Manag. **27**, 7–8 (2011)
25. B. Culiberg, H. Cho, M. Kos Koklic, V. Zabkar, J. Bus. Ethics. **182** (2022)
26. M. F. Chen, Bus. Strateg. Environ. **29**, 6 (2020)
27. I. Ajzen, Organ. Behav. Hum. Decis. Process. **50**, 2 (1991)
28. R. Lülfs, R. Hahn, Organ. Environ. **27**, 1 (2014)
29. I. Ajzen, Human Behavior and Emerging Technologies **2**, 4 (2020)
30. K. Zhang, Int. J. Humanit. Soc. Sci. Invent. **7**, 5 (2018)
31. A. Rivis, P. Sheeran, Curr. Psychol. **22**, 3 (2003)
32. M. T. Liu, Y. Liu, Z. Mo, Asia Pacific J. Mark. Logist. **32**, 8 (2019)
33. S. Karimi, A. Saghaleini, Glob. Ecol. Conserv. **26** (2021)
34. I. Ajzen, Organ. Behav. Hum. Decis. Process. **50** (1991)
35. J. Langemeyer, C. Madrid-Lopez, A. Mendoza Beltran, G. Villalba Mendez, Landsc. Urban Plan. **210** (2021)
36. D. K. Hassan, M. Hewidy, M. A. El Fayoumi, Ain Shams Eng. J. **13**, 3 (2022)
37. J. P. Newell, A. Foster, M. Borgman, S. Meerow, Cities **125** (2021)
38. DKPP Surabaya, Potensi Daerah Hasil Produksi Sektor Pertanian di Kota Surabaya (Dinas Ketahanan Pangan dan Pertanian Kota Surabaya, Surabaya, 2022)
39. Y. H. Al-Mamary, M. Al-nashmi, Y. A. G. Hassan, A. Shamsuddin, Int. J. Hybrid Inf. Technol. **9** 6 (2016)
40. E. Ulker-Demirel, G. Ciftci, J. Hosp. Tour. Manag. **43** (2019)
41. I. Ajzen, Psychol. Heal. **26**, 9 (2011)
42. E. L. Bird, J. Panter, G. Baker, T. Jones, and D. Ogilvie, J. Transp. Heal. **10** (2018)
43. F. F. Sniehotta, J. Presseau, and V. Araújo-Soares, Health Psychol. Rev. **8**, 1 (2014)
44. L. Gao, S. Wang, J. Li, and H. Li, Resour. Conserv. Recycl. **127** (2017)
45. M. Savari, H. Gharechaee, J. Clean. Prod. **263** (2020)
46. K. Shalender, N. Sharma, Environ. Dev. Sustain. **23**, 1 (2021)
47. N. López-Mosquera, T. García, R. Barrena, J. Environ. Manage. **135** (2014)
48. D. L. Hoeksma, M. A. Gerritzen, A. M. Lokhorst, P. M. Poortvliet, Meat Sci. **128** (2017)
49. M. F. Chen, P. J. Tung, Int. J. Hosp. Manag. **36** (2014)
50. M. P. Singh, A. Chakraborty, M. Roy, Resour. Conserv. Recycl. **135** (2018)
51. W. Zhao, M. Quddus, H. Huang, Q. Jiang, K. Yang, Z. Feng, Accid. Anal. Prev. **151** (2020)
52. R. Watakakosol et al., Subst. Use Misuse **56**, 1 (2021)
53. K. S. Courneya, T. M. Bobick, Psychol. Sport Exerc. **1**, 1 (2000)
54. N. K. N. Mat, *I. Sentosa, The integration of technology acceptance model and technology of puchasing behavior (A Structural Equation Modeling Approach)*, in Proc. Proceeding Asia Pacific Conf. Manag. Technol. Technol. Entrep., January (2008)
55. S. Ozkan, I. E. Kanat, Gov. Inf. Q. **28**, 4 (2011)

56. S. Chen, Transp. Res. PART A **88**, (2016)
57. B. Setiawan, A. Z. Afiff, I. Heruwasto, Soc. Mar. Q. **26**, 3 (2020)
58. Y. H. Shin, J. Im, S. E. Jung, K. Severt, Int. J. Hosp. Manag. **69** (2017)
59. J. J. Kim, J. Hwang, J. Hosp. Tour. Manag. **42** (2020)
60. L. H. Tran, N. A. Nguyen, T. D. Tran, T. P. L. Nguyen, Data Br. **42** (2022)
61. N. López-Mosquera and M. Sánchez, J. Environ. Manage. **113** (2012)
62. X. Qi, A. Ploeger, Food Qual. Prefer. **92** (2021)
63. I. G. M. Yuda Bakti, T. Rakhmawati, S. Sumaedi, T. Widianti, M. Yarmen, N. J. Astrini, Transp. Res. Procedia **48** (2018)
64. M. C. Onwezen, G. Antonides, and J. Bartels, J. Econ. Psychol. **39** (2013)
65. U. Liebe, P. Preisendörfer, and J. Meyerhoff, Environ. Behav. **43**, 1 (2011)
66. I. Ajzen, Personality & Behavior **8** (2005)
67. X. Fu, Z. Juan, Transportation (Amst). **44**, 5 (2017)
68. S. H. Schwartz, Advances in Experimental Social Psychology **10** (1977)
69. E. van der Werff, L. Steg, K. Keizer, Glob. Environ. Chang. **23**, 5 (2013)
70. O. Kácha, S. van der Linden, Curr. Res. Ecol. Soc. Psychol. **2** (2021)
71. S. R. H. Hati, R. Fitriasih, A. Safira, J. Information, Commun. Ethics Soc. **18**, 1 (2020)
72. X. Zhang, G. Geng, P. Sun, J. Clean. Prod. **166** (2017)
73. M. T. Liu, Y. Liu, Asia Pacific J. Mark. Logist. **31**, 4 (2019)
74. B. Chen, J. Lee, Int. Trade, Polit. Dev. **4**, 1 (2020)
75. H. Shi, S. Wang, D. Zhao, J. Clean. Prod. **147** (2017)
76. A. Kumar, S. Shabir, N. Mukund, N. R. Velaga, IATSS Research **46**, 4 (2022)
77. M. M. Cahigas, Y. T. Prasetyo, S. F. Persada, A. K. S. Ong, R. Nadlifatin, Res. Transp. Bus. Manag. **45** (2022)
78. X. Ru, S. Wang, Q. Chen, S. Yan, Journal of Cleaner Production, **197** (2018)
79. A. Tommasetti, P. Singer, O. Troisi, G. Maione, Sustain. **10**, 7 (2018)
80. A. K. Yadav, S. S. Choudhary, N. M. Pawar, N. R. Velaga, IATSS Res. **46**, 4 (2022)
81. P. Ataei, S. Gholamrezai, R. Movahedi, V. Aliabadi, J. Rural Stud. **81** (2020)
82. V. Aliabadi, P. Ataei, S. Gholamrezai, Weather, Climate, and Society **14**, 2 (2022)
83. G. P. P. Govindharaj et al., Sustain. Prod. Consum. **26** (2021)
84. K. Esfandiar, R. Dowling, J. Pearce, E. Goh, J. Hosp. Tour. Manag. **46** (2021)
85. A. Nayum, J. Th, Energy Res. Soc. Sci. **89** (2022)
86. A. Nayum, J. Thøgersen, Energy Res. Soc. Sci. **89** (2022)
87. C. Wan, G. Q. Shen, S. Choi, Environmental Science & Policy **75** (2017)
88. H. Shi, J. Fan, and D. Zhao, J. Clean. Prod. **145** (2017)
89. Y. Zhou, A. M.-W. Loi, G. W.-H. T. A, P.-S. Lo, W. Lim, Data Br. **42** (2020)
90. S. Gu,Y. Wu, World Scientific Research Journal **5**, 9 (2019)
91. X. Ru, S. Wang, and S. Yan, Resour. Conserv. Recycl. **134**, 96 (2018)
92. A. Y. J. Tsai, A. Y. K. Tan, Int. J. Environ. Res. Public Health **19**, 15 (2022)
93. F. G. Kaiser, Pers. Individ. Dif. **41**, 1 (2006)
94. C. A. Klöckner, Glob. Environ. Chang. **23**, 5 (2013)
95. W. Wei, C. Gu, C. Yang, Systems **10**, 1 (2022)
96. S. Bamberg, G. Möser, J. Environ. Psychol. **27**, 1 (2007)
97. Y. Liu, H. Sheng, N. Mundorf, C. Redding, Y. Ye, Int. J. Environ. Res. Public Health **14**, 12 (2017)
98. M. Savari, H. E. Damaneh, H. E. Damaneh, M. Cotton, Sci. Rep. **13**, 1 (2023)

99. J. F. J. Hair, J. W. C. Black, B. J. Babin, R. E. Anderson, Multivariate Data Analysis, 7th ed. (USA, 2014)
100. C. C. Wang, K. S. Liu, C. L. Cheng, Y. Y. Cheng, Comput. Human Behav. **29**, 3 (2013)
101. G. Gravili and M. Fait, Social recruitment in HRM: a theoretical approach and empirical analysis. (Emerald Group Publishing Limited, 2016)
102. Y. G. Kim, Procedia - Soc. Behav. Sci. **159** (2014)
103. W. Liao, C. Fang, Sustainability **11**, 18 (2019)
104. Y. G. Kim, Procedia - Social and Behavioral Sciences **159** (2014)
105. J. F. J. Hair, G. T. M. Hult, C. M. Ringle, M. Sarstedt, A Primer on Partial Least Squares Structural Equation Modeling (SAGE Publications, California, 2014)
106. Y. Liu, M. T. Liu, W. Chan, Asia Pacific Journal of Marketing and Logistics **33**, 5 (2022).
107. Y. Liu, M. T. Liu, A. Pérez, W. Chan, J. Collado, and Z. Mo, Asia Pacific J. Mark. Logist. **33**, 5 (2021)
108. R. Blundel, A. Monaghan, C. Thomas, Bus. Ethics **22**, 3 (2013)
109. S. Koirala, SMEs: Key Drivers of Green and Inclusive Growth (2018)
110. D. R. Wijayanti and S. Suryani, Procedia - Soc. Behav. Sci. **184** (2014)
111. N. B. Ryder, Am. Sociol. Rev. **30**, 6 (1965)
112. A. Singh-Manoux, M. Marmot, Soc. Sci. Med. **60**, 9 (2005)
113. R. Basu and N. Sondhi, J. Retail. Consum. Serv. **21**, 5 (2014)
114. V. Quintal, I. Phau, D. Sims, I. Cheah, J. Retail. Consum. Serv. **30** (2016)
115. A. F. Zillich, C. Riesmeyer, Soc. Media Soc. **7**, 3 (2021)
116. D. Pardana et al., *Attitude analysis in the theory of planned behavior: Green marketing against the intention to buy environmentally friendly products*, in IOP Conf. Ser. Earth Environ. Sci. (2019)
117. L. Wang and Y. Zhang, J. Sports Sci. **34**, 7 (2016)
118. I. Ahmed, Front. Psychol. **13** (2022)
119. S. Kabir, H. Ahasanul, A. Sarwar, Int. J. Eng. Inf. Syst. **1**, 6 (2017)
120. S. Moghavvemi, N. A. M. Salleh, A. Sulaiman, and M. Abessi, Behav. Inf. Technol. **34**, 12 (2015)

Juridical analysis of land asset utilization on legal entity college State University of Surabaya

Lusy Andriani[1*], *Yuniarto Wiryo* Nugroho[2], *Mohamad Aldy* Firdaus[3], and Kholidah[3]

[1]Faculty of Economics and Business, State University of Surabaya, Surabaya, Indonesia.
[2]Faculty of Social Sciences Education, IKIP Widya Darma Surabaya, Surabaya, Indonesia.
[3]Faculty of Law, Airlangga University, Surabaya, Indonesia.

Abstract. Surabaya State University is one of the Universities that has changed its campus autonomy from a Public Service Agency State University to a Legal Entity State University. These changes are regulated in Government Regulation Number 37 of 2022 concerning State Universities Legal Entities of Surabaya State University. The change in status has a significant impact in various ways and one of them is on the management of Unesa's Wealth, as stipulated in [1] Article 88 paragraph (1) of PP No.37 of 2022 concerning PTNBH Surabaya State University stated that "Unesa's initial wealth as referred to in [1] Article 87 paragraph (1) point a is in the form of state wealth separated except land. This research arises because specifically the use of land assets is regulated separately unlike other assets of Surabaya State University, so further study is needed in this study. The problem to be studied is, first, how is the juridical analysis of land assets at Surabaya State University as PTNBH? Second, how is the management and utilization of land assets at Surabaya State University as PTNBH? The problems in this study will be analyzed using normative legal research methods, namely studies based on legal philosophy, legal theories, legal principles and applicable legal norms using a normative juridical approach.

1 Introduction

Higher Education as regulated in Law Number 12 of 2012 concerning Higher Education (Law on Higher Education) has a definition described in [1] Article 1 number 2, namely, the level of education after secondary education which includes diploma programs, undergraduate programs, master programs, doctoral programs, and professional programs, as well as specialist programs organized by universities based on Indonesian culture. From this definition, higher education must have a teaching basis from Pancasila, the Constitution of the Republic of Indonesia Year 1945, the Unitary State of the Republic of Indonesia, and Bhinneka Tunggal Ika.

This basis makes the functions of higher education as stipulated in [1] Article 4 of the Higher Education Law, among others: (a) Develop abilities and shape the character of

* Corresponding author: lusyandriani@unesa.ac.id

dignified national civilization in order to educate the nation's life; (b) Developing an innovative, responsive, creative, skilled, competitive, and cooperative academic community through Tridharma; and (c) Developing Science and Technology by paying attention to and applying Humanities values.

The function of education is also accompanied by the objectives as stipulated in [1] Article 5 of the Higher Education Law, namely: (a) the development of students potential to become human beings who believe and fear God Almighty and have noble character, healthy, knowledgeable, capable, creative, independent, skilled, competent, and cultured for the benefit of the nation; (b) the production of graduates who master the branches of science and / or technology to meet national interests and increase nation's competitiveness; (c) the production of Science and Technology through Research that pays attention to and applies Humanities values to benefit the progress of the nation, as well as the progress of civilization and the welfare of mankind; and (d) the realization of community service based on tools and research work that is useful in advancing general welfare and educating the life of the nation.

Seeing from the functions and objectives of higher education cannot be realized without the implementation and management itself, so in this case, the government realizes in the form of Government Regulation Number 4 of 2014 concerning the Implementation of Higher Education and Management of Higher Education (PP Management of PT). Through [1] Article 21 of the PT Management PP, it stipulates that Higher Education Management includes: a. Higher Education autonomy, b. Higher Education Management patterns; c. governance of Higher Education, and d. public accountability.

Referring to [1] Article 22 paragraph (1) of the PT Management PP the form of Higher Education autonomy is the management of its institution as the center of implementing the Tridharma of Higher Education. Then the autonomy of the university is divided into 3 (three) as stipulated in [1] Article 22 paragraph (2) of the PT Management PP, namely as follows: (a) State Universities (PTN); (b) Legal Entity State University (PTNBH); and (c) Private Universities (PTS). The focus of this writing is to discuss Legal Entity State Universities (PTNBH). PTNBH is a State University established by the Government which has the status of a public legal entity and is given full autonomy in managing the campus. This form of autonomy is divided into two, namely academic and non-academic autonomy.

The existence of this form of autonomy can be seen in that legal entity-based higher education is one application to build the world of education by making a legal subject acceptable, considering that the goal is to carry out the implementation of strong and authoritative education [2]. With the existence of Higher Education as a legal entity, the implementation of education will be supported in an organization that has rights such as humans, not easily used, deceived or made slaves by parties or legal entities that have no interest in education.

Discussing PTNBH that the existence of this autonomy is in the financial sector, the government through Government Regulation Number 26 of 2015 Jo. Regulation Number 8 of 2020 concerning the Form and Mechanism of Funding for State Universities Legal Entities. Therefore, the source of funding for PTNBH comes from two sources, namely the State Budget (APBN) and other than the state budget as stipulated in [1] Article 2 of PP No. 26 of 2015.

The development of PTNBH to date, it is known that President Joko Widodo has established 5 PTNs to become PTNBH. Including Surabaya State University which has been outlined in Government Regulation Number 37 of 2022 concerning State Universities Legal Entities of Surabaya State University (PP PTNBH Unesa). Please note that the issuance of PP PTNBH Unesa makes Surabaya State University one of 21 PTNBH throughout Indonesia. And as it should be Surabaya State University (Unesa) has funding sources other than the State Budget.

The income that Unesa may have as PTNBH is the utilization of assets divided into 2 parts, as stipulated in [1] Article 88 paragraph (1) of PP PTNBH Unesa stated that, "Unesa's initial wealth as referred to in [1] Article 87 paragraph (1) point a is in the form of state wealth separated except land. That implicitly the [1] Article divides assets into land assets and assets other than land. The difference in ownership status is certainly the cause of different rules in utilizing the two types of assets. Land assets with the status of State Property (BMN) follow the provisions for the utilization of BMN, while the utilization of non-BMN land assets and assets other than land is regulated separately by PTNBH.

The distribution of these assets also affects the regulation regarding the utilization of these assets. Focusing on land assets the use of land assets causes problems due to differences in regulations as mentioned earlier. Related to land assets in PTNBH Unesa in particular, in this case, further study is needed on how the juridical analysis of land assets at Surabaya State University as PTNBH, and how to manage and utilize land assets at Surabaya State University as PTNBH.

2 Methods

2.1 Research approaches

This research uses a normative legal research method approach, namely a study based on legal philosophy, legal theories, legal principles and applicable legal norms. This research also uses normative juridical, empirical juridical and comparative approaches. This research is legal research, where this research aims to find the truth of coherence, namely whether legal rules are in accordance with legal norms [2] and whether these norms are in accordance with legal principles, as well as a person's actions in accordance with legal norms. The method used in writing this research is a normative legal research method, namely a study based on legal philosophy, legal theories, legal principles and applicable legal norms. This research uses a normative juridical approach.

Researchers also use comparative because it aims to compare with other state universities by analyzing the respective laws between Surabaya State University and other state universities which are used as comparisons. Another research consideration in using comparative research is as a benchmark for the management and utilization of land assets at the University in order to exchange opinions regarding ways or tips for handling a university asset in a systematic or comprehensive and systematic manner.

2.2 Research focus

The focus of this research is to discuss the juridical analysis of land assets at Surabaya State University as a Legal Entity State University (PTN-BH) as regulated in [1] Article 88 paragraph (1) of PP PTNBH Unesa, which states that, "Unesa's initial wealth as referred to in [1] Article 87 paragraph (1) letter a constitutes separate state assets except land. Management of land assets at Surabaya State University PTN BH. Utilization of land assets at Surabaya State University PTN BH.

2.3 Techniques for collecting and processing legal materials

There are 4 data collection techniques in qualitative research, including observation, interviews, documentation and a combination of the three [3]. In this research, the data collection techniques used were focused group discussions, interviews and empirical research. This is done in order to obtain complete information regarding the analysis of the position, function and content of the Chancellor's Regulations and Senate Regulations based on Law Number 12 of 2011 concerning the Formation of Legislative Regulations.

1. Focus Group Discussion (FGD) between research members. This technique is used to obtain basic data on laws and regulations that are still in force regarding the accuracy of competence and material in the content of the discussion.
2. Interview. The interview method used in this research is in-depth interviews, namely by extracting information in depth, openly, and freely with the problem and focus of the research. Informants were selected who had competence in the field of Constitutional Law or Governance Law.
3. Empirical research is a research method carried out using empirical evidence. Empirical evidence is information obtained through observation or experimentation. Empirical evidence is done by analyzing the data and then collecting this empirical evidence using quantitative and qualitative research methods. According to the Big Indonesian Dictionary (KBBI), empirical is something based on experience, especially experience obtained through discovery, experimentation or observation. So, empirical research can be interpreted as social science, namely a group of sciences that focuses on research on human behavior and the environment. However, it is different from natural science or science which examines nature and its symptoms. Several opinions from experts regarding the meaning of empirical research: According to Amiruddin and Zainal [4] Asikin, empirical research focuses on examining a phenomenon or situation of the research object in detail by collecting the facts that occur and developing existing concepts. Anwar and Adang [5] said that empirical research is a science based on common sense, not speculative and based on observations of reality. According to Hilman [6], empirical research is research that is exploratory, descriptive and explanatory.

2.4 Analysis of legal materials

At this stage, data is collected through focused group discussions, interviews and empirical research. By collecting all data related to the analysis of PTN BH's land assets. Data reduction is a selection process, focusing attention on rough simplifications that emerge from written notes during data collection through focused group discussions, interviews, and empirically, where data reduction takes place continuously throughout the research. The selection of rough data was obtained from the results of focused group discussions, interviews and empirical research. This selection is done by summarizing data that is appropriate to what is being researched, while those that are not appropriate are not used.

Data presentation (data display) is a collection of information that is structured and provides several possibilities for drawing conclusions and making decisions that continue to develop into a presentation of data, which can be presented not only in the form of photos, but also in the form of descriptions of conversations from interviews conducted by researchers with the sources used. research subject. Next, the data will be presented in the form of sentences supported by documentation in the form of photos so that the data presented from the information obtained becomes valid data. In this research, the data presented is data relating to Land Assets at Surabaya State University.

The final stage of data analysis is drawing conclusions and verifying the data. Drawing conclusions is part of an activity and a complete form. Data is presented in descriptive form about Land Assets at Surabaya State University along with their management and utilization.

The data collected is supporting data about Assets at Surabaya State University along with management and utilization obtained from focused group discussions, interviews and empirical research. So that the conclusions from the results of this research can connect theory with existing data. The final stage of analysis needs to be verified so that it is stable enough and truly accountable.

3 Result and discussion

3.1 Juridical analysis of land assets at Surabaya State University as PTNBH.

PTNBH is a form of implementing higher education autonomy as regulated in [1] Article 65 of Law Number 12 of 2012 concerning Higher Education (hereinafter referred to as the Higher Education Law). [1] Article 65 paragraph (3) of the Higher Education Law regulates that the implementation of autonomy in question is as follows: (1) Initial wealth consists of separate state assets except land; (2) Governance and independent decision making; (3) Units that carry out accountability and transparency functions; (4) The right to manage funds independently, transparently and accountably; (5) Authority to appoint and dismiss lecturers and educational staff; (6) Authority to establish business entities and develop endowment funds; And (7)Authority to open, organize and close Study Programs.

The implementation of higher education autonomy mandated by PTNBH is certainly very broad, if seen from these regulations. In addition, there are regulations that give PTNBH the authority to be given the right to manage state assets as regulated in [1] Article 13 of Government Regulation Number 26 of 2015 concerning Forms and Mechanisms of Legal Entity State Higher Education Funding. Seeing the freedom for

PTNBH in carrying out autonomy, the research focuses on point number one, namely initial wealth in the form of state assets which are separated except for land.

Surabaya State University as PTNBH is obliged to follow the rules for separating initial assets as regulated in Minister of Finance Regulation Number 108/PMK.06/2017 concerning Procedures for Determining Initial Asset Value Legal Entity State Universities (PMK 108/2017). However, these regulations clearly exclude land assets, the question that arises is how is land regulated as a separate PTNBH asset? Of course, [1] Article 19 PMK 108/2017 regulates the determination of the initial asset value for PTNBH. Where the Minister of Finance in this case determines the initial wealth value by decree. Then, in this regulation, the handover of assets is carried out which is stated in the handover minutes which are signed by the relevant echelon I officials at the Technical Ministry and the PTNBH leadership. The handover becomes PTNBH's initial assets, but in this case land is excluded. This means that land remains state property whose management authority remains with the Ministry of Finance.

The land that is separated in this case falls under the authority of the Ministry of Finance, so the status or rights to land managed by central government agencies are Management Rights as regulated in Government Regulation Number 18 of 2021 concerning Management Rights for Land Rights, Flats and Land Registration. The Ministry of Finance as one of the subjects is regulated in [1] Article 5 paragraph (1) PP Number 18 of 2021. This regulation is supported by [1] Article 88 PP UNESA Statutes where land is state property which is administered by the Minister who handles government affairs in the financial sector.

Management Rights as regulated in PP Number 18 of 2021 are control rights from the state whose implementation authority is partly delegated to the management rights holder, where in accordance with [1] Article 7, the management rights holder is given the authority to: (1) Prepare a plan for the allocation, use and use of land in accordance with the spatial plan; (2) Use and utilize all or part of the management rights land for your own use or in collaboration with other parties; And (3) Determine rates and/or annual mandatory fees from other parties in accordance with the agreement.

The authority is limited only to utilization in the form of cooperation, as evidenced by [1] Article 12 PP Number 18 of 2021 where management rights cannot be used as collateral for debts burdened with mortgage rights and cannot be transferred to other parties [7]. This is in line with [1] Article 90 paragraph (1) PP UNESA Statutes where land assets become separate assets that cannot be transferred and cannot be used as collateral to other parties.

3.2 Utilization and management of land assets at Surabaya State University with PTNBH status.

Indonesia's legal system is an inheritance of the Netherlands colonial era. During colonial, the legal system that followed civil law is applied in Indonesia based on the principle of concordance. But Indonesia also adheres to two other legal systems, namely the customary law system and Law. No. 5 of 1960 as an example.

In law number 5 of 1960 about agrarian principles (UUPA), there are eleven principles as basic which must automatically animate from the implementation of UUPA and all its implementing regulations [2]:

1. Nationality Principle, found in [1] Article 1 paragraph (1), (2), (3).
2. Principle that at highest level, earth, water, space and natural resources contained therein are controlled by state is found in [1] Article 2 paragraph (1).
3. Prioritizing Principle for national and state interests, is found in [1] Article 3.
4. Principle that all land rights have a social function is found in [1] Article 6.
5. Principle that only Indonesian citizens have holders' rights to land, is found in [1] Article 9
6. Principle of equality for every Indonesian citizen is found in [1] Article 9 paragraph (2).
7. Principle that agricultural land must be actively worked by holder himself, is found in [1] Article 10 paragraph (1).
8. Principle of land use by planned is found in [1] Article 2 paragraph (2) letter a.
9. Principle of legal unity is found in words repeal Dutch land law since UUPA applies
10. Principle of guarantee legal certainty and legal protection, is found in [1] Article 19 paragraph (1)
11. Horizontal separation principle, found in [1] Article 44 paragraph (1)

Important to notice, that infrastructure is essential to an operational organization, society including university. Assets are of relevance in the day to providing substantial social, economic and environmental impacts in terms of investment, quality of life, resources, and energy. Assets have been designed and built with independent purposes that are combined with specific business and user needs, including the land assets.

The horizontal separation principle is one of the principles sense in UUPA [8]. The horizontal separation principle in land law indicates that the land holder is different or separate from the objects that stand on it. Understanding of "control" can be used in both physical and juridical sense [9]. Juridical control-based on rights, which are protected by law and generally give rights holders authority to control land that is owned physically, for example, the land holder uses or takes advantage of land that is owned, not handed over to another party [10]. However, there is also juridical control which, even though it gives authority to physically control land that is owned, in reality physical control is carried out by another party, for example landholder who does not use his own land but rents it to another party, in this case that land is legally owned by landholder but physically controlled by the land tenant. In this case, the landholder based on his legal control rights, has right to take back that land from a tenant under certain conditions. There is also juridical control which does not give authority to physically control that land, for example, a creditor holding guarantee rights to land has juridical control rights over land used as collateral, but physical control remains with landholder.

The horizontal separation principle is found in [1] Article 44 paragraph (1) of UUPA, namely a person or corporation has the right to lease land, if he has the right to use others' land for building purposes, by paying the holder a certain amount of money as rent. Implementation of the horizontal separation principle is the rights to rent a building, that is a person or corporation rents land belonging to another person which is not being used or has no building by paying a certain amount of money as rent, the amount of which is determined on the basis from an agreement, for a certain period of time, and tenant is given right to construct a building to be used for a certain period of time agreed by both parties [11]. It can be concluded that in rental rights for buildings, there is a horizontal separation between the holder of land and the building above it, namely land belongs to landholder, while the building belongs to the land tenant. The horizontal separation principle is not clearly explained in UUPA, this legal principle is hidden vaguely in several [1] Articles found such as [1] Article 44 of UUPA which is explained above, in [1] Article 46. Paragraph (2): "By exploit right to legally collect forest products, do not automatically obtained holder right to that land."

It can be concluded from words [1] Article 46 paragraph (2) above, that if someone obtains by right to legally collect forest products, this does not automatically mean that land used becomes property from forest product collectors. Forest products belong to the person who obtains these rights, while land belongs to the state. The horizontal separation principle originates from customary law, where this principle is applied to land law in Indonesia, so that buildings and/or plants located on them do not automatically become part of land, this condition is excluded if: (a) Building and or plants are physically an integral part of relevant land, meaning that building and or plants are based on perennials; (b) Building and or plants belong to land holder; and (c) This intention is stated expressly in deed prove commission through relevant legal act.

With this horizontal separation principle, the subject of the land rights holder can be different from the subject of the building holder, so that land and buildings will be submitted to different laws, land will be subject to land law, while building will be subject to debt law which regulates the power of rights over non-land objects [12]. Principles currently applied in National Land Law, override the principle of attachment found in [1] Articles 500, 571 and 601 BW. In the horizontal separation principle, buildings and plants on land are not part of the land so buildings and plants holder on a plot of land does not necessarily fall to landholder. Legal acts regarding land do not necessarily cover buildings and plants belonging to the holder of land on it. If legal actions regarding land are intended to include buildings and plants, then this must be explicitly stated in a deed that proves the implementation of relevant legal action.

PTNBH's initial assets as regulated in PMK 108/2017, land assets are included in State Property (BMN). In the case of land assets that become BMN, there is Minister of Finance Regulation Number 115/PMK.06/2020 concerning the Utilization of State Property, hereinafter referred to as PMK on the Utilization of BMN. So, from the previous information, there are differences in the status of land assets in these Legal Entity Universities. The first status for land assets that have been separated from the initial assets of a PTNBH falls under the authority of the Ministry of Finance by entering it as BMN. The second status is obtained from PTNBH's own assets, so its management is not included in PTNBH, this is in accordance with [1] Article 91 PP UNESA Statute.

Differences in regulations regarding land assets result in different rights to the land, and also different types of use of the land. However, in principle, the profits from this utilization are entirely the income of PTNBH itself. 12 In answering problems related to the utilization and management of land assets, they are divided into two which will be explained as follows.

1. Utilization and management of land assets whose assets have been separated
   It is often mentioned that land assets that have been separated from PTNBH's assets have become BMN whose management is under the Ministry of Finance. So, the right rule to answer this research is PMK on BMN Utilization. In accordance with [1] Article 8 there are forms of BMN utilization in the form of:
   a. Rental, is the use of BMN by another party for a certain period of time and receives compensation in the form of cash;
   b. Borrow and Use, is the utilization of BMN through handing over the use of BMN from the Central Government to the Regional Government or Village Government within a certain period of time without receiving compensation and after the certain period ends it is handed back to the BMN manager/User;
   c. Utilization Cooperation (KSP) is the utilization of BMN by other parties within a certain period of time in order to increase non-tax state revenues and other financing;
   d. Built, Operate, Transerv (BOT) is the use of BMN in the form of land by another party by constructing buildings and/or facilities and facilities, then utilized by the other party within a certain agreed period of time, to then be handed back the land and buildings and/or facilities and facilities after the period ends;
   e. Built, Transerv, Operate (BTO) is the utilization of BMN in the form of land by another party by constructing buildings and/or facilities and their facilities and after completion of the construction they are handed over for use by the other party within the agreed period of time;
   f. Infrastructure Provision Cooperation (KSPI), is the utilization of BMN through cooperation between the government and business entities in infrastructure provision activities in accordance with the provisions of statutory regulations;
   g. Limited Cooperation for Infrastructure Financing (KETUPI), is the utilization of BMN through optimization in improving BMN's operational functions in order to obtain funding to finance the provision of other infrastructure.
   h. That from these forms of utilization of BMN, Surabaya State University as one of the PNTBH has accommodated the PMK on BMN Utilization through the Chancellor's Regulation as its technical implementing regulation. These technical regulations are still in the design process and will soon be published and implemented at Surabaya State University [13]. The regulation is the Chancellor's Regulation concerning Mechanisms and Procedures for Management of UNESA Property, as stated in the picture below.

UNESA
PERATURAN REKTOR UNIVERSITAS NEGERI SURABAYA
NOMOR 23 TAHUN 2023
TENTANG
MEKANISME DAN TATA CARA PENGELOLAAN SARANA DAN PRASARANA
BARANG MILIK UNESA
UNIVERSITAS NEGERI SURABAYA

DENGAN RAHMAT TUHAN YANG MAHA ESA

REKTOR UNIVERSITAS NEGERI SURABAYA,

Menimbang : a. bahwa untuk melaksanakan ketentuan Pasal 92 Peraturan Pemerintah Republik Indonesia Nomor 37 Tahun 2022 tentang Perguruan Tinggi Negeri Badan Hukum Universitas Negeri Surabaya;
b. bahwa berdasarkan pertimbangan sebagaimana dimaksud dalam huruf a, perlu menetapkan Peraturan Rektor Universitas Negeri Surabaya tentang Pengelolaan Barang Milik Unesa Universitas Negeri Surabaya;

https://doi.org/10.1051/e3sconf/202345003008

Fig. 1. Chancellor's regulations regarding the mechanism for using facilities and infrastructure

Starting from the PMK on the Utilization of BMN to the Regulation of the Chancellor of Surabaya State University, it has been explained that the use of land assets whose assets are separated must obtain approval from The Ministry of Finance comes. These land assets are under the authority of the Ministry of Finance as the manager of BMN, while UNESA as PTNBH is only the manager of BMN. This has been carried out by UNESA in relation to the mapping of the land assets that will be used, which will be described in a City Plan Certificate issued by the Surabaya City Public Housing and Settlement Area and Land Service.

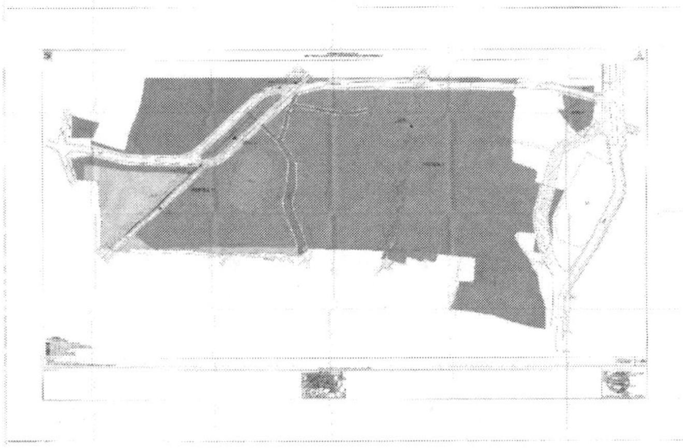

Fig. 2. Source: Department of Public Housing and Settlement Areas and Land City of Surabaya

The description of the image consists of zones and subzones that are used in their designation for carrying out activities. Zones are used in carrying out activities consisting of, Trade and Services Zones, Local Protection Zones, Green Open Space Zones, and Public Service Facilities Zones. Then these zones are further divided into sub zones, namely, Regional Service Scale Trade and Services, Local Service Scale Trade and Services, SUTT/SUTET Opportunities, Low Density Housing, Green Open Spaces, and Public Service Canters-Education.

The distribution of the use of land assets helps in managing which can be used as BMN which adds to UNESA's own income, and this use can be carried out according to the mechanisms and forms in the laws and regulations mentioned previously. Apart from land assets that have been separated, the use and management of land assets after becoming PTNBH is different from what is described above, as explained as follows.

3.3 Utilization and management of Surabaya State University land assets which have become PTNBH

UNESA as PTNBH certainly in its development has not only received funding sources other than the State Revenue and Expenditure Budget. In accordance with [1] Article 11 of Government Regulation Number 8 of 2020 Jo. 26 of 2015 concerning Forms and Mechanisms of Funding for State Universities as Legal Entities, UNESA has the right to funding sourced from the community, education costs, endowment fund management, PTN Legal Entity business, cooperation with the Tridharma of Higher Education, Management of PTN Legal Entity assets, budget regional income and expenditure, and loans.

The flexibility of PTNBH funding has been proven by the collaboration between UNESA and the Magetan Regency Government regarding the Opening of the Study Program Outside the Main Campus (PSDKU), in this case, the Magetan Regency Government provided a grant in the form of land to be used by UNESA.

4 Conclusion

The focus of this research is that land grants made on the basis of cooperation must be managed and utilized well by UNESA. Talking about UNESA as PTNBH, of course, the land assets obtained after the assets are separated will be the profits obtained. As expressly regulated in [1] Article 91 PP UNESA Statute which stipulates that:

1. Wealth in the form of land sourced from the development of UNESA funds after the initial determination of wealth is UNESA property.
2. (2) Land as referred to in paragraph (1) is recorded as wealth in UNESA's balance sheet and administered by UNESA.
3. (3) Land acquired and owned by UNESA after the determination of UNESA's initial assets can be transferred to other parties after obtaining approval from the Council of Trustees (MWA).

The land assets obtained by UNESA after becoming PTNBH can be managed and utilized as widely as possible for the benefit of the Tridharma of Higher Education, because PTNBH's freedom as regulated in statutory regulations extends to management rights over state assets. It is necessary to pay further attention, that the legal entity form of PTNBH is still a question. Of course, this influences the extent of PTNBH's authority in managing assets as a form of autonomy in education. So, this research does not just stop here, considering that PTNBH's large authority also includes the right to manage state assets.

References

1. R. I. Goverment, Law Number 12 of 2012 concerning Higher Education (Indonesian republican government, Indonesia, 2012)

2. P. Langbroek, K. Van Den Bos, M. S. Thomas, M. Milo, W. Van Rossum, Utrecht L. Rev.**13**, 1 (2017)
3. A. L. Hall, R. C. Rist, Psychology & Marketing **16**, 4 (1999)
4. A. A. H. Zainal, A. Amiruddin, Introduction to Legal Research Methods (Rajawali Press, Surabaya, Indonesia, 2013)
5. Y. Anwar, Adang, Introduction to Legal Sociology (Grasido, Jakarta, Indonesia, 2008)
6. H. Hilman Methods for Making Legal Science Working Papers or Theses (Mandar Maju, Bandung, Indonesia, 2013)
7. A. Suprihanto, *The Legal Protection for the Creditors as the Holders of the Mortgage Rights Towards the Buildings Rights Over the Managements Rights*, In The 2nd International Conference of Law, Government and Social Justice, ICOLGAS, Atlantis Press (2020)
8. M. Farida, Legislative Science (Kanisius. Yogyakarta, Indonesia, 2007)
9. M. Farida, Compedium Report on Legislation, Department of Law and Human Rights of the Republic of Indonesia, National Legal Development Agency (National Legal System Research and Development Center, Jakarta, Indonesia, 2008)
10. W. Serrano, Journal Buildings, **13** (2023)
11. B. Manan, Bagir, The role of statutory regulations in developing national law (Armico, Bandung, Indonesia, 1987)
12. R. E. Armen, E. R. Adawiah, G. Dewi, International Journal of Law and Management **65**, 6 (2023)
13. Rector of State university of Surabaya, Surabaya state university rector's regulation number 23 of 2023 concerning mechanisms and procedures for managing facilities and infrastructure (State University of Surabaya, Surabaya, 2023)

Use of bibliometric software to explore the relationship between scientific literacy and socio-scientific issues

Arie Widya Murni[1*], *Suryanti*[2], and *Nadi* Suprapto[3]

[1,2]Doctoral Study Program in Basic Education, State University of Surabaya, Surabaya, Indonesia
[3]Department of Physics, State University of Surabaya, Surabaya, Indonesia

Abstract. This article aims to provide a bibliometric analysis of relevant literature to explore the relationship between scientific culture and social science issues in research trends or publications indexed by the Scopus database. This analysis can reveal the extent of the relationship between scientific literacy and socio-scientific issues. The study methodology used was the Preferred Reporting Items for Systematic Reviews and Meta-Analyses explained in a four-step usage flow diagram including identification, selection, eligibility and inclusion, which will then be applied when carrying out the review process to assist reporting authors. This search examined all journal articles related to scientific literacy and socio-scientific issues. All data were collected in the Scopus database from 106 articles published between 2014 and 2022. Much research has been conducted on issues of scientific knowledge and social sciences in a variety of fields. However, research that specifically addresses scientific literacy and socio-scientific issues remains rare. This is clearly shown through the analysis results performed using VOSviewer software. This article uses bibliometric analysis to identify the main subject areas of each of the studies undertaken and to explore the relationship between scientific literacy and socio-scientific issues in the fields.

1 Introduction

The development of human civilization has now entered the 21st-century, where there have been very significant developments in various fields, including developments in information, industry, technology, education, to needs that refer to human resource skills [1]. The various demands of the 21st-century require every individual to be able to survive and compete with other individuals in the wider community [2]. 21st-century skills are the key skills that each individual must master to face the challenges of the future [3]. In 21st-century skills, it is explained that humans must be ready to become lifelong learners. When someone is ready to become an independent lifelong learner,

* Corresponding author: arie.21011@mhs.unesa.ac.id

each individual needs awareness to maximize themselves in developing abilities according to their field.

Applying skills in daily work is a key requirement for acquiring basic knowledge, including literacy, numeracy, technological literacy, financial literacy, cultural literacy, as well as civic and scientific literacy. Scientific literacy is considered a skill that students need to acquire in the 21st-century [4], [5], in which scientific literacy skills require understanding and applying knowledge. knowledge to solve problems related to science and technology in everyday life [6]. On the other hand, scientific literacy is also necessary in the ability to make decisions [7] and participate in the socio-cultural environment, so that students can promote an active role in social life from an early age.

Rychen and Salganik [8] write that scientific knowledge is about understanding how to change the way one interacts with the world so that it can be used to achieve broader goals. On the other hand, scientific literacy can enable someone to use the knowledge they have to identify problems, obtain new knowledge, explain scientific phenomena, and draw conclusions based on evidence relating to scientific issues [9]. So that students who are familiar with scientific literacy skills can make scientifically sound decisions [10–12].

Even though the results of the research conducted [13] show that reasoning and argumentation skills can explain scientific phenomena and can be stimulated and trained through the argumentation process in debates on social scientific issues, unfortunately not all educators are motivated to get used to presenting real problems that are close to their daily lives to discuss together with his students.

This article aims to provide a bibliometric analysis of relevant literature to explore the relationship between scientific literacy and socio-scientific issues. As well as research or publication trends indexed by the Scopus database. This analysis can reveal the extent of the relationship between scientific literacy and socio-scientific issues.

2 Research methods

The PRISMA statement [14], [15] is contained in a flow diagram using four phases consisting of identification, screening, eligibility, and inclusion *which* are then adopted when carrying out the review process to help authors report systematic reviews more effectively. The research method used is prism.

2.1 Eligibility criteria

All original studies investigated the relationship between scientific literacy and socio-scientific problems. Studies were identified by searching for relevant papers between 2014 and 2022 via the Scopus electronic database. All articles covering the topic are written in English and are empirical studies. Articles published in scientific peer-reviewed journals. The following are some examples of research journal names used: Educational Psychology Review, Chemistry Education Research and Practice, Education Sciences, Journal of Education and Learning, Journal of Technology and Science Education, International Journal of Instruction, International Journal of Science Education, Journal of Social Sciences Kasetsart, etc.

2.2 Search procedure

The search process uses keywords such as "Scientific Literacy" AND "Socio-scientific Issues"; and "Scientific Literacy" OR "Socio-scientific Issues". This process is carried out using the Scopus electronic database. The reason why the focus is on papers published in Scopus journals is based on the assumption that papers go through strict screening procedures and the results are more objective.

2.3 Study selection and data collection process

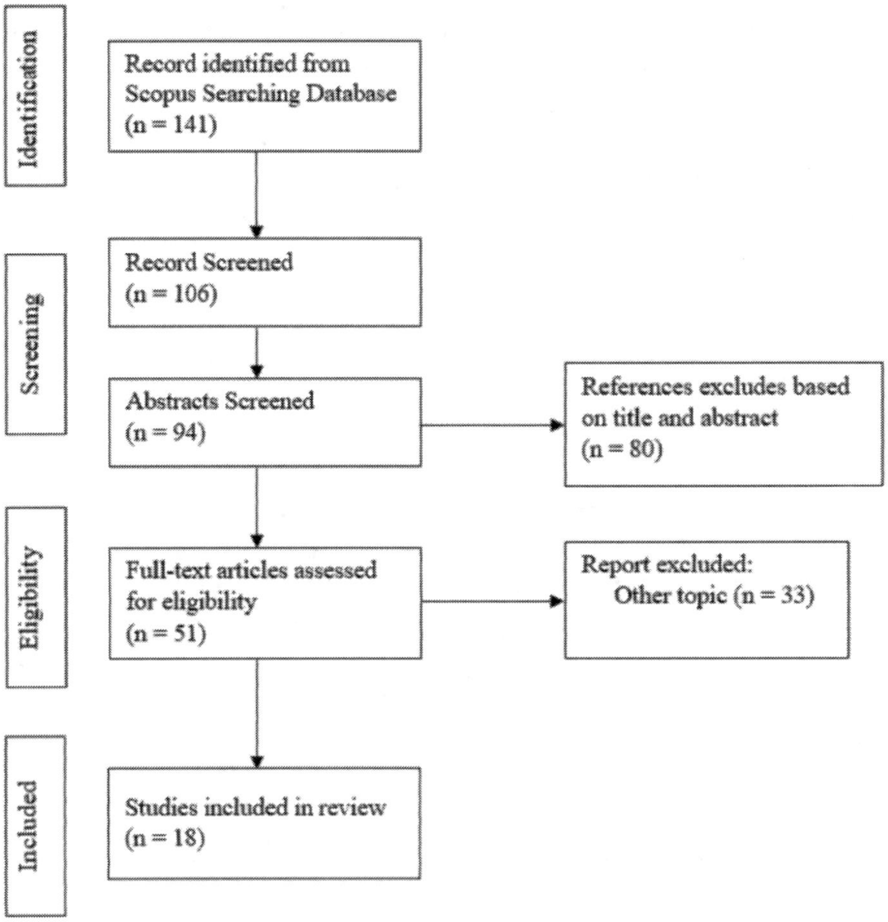

Fig. 1. PRISMA Flow Diagram of Study Selection Process

2.4 Data items

The data extracted from each study includes: background to research problems, the concept of scientific literacy, examples of social scientific problems, and the relationship between social scientific problems and scientific literacy. Data mining, mapping and grouping of articles taken from the database were carried out using VOSviewer software.

3 Results and discussion

3.1 Performance analysis

The results of the bibliometric analysis of this study are referred to [16]. There are two types of analytics, which are performance analytics in the form of the number of publications per year, most cited articles, journal with the most articles, journal ranking and country with most articles; and scientific map in the form: Circle Network Visualization, Frame Overlay Visualization, and Density Visualization.

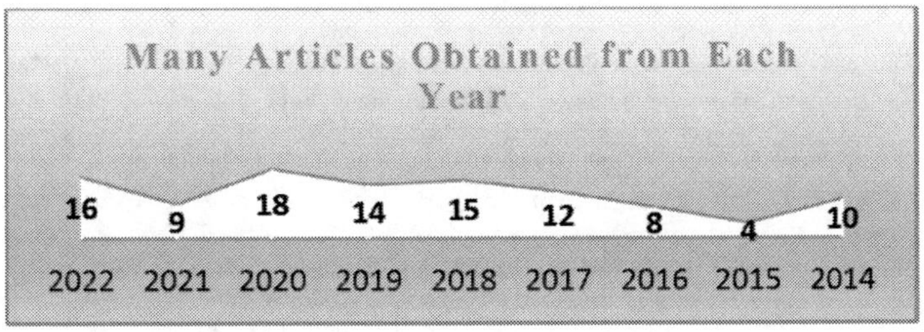

Fig. 2. Number of article publications obtained from 2014 to 2022 according to keywords

Articles in Scopus-indexed journals are often used as reference material in other research. The greater the number of citations or citations for an article, means that the research results can be used as a reference in other research. The search results for these articles show a total of 820 citations from 106 articles in 2014-2022.

Table 1. Top Writer

No	Number of Quotes	Authors	Article Title	Year	Journal Name
1	137	Sadler	Exploring Undergraduates' Breadth of Socio-scientific Reasoning Through Domains of Knowledge; Developing and Using Multiple Models to Promote Scientific Literacy in the Context of Socio-Scientific Issues; Students' perceptions of socio-scientific issue-based learning and their appropriation of epistemic tools for systems thinking; Student development of model-based reasoning about carbon cycling and climate change in a socio-scientific issues unit; Assessment of scientific literacy: Development and validation of the Quantitative Assessment of Socio-Scientific Reasoning (QuASSR); Thai pre-service science teachers' struggle in using Socio-scientific Issues (SSIs) during practicum.	2022; 2021; 2020; 2017; 2017; 2016	Research in Science Education; Science and Education: International Journal of Science Education; Journal of Research in Science Teaching; Asia-Pacific Forum on Science Learning and Teaching
2	85	Zangori	Developing and Using Multiple Models to Promote Scientific Literacy in the Context of Socio-Scientific Issues; Students' perceptions of socio-scientific issue-based learning and their appropriation of epistemic tools for systems thinking; Student development of model-based reasoning about carbon cycling and climate change in a socio-scientific issues unit.	2021; 2021; 2017	Science and Education; International Journal of Science Education; Journal of Research in Science Teaching
3	13	Forbes	Exploring Undergraduates' Breadth of Socio-scientific Reasoning Through Domains of Knowledge; Cultivating water literacy in stem education: Undergraduates' socio-scientific reasoning about socio-hydrologic issues; Application of construal level and value-belief norm theories to undergraduate decision-making on a wildlife socio-scientific issue.	2022; 2020; 2018	Research in Science Education; Water (Switzerland); International Journal of Science Education

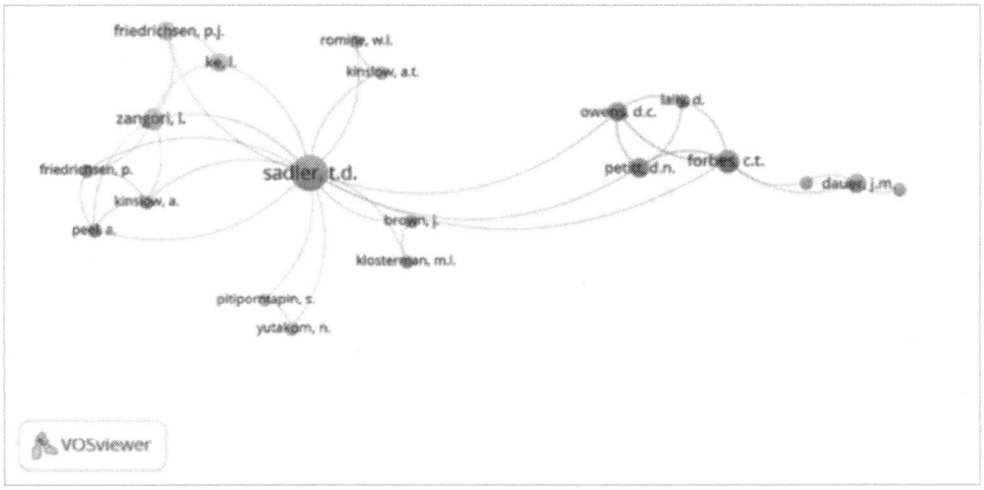

Fig. 3. Visualization of the network of top writers related to scientific literacy and socio-scientific issues

It appears that Sadler is in first place for top writer, the reason is that apart from having the highest number of citations, it turns out that Salder is an active writer every year. However, it turns out that Salder was not always and never even the first author, but rather he was the second, third, and even fifth author. So, this data shows that a person's existence may not always be first, but writing regularly and consistently maintaining the quality of writing can increase public trust so that a person's citations will increase.

Zangori came in second as best author, showing that his research with Sadler in 2017 titled Developing Students' Model-Based Reasoning on the Carbon Cycle and Climate Change in the social issues unit whose scientists received the most citations, specifically 46 citations and Zangori's. The study is his first on the topic of social science issues. The next writer who ranks third in the top is Forbest. Forbest may be a little different from Zangori, if Zangori started out as the first writer with Sadler, while Forbest in his final year started writing with Sadler so it appears in the visualization that his chain network is longer than Zangori's.

3.2 Scientific literacy topic areas and socio-scientific issues using vosviewer

In bibliometric research, thematic mapping is essential [17]. All thematic areas related to the keyword's scientific literacy and social scientific issues can be viewed in Figure 4. In bibliometric analysis, VOSviewer can display three images of different map visualizations, which are Figure 4. related to the visualization network, Figure 5. related to overlay visualization, and Figure 6. related to density visualization. The minimum number of relationships between subjects is limited to two events or conditions.

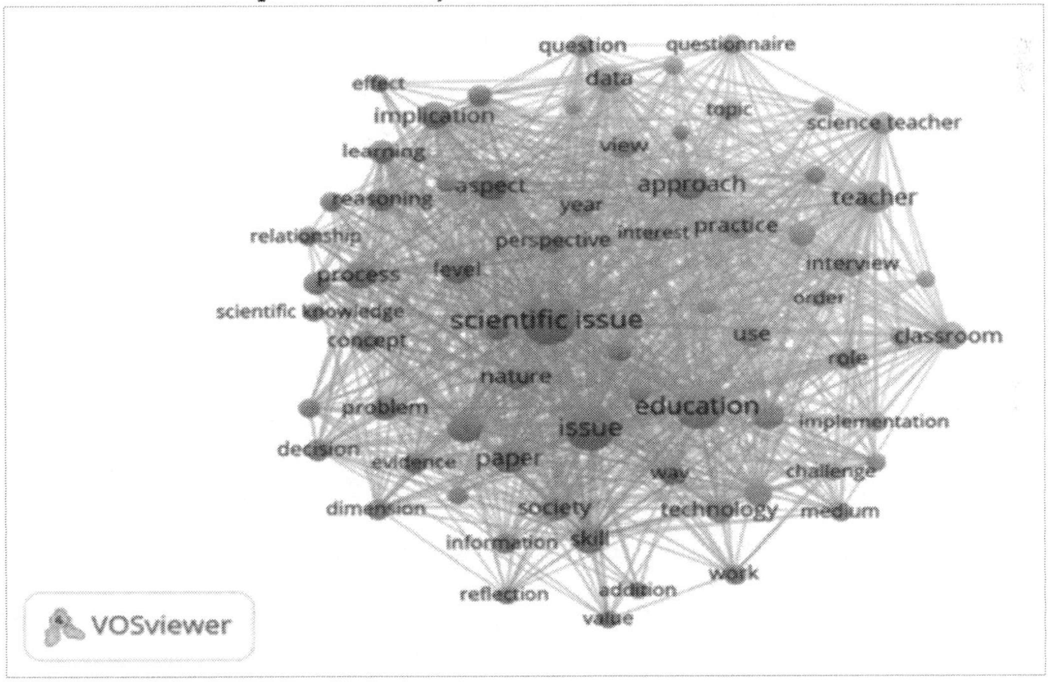

Fig. 4. Network visualization of scientific literacy topics and socio-scientific issues

After analysis with VOSviewer software, three clusters were obtained in the mapping of all topics. The color that appears in each cluster has a meaning, whereas red means that the data in the research is empirical or the research in this cluster has a strong empirical basis. Research in the red cluster bases findings on empirical data collected directly from observations or experiments, so the cluster in red indicates that research results tend to be more reliable and trustworthy because they are supported by strong empirical data. Furthermore, the green color indicates that research in this cluster encourages more in-depth analysis or interpretation of data. This means that research in this cluster tends to focus on an in-depth understanding of existing data and may produce deeper insights or understanding related to the topic discussed. Lastly, the blue color indicates that this cluster has a lower impact or relevance compared to other clusters because perhaps the number of publications or the number of citations is lower. In addition, the blue cluster provides an indication that research in this cluster may receive less attention or be considered less significant in the context of bibliometric research.

The thickness of the connecting lines indicates the strength of the topic area or keyword pair. In addition to groups and rows, the size of the circle indicates the frequency of a keyword or topic. In addition, buttons or keywords that have no network with other keywords and appear to be buttons but do not display information will also have the potential to become new research topics in the future.

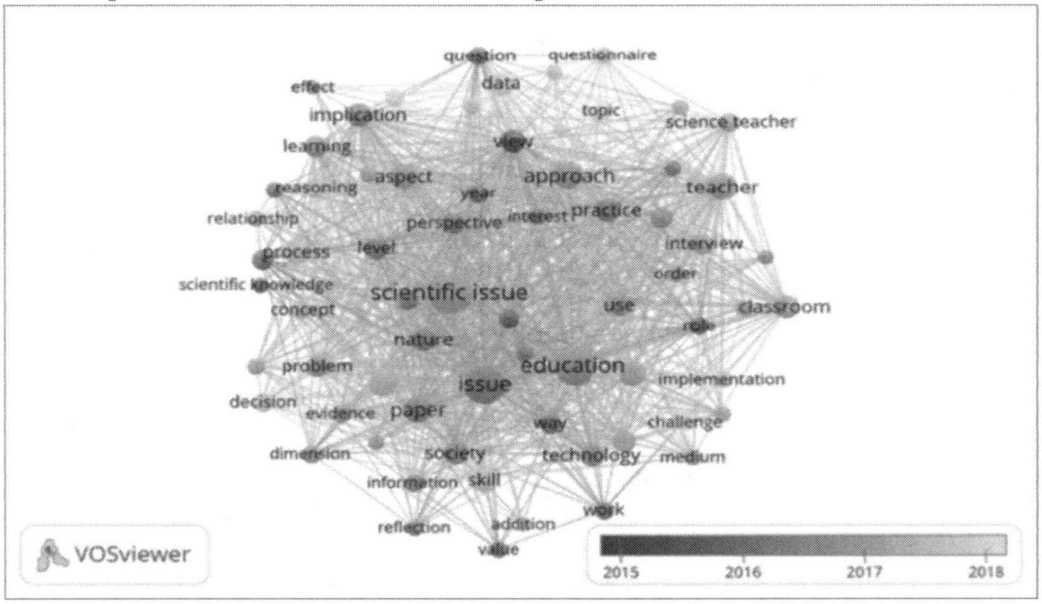

Fig. 5. Overlay visualization of scientific literacy topics and socio-scientific issues

Figure 5 above shows year-over-year trends related to search areas or keywords. The color in the keyword indicates the search period. The brighter the color seen, the more recent the year the research was conducted. It can be seen that the oldest year is 2015 in purplish blue and the maximum year obtained is 2018 which is marked in yellow. Even though the Scopus database was obtained from 2014 to 2022.

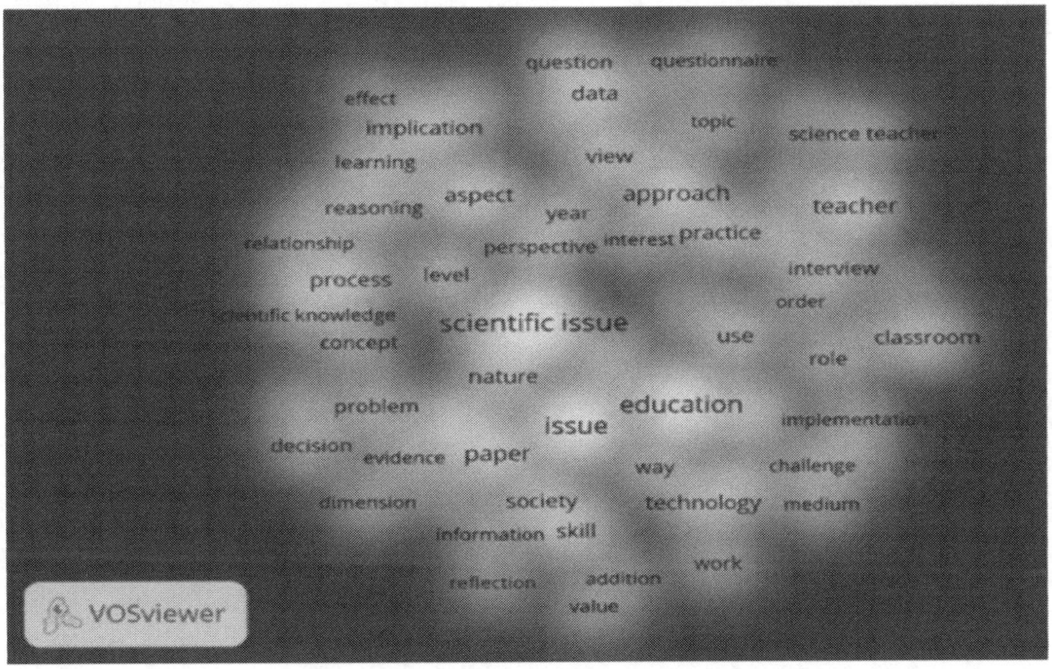

Fig. 6. Visualization of the density of scientific literacy topics and cosio scientific issues

In research related to the field of scientific literacy and socio-scientific issues, this can be seen in Figure 6. The lighter the color, the more researchers are conducting research related to this field and conversely, the more color is seen. dark. Fewer people conduct research. From Figures 4, 5, and 6 (topic network visualization, overlay visualization, and density visualization), no major topics or keywords are shown. Therefore, the researcher reduced the keywords to focus on the relationship between scientific culture and social science issues. The topic area is narrowed because the buttons are visible but the information is not yet available. Therefore, the researcher believes that this topic to be a potential new research topic in the future.

3.3 Scientific literacy topic areas and socio-scientific issues

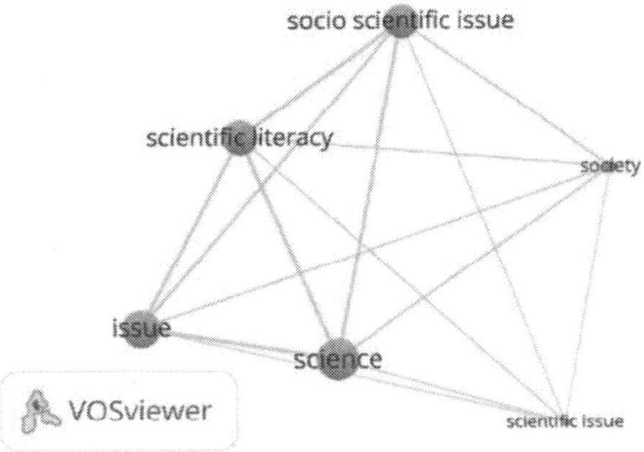

Fig. 7. Overlay network of the narrowing of socio-scientific issues topic areas and scientific literacy

Based on the results of keyword reduction focus on the relationship between scientific culture and social science issues. Figure 7 presents a comprehensive visualization of the narrowing of thematic areas of scientific literacy and socio-scientific issues. We see that there is a close relationship between scientific culture and socio-scientific issues, as evidenced by the thickness of this line and the fact that this line is the shortest among the others.

The results of this analysis show that the relationship between scientific literacy and social science issues is congruent, which is indicated by an upright prism-shaped shape. The points that make up the shape also appear strong as evidenced by the very significant area of the points. This congruence shows parallelism, congruence, and the same forces that influence each other.

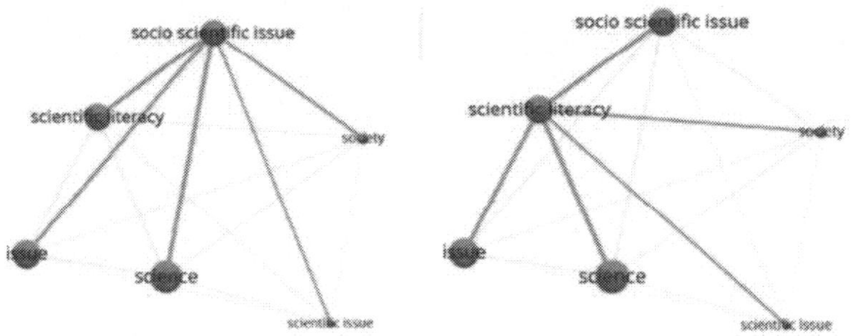

Fig. 8. Overlay visualization of congruent topics of scientific literacy and socio-scientific issues

Scientific literacy is one of the most important skills to prepare the next generation, and using scientific knowledge and information helps meet life's challenges [18–20]. Therefore, due to the importance of scientific knowledge, the amount of research on this topic must increase. This is a research opportunity for future researchers looking for related keywords. Keywords that do not have a strong network with other keywords can become new research topic opportunities. Figure 8 shows that social science issues, scientific culture, issues, and science have an important connection, but science and social issues are related to each other, although not very strongly, so this could be an opportunity for research as research related to this area is still very new.

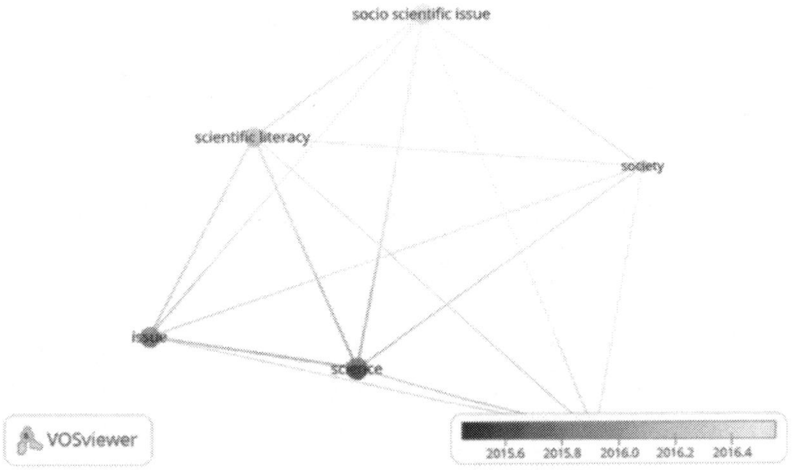

Fig. 9. Overlay visualization of the narrowing of socio-scientific issues topic areas and scientific literacy

Figure 9. shows an overlay visualization of the narrowing of the topic areas for socio-scientific issues and scientific literacy. In the modern information age, solving social science problems is becoming increasingly important [21]. However, it should be understood that until mid-2016, the results of overlay visualization using VOS Viewer in the context of socio-scientific issues could not yet be realized. This discovery is an important step, marking a change in the way we approach the analysis of social science questions. With more advanced visualization capabilities, we can sometimes analyze complex data to get a clearer, more focused picture. This will not only increase public understanding of these issues but also provide a strong basis for making more informed and responsible decisions regarding issues that affect people's daily lives [22].

The density of visualization narrows the thematic areas of social science and scientific culture issues, expressed through bright colors, containing profound meanings. The bright colors in this visualization indicate a significant increase in attention and research on social science issues as well as a significant increase in scientific literacy [23]. The presence of this prominent color also reflects the growing effort and investment devoted to understanding and solving complex problems that combine scientific and social aspects [24]. Furthermore, it reflects a growing awareness of the importance of scientific literacy in society, as it equips individuals with the ability to make informed decisions when faced with challenges and important decisions in everyday life.

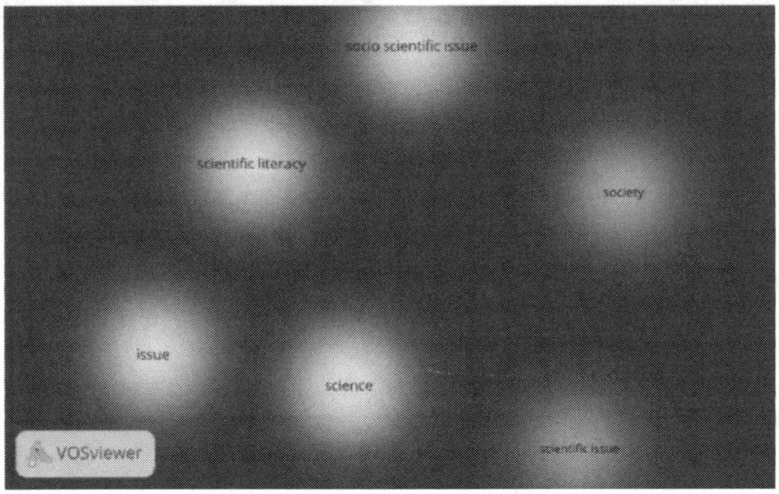

Fig. 10. Visualization density of the narrowing of socio-scientific issues topic areas and scientific literacy

As the density of visualizations shows increasing attention to socio-scientific issues and scientific knowledge in bright colors, there is a need to expand educational and research efforts in this area. This development shows that society is increasingly aware of the complexity of issues including scientific and social aspects, as well as the great importance of scientific culture [25]. Therefore, an important first step involves increasing public access to scientific information. Educational programs that prioritize scientific knowledge and socio-scientific issues need to be improved, both academically and through easily accessible online educational resources.

Through these steps, we ensure that the increased focus on socio-scientific issues and scientific literacy is not just a temporary trend. Rather, it is an inseparable part of education and knowledge of society as a whole. This will bring great benefits in forming a society that is more informed, critical and able to face the complex challenges of this modern era [26].

The rapid progress of scientific knowledge will make classical memorization of material increasingly meaningless, so it is necessary to present real-life problems or social scientific issues. Environmental problems, social problems, and other problems that are often encountered in everyday life, whether real, printed or digital, can be used as a study with students [27–29]. Problems that basically do not reflect good things should be studied together with students in learning so that from an early age students are inspired not to do the wrong thing, can make the right decisions, act according to norms, and are able to spread goodness through actions or invitations that are corrective in nature. which is not good.

This is in line with socio-scientific issues (SSI) based teaching which was chosen because it aims to invite students to be actively involved in learning [30], both inside and outside the classroom. This is related to the opinion [31] that the main aim of SSI-based learning is for students to develop a sense of having something to say truthfully about the problems they encounter so that students develop a sensitivity to care about the surrounding environment, both natural, and social.

4 Conclusion

In summary, this study conducted an in-depth analysis of 106 journal articles related to scientific culture and social sciences issues, all from the Scopus database. Although there is a wealth of research in various fields on these topics, there is a lack of research that focuses specifically on the connection between scientific knowledge and socio-scientific issues. This difference can be seen from the results of the analysis performed using VOSviewer. Through a bibliographic review, this article successfully describes the main thematic areas of each study, providing insight into the dynamic relationship between scientific culture and socio-scientific issues Associations in currently popular fields. We hope that this comprehensive review of current research will provide a valuable foundation for future research, providing updated perspectives and opportunities for further exploration in this important area.

References

1. J. A. Bellanca and R. S. Brandt, *21st-century skills : rethinking how students learn.* (Solution Tree Press, 2010)
2. H. Lee, H. Chang, K. Choi, S. W. Kim, D. L. Zeidler, Int J Sci Educ. **34**, 6 (2012)
3. PISA., Organisation for Economic Co-operation and Development., Source OECD (Online service), *Assessing scientific, reading and mathematical literacy : a framework for PISA 2006.* OECD, (2006)
4. R. C. Artaga, Jurnal Ilmiah Sekolah Dasar **5**, 4 (2021)
5. OECD, PISA 2015 Assessment and Analytical Framework science, reading, mathematic, financial literacy and collaborative problem solving (2017)
6. Syahmani, Sauqina, E. Hafizah, *Correlation of students' environmental literacy and scientific literacy after students' involvement in wetlands-based stem educational approach,* in IOP Conference Series: Earth and Environmental Science, IOP Publishing Ltd, May (2021)
7. E. Pratiwi, Suryanti, and E. Sudibyo, Jurnal Education and Development **9**, 1 (2021)
8. D. S. Rychen, L. H. Salganik, The definition and selection of key competencies executive summary (Organization for Economic Cooperation and Development, 2003)
9. N. Shofiyah, I. Afrilia, F. E. Wulandari, S*cientific approach and the effect on students' scientific literacy*, in Journal of Physics: Conference Series, Institute of Physics Publishing, Aug. (2020)
10. W. H. Bingle, P. J. Gaskell, *Sci Educ*. **78**, 2 (1994)
11. A. Uskola, B. Burgoa, G. Maguregi, *Revista Eureka* **18**, 1 (2021)
12. H. P. Rizal, P. Siahaan, G. Yuliani, *Implementation of socioscientific issues instruction to fostering students' decision-making based gender on environmental pollution*, in Journal of Physics: Conference Series, Institute of Physics Publishing, Mar. (2017)
13. S. Simon, S. Erduran, J. Osborne, J Res Sci Teach. **41**. 10 (2002)
14. A. Liberati et al., J. Clin. Epidemiol. **151**. 4 (2009)
15. D. Moher et al., PLoS Medicine **6**, 7 (2009)
16. N. Donthu, S. Kumar, D. Mukherjee, N. Pandey, W. M. Lim, J Bus Res. **133** (2021)
17. J. Ye, D. Chen, and L. Kong, J. Balt. Sci. Educ. **18**, 5 (2019)

18. R. C. Laugksch, Sci Educ. **15**. (2017)
19. T. D. Sadler, D. L. Zeidler, J Res Sci Teach. **46**, 8 (2009)
20. E. A. Berman, J. L. Kuden, Scientific literacy in agriculture to zoology: information literacy in the life sciences (Elsevier Inc., 2017)
21. T. Stouthart, D. Bayram, J. van der Veen, Sustainability **15**, 14 (2023)
22. Suryanti, M. Nursalim, N. Lutfi, I. Yuliana, Eur. J. Educ. Res. **12**, 1(2024)
23. N. Azizah, N. Suprapto, Jurnal Pendidikan **14**, 22 (2022)
24. A. Shukla-Jones, S. Friedrichs, and D. E. Winickoff, Gene editing in an international context: Scientific, economic and social issues across sectors (OECD, 2018)
25. S. Marmoah, J. I. S. Poerwanti, Suharno, Heliyon **8**, 4 (2022)
26. M. N. Fita, B. Jatmiko, E. Sudibyo, Stud. Learn. Teach. **2**, 3 (2021)
27. H. Hilman, B. Musthafa, and M. Agustin, *The design of literacy environment model in primary school*, in Journal of Physics: Conference Series, Institute of Physics Publishing, Mar. (2020)
28. S. Techakosit, P. Wannapiroon, Procedia Soc Behav Sci. **174**, (2015)
29. J. Simonneaux, L. Simonneaux, Res Sci Educ. **42**, 1 (2012)
30. M. L. Presley et al., Science Educator **22**, 1 (2013)
31. T. S. Hancock, P. J. Friedrichsen, A. T. Kinslow, and T. D. Sadler, Sci Educ. **28**, 7 (2019)

AUTHOR INDEX

AUTHOR INDEX